Linux 设备驱动程序开发(影印版)
Linux Device Drivers Development

John Madieu 著

南京　东南大学出版社

图书在版编目(CIP)数据

Linux 设备驱动程序开发:英文/(法)约翰·马杜(John Madieu)著. —影印本. —南京:东南大学出版社,2018.8
书名原文:Linux Device Drivers Development
ISBN 978-7-5641-7753-9

Ⅰ.①L… Ⅱ.①约… Ⅲ.①Linux 操作系统-驱动程序-程序设计-英文 Ⅳ.①TP316.89

中国版本图书馆 CIP 数据核字(2018)第 099725 号
图字:10-2018-097 号

© 2017 by PACKT Publishing Ltd.

Reprint of the English Edition, jointly published by PACKT Publishing Ltd and Southeast University Press, 2018. Authorized reprint of the original English edition, 2018 PACKT Publishing Ltd, the owner of all rights to publish and sell the same.

All rights reserved including the rights of reproduction in whole or in part in any form.

英文原版由 PACKT Publishing Ltd 出版 2017。

英文影印版由东南大学出版社出版 2018。此影印版的出版和销售得到出版权和销售权的所有者——PACKT Publishing Ltd 的许可。

版权所有,未得书面许可,本书的任何部分和全部不得以任何形式重制。

Linux 设备驱动程序开发(影印版)

出版发行:东南大学出版社
地　　址:南京四牌楼 2 号　　邮编:210096
出 版 人:江建中
网　　址:http://www.seupress.com
电子邮件:press@seupress.com
印　　刷:常州市武进第三印刷有限公司
开　　本:787 毫米×980 毫米　　16 开本
印　　张:36.5
字　　数:715 千字
版　　次:2018 年 8 月第 1 版
印　　次:2018 年 8 月第 1 次印刷
书　　号:ISBN 978-7-5641-7753-9
定　　价:108.00 元

本社图书若有印装质量问题,请直接与营销部联系。电话(传真):025-83791830

Credits

Author
John Madieu

Reviewer
Jérôme Pouiller

Commissioning Editor
Gebin George

Acquisition Editor
Gebin George

Content Development Editor
Devika Battike

Technical Editor
Swathy Mohan

Copy Editors
Juliana Nair
Safis Editing

Project Coordinator
Judie Jose

Proofreader
Safis Editing

Indexer
Rekha Nair

Graphics
Kirk D'Penha

Production Coordinator
Arvindkumar Gupta

About the Author

John Madieu is an embedded Linux and kernel engineer living in France, in Paris. His main activities consist of developing drivers and Board Support Packages (BSP) for companies in domains such as automation, transport, healthcare, energy, and the military. John works at EXPEMB, a French company that is a pioneer in electronical board design based on computer-on-module, and in embedded Linux solutions. He is an open source and embedded systems enthusiast, convinced that it is only by sharing knowledge that one learns more.

He is passionate about boxing, which he practised for 6 years professionally, and continues to transmit this passion through sessions of training that he provides voluntarily.

> *I would like to thank Devika Battike, Gebin George, and all the Packt team for their efforts to release this book on time. They are the people without whom this book would probably never have seen the light of day. It was a pleasure to work with them.*

> *Finally, I would like to thank all the mentors I have had over the years, and who still continue to accompany me. Mentors such as Cyprien Pacôme Nguefack for his programming skills that I have learned over the years, Jérôme Pouillier and Christophe Nowicki for introducing me buildroot and leading me to kernel programming, Jean-Christian Rerat and Jean-Philippe DU-Teil of EXPEMB for their coaching and accompaniment in my professional career; to all those I could not mention, I wish to thank them for having transmitted these connoises to me, which I have tried to disseminate through this book.*

About the Reviewer

Jérôme Pouiller is a true geek and fascinated by understanding how things do work.

He was an early adopter of Linux. He found in Linux a system with no limits, where everything could be changed. Linux has provided an excellent platform to hack anything.

He graduated in machine learning at Ecole Pour l'Informatique et les Technologies Avancées (EPITA). Beside his studies, he learned electronics by himself. He quickly turned his attention to the piece of software at crossroad of all advanced systems: the operating system. It is now one of his favorite subjects.

For 15 years now, Jérôme Pouiller has designed (and often debugged) Linux firmware for a variety of industries (multimedia, healthcare, nuclear, military).

In addition to his consulting activities, Jérôme Pouiler is professor of operating systems at Institut National des Sciences Appliquées (INSA). He has written many course materials about system programming, operating system design, realtime systems, and more.

www.PacktPub.com

For support files and downloads related to your book, please visit `www.PacktPub.com`.

Did you know that Packt offers eBook versions of every book published, with PDF and ePub files available? You can upgrade to the eBook version at `www.PacktPub.com` and as a print book customer, you are entitled to a discount on the eBook copy. Get in touch with us at `service@packtpub.com` for more details.

At `www.PacktPub.com`, you can also read a collection of free technical articles, sign up for a range of free newsletters and receive exclusive discounts and offers on Packt books and eBooks.

`https://www.packtpub.com/mapt`

Get the most in-demand software skills with Mapt. Mapt gives you full access to all Packt books and video courses, as well as industry-leading tools to help you plan your personal development and advance your career.

Why subscribe?

- Fully searchable across every book published by Packt
- Copy and paste, print, and bookmark content
- On demand and accessible via a web browser

Customer Feedback

Thanks for purchasing this Packt book. At Packt, quality is at the heart of our editorial process. To help us improve, please leave us an honest review on this book's Amazon page at https://www.amazon.com/dp/1785280007.

If you'd like to join our team of regular reviewers, you can e-mail us at customerreviews@packtpub.com. We award our regular reviewers with free eBooks and videos in exchange for their valuable feedback. Help us be relentless in improving our products!

I would like to thank my girlfriend for her support and all the sleepless nights accompanying the writing of this book, as well as Brigitte and François, my dear parents, for whom I have a thought and to whom I dedicate this book entirely.

- John Madieu

I would like to dedicate this book in the memory of my father, who left too.

- Jérôme Pouiller

Table of Contents

Preface   1
Chapter 1: Introduction to Kernel Development   9
   Environment setup   10
      Getting the sources   10
         Source organization   11
      Kernel configuration   12
      Build your kernel   13
   Kernel habits   14
      Coding style   14
      Kernel structures allocation/initialization   15
      Classes, objects, and OOP   16
   Summary   17
Chapter 2: Device Driver Basis   19
   User space and kernel space   20
      The concept of modules   21
      Module dependencies   21
         depmod utility   21
      Module loading and unloading   22
         Manual loading   22
            modprobe and insmod   22
            /etc/modules-load.d/<filename>.conf   22
         Auto-loading   23
         Module unload   23
   Driver skeletons   24
      Module entry and exit point   25
         __init and __exit attributes   25
      Module information   27
         Licensing   29
         Module author(s)   30
         Module description   31
   Errors and message printing   31
      Error handling   31
      Handling null pointer errors   34
      Message printing – printk()   35
   Module parameters   37
   Building your first module   39

The module's makefile	39
In the kernel tree	41
Out of the tree	44
Building the module	44
Summary	45
Chapter 3: Kernel Facilities and Helper Functions	**47**
Understanding container_of macro	47
Linked lists	50
Creating and initializing the list	51
Dynamic method	51
Static method	52
Creating a list node	52
Adding a list node	53
Deleting a node from the list	54
Linked list traversal	54
Kernel sleeping mechanism	55
Wait queue	55
Delay and timer management	58
Standard timers	59
Jiffies and HZ	59
Timers API	59
Timer setup initialization	60
Standard timer example	61
High resolution timers (HRTs)	62
HRT API	62
HRT setup initialization	62
Dynamic tick/tickless kernel	64
Delays and sleep in the kernel	64
Atomic context	64
Nonatomic context	65
Kernel locking mechanism	65
Mutex	66
Mutex API	66
Declare	66
Acquire and release	67
Spinlock	68
Spinlock versus mutexes	70
Work deferring mechanism	70
Softirqs and ksoftirqd	70
ksoftirqd	71
Tasklets	72
Declaring a tasklet	72

Enabling and disabling a tasklet	73
Tasklet scheduling	73
Work queues	75
Kernel-global workqueue – the shared queue	75
Dedicated work queue	78
Programming syntax	78
Predefined (shared) workqueue and standard workqueue functions	81
Kernel threads	82
Kernel interruption mechanism	**82**
Registering an interrupt handler	82
Interrupt handler and lock	85
Concept of bottom halves	86
The problem – interrupt handler design limitations	86
The solution – bottom halves	86
Tasklets as bottom halves	87
Workqueue as bottom halves	88
Softirqs as bottom half	89
Threaded IRQs	**89**
Threaded bottom half	91
Invoking user-space applications from the kernel	**92**
Summary	**93**

Chapter 4: Character Device Drivers — 95

The concept behind major and minor	**96**
Device number allocation and freeing	97
Introduction to device file operations	**98**
File representation in the kernel	99
Allocating and registering a character device	**101**
Writing file operations	**102**
Exchanging data between kernel space and user space	102
A single value copy	103
The open method	104
Per-device data	104
The release method	105
The write method	106
Steps to write	106
The read method	108
Steps to read	109
The llseek method	110
Steps to llseek	111
The poll method	112
Steps to poll	113
The ioctl method	116

Generating ioctl numbers (command)	117
Steps for ioctl	118
Filling the file_operations structure	120
Summary	120

Chapter 5: Platform Device Drivers — 121

Platform drivers — 122
Platform devices — 126
Resources and platform data — 126
Device provisioning - the old and depreciated way — 126
Resources — 127
Platform data — 129
Where to declare platform devices? — 131
Device provisioning - the new and recommended way — 131
Devices, drivers, and bus matching — 132
How can platform devices and platform drivers match? — 134
Kernel devices and drivers-matching function — 135
OF style and ACPI match — 136
ID table matching — 136
Name matching - platform device name matching — 140
Summary — 140

Chapter 6: The Concept of Device Tree — 141

Device tree mechanism — 141
Naming convention — 142
Aliases, labels, and phandle — 143
DT compiler — 144
Representing and addressing devices — 145
SPI and I2C addressing — 145
Platform device addressing — 147
Handling resources — 148
Concept of named resources — 149
Accessing registers — 150
Handling interrupts — 151
The interrupt handler — 151
Interrupt controller code — 152
Extract application-specific data — 153
Text string — 153
Cells and unsigned 32-bit integers — 154
Boolean — 155
Extract and parse sub-nodes — 155
Platform drivers and DT — 156
OF match style — 156
Dealing with non-device tree platforms — 159

Support multiple hardware with per device-specific data	160
Match style mixing	162
Platform resources and DT	164
Platform data versus DT	166
Summary	167

Chapter 7: I2C Client Drivers — 169

The driver architecture	170
The i2c_driver structure	170
The probe() function	171
Per-device data	172
The remove() function	173
Driver initialization and registration	174
Driver and device provisioning	174
Accessing the client	175
Plain I2C communication	175
System Management Bus (SMBus) compatible functions	177
Instantiating I2C devices in the board configuration file (old and depreciated way)	178
I2C and the device tree	179
Defining and registering the I2C driver	180
Remark	181
Instantiating I2C devices in the device tree - the new way	182
Putting it all together	182
Summary	183

Chapter 8: SPI Device Drivers — 185

The driver architecture	186
The device structure	186
spi_driver structure	189
The probe() function	189
Per-device data	190
The remove() function	191
Driver initialization and registration	191
Driver and devices provisioning	192
Instantiate SPI devices in board configuration file – old and depreciated way	193
SPI and device tree	194
Instantiate SPI devices in device tree - the new way	196
Define and register SPI driver	196
Accessing and talking to the client	197
Putting it all together	202
SPI user mode driver	202
With IOCTL	204

Summary	207
Chapter 9: Regmap API – A Register Map Abstraction	**209**
Programming with the regmap API	210
regmap_config structure	211
regmap initialization	214
SPI initialization	214
I2C initialization	215
Device access functions	216
regmap_update_bits function	217
Special regmap_multi_reg_write function	218
Other device access functions	219
regmap and cache	219
Putting it all together	221
A regmap example	221
Summary	224
Chapter 10: IIO Framework	**225**
IIO data structures	227
iio_dev structure	227
iio_info structure	231
IIO channels	232
Channel attribute naming conventions	234
Distinguishing channels	236
Putting it all together	238
Triggered buffer support	241
IIO trigger and sysfs (user space)	245
Sysfs trigger interface	245
add_trigger file	245
remove_trigger file	246
Tying a device with a trigger	246
The interrupt trigger interface	246
The hrtimer trigger interface	247
IIO buffers	248
IIO buffer sysfs interface	248
IIO buffer setup	249
Putting it all together	251
IIO data access	257
One-shot capture	258
Buffer data access	258
Capturing using the sysfs trigger	258
Capturing using the hrtimer trigger	260
IIO tools	261

Summary	261
Chapter 11: Kernel Memory Management	**263**
System memory layout - kernel space and user space	265
Kernel addresses – concept of low and high memory	267
Low memory	268
High memory	268
User space addresses	269
Virtual Memory Area (VMA)	272
Address translation and MMU	274
Page look up and TLB	280
How does TLB work	280
Memory allocation mechanism	282
Page allocator	283
Page allocation API	283
Conversion functions	285
Slab allocator	286
The buddy algorithm	286
A journey into the slab allocator	289
kmalloc family allocation	291
vmalloc allocator	294
Process memory allocation under the hood	296
The copy-on-write (CoW) case	297
Work with I/O memory to talk with hardware	298
PIO devices access	298
MMIO devices access	299
__iomem cookie	300
Memory (re)mapping	302
kmap	302
Mapping kernel memory to user space	303
Using remap_pfn_range	303
Using io_remap_pfn_range	305
The mmap file operation	305
Implementing mmap in the kernel	307
Linux caching system	308
What is a cache?	308
CPU cache – memory caching	309
The Linux page cache – disk caching	310
Specialized caches (user space caching)	310
Why delay writing data to disk?	310
Write caching strategies	311
The flusher threads	312
Device-managed resources – Devres	312

Summary	314
Chapter 12: DMA – Direct Memory Access	**315**
Setting up DMA mappings	316
Cache coherency and DMA	316
DMA mappings	317
Coherent mapping	317
Streaming DMA mapping	318
Single buffer mapping	319
Scatter/gather mapping	319
Concept of completion	322
DMA engine API	324
Allocate a DMA slave channel	325
Set slave and controller specific parameters	326
Get a descriptor for transaction	329
Submit the transaction	330
Issue pending DMA requests and wait for callback notification	331
Putting it all together – NXP SDMA (i.MX6)	332
DMA DT binding	337
Consumer binding	337
Summary	339
Chapter 13: Linux Device Model	**341**
LDM data structures	342
The bus	342
Bus registration	347
Device driver	348
Device driver registration	349
Device	350
Device registration	351
Deep inside LDM	352
kobject structure	352
kobj_type	354
ksets	356
Attribute	357
Attributes group	358
Device model and sysfs	359
Sysfs files and attributes	361
Current interfaces	362
Device attributes	362
Bus attributes	364
Device drivers attributes	365
Class attributes	366

Allow sysfs attribute files to be pollable	367
Summary	369

Chapter 14: Pin Control and GPIO Subsystem — 371

Pin control subsystem	371
Pinctrl and the device tree	372
The GPIO subsystem	376
The integer-based GPIO interface: legacy	377
Claiming and configuring the GPIO	377
Accessing the GPIO – getting/setting the value	378
In atomic context	379
In a non-atomic context (that may sleep)	379
GPIOs mapped to IRQ	379
Putting it all together	380
The descriptor-based GPIO interface: the new and recommended way	382
GPIO descriptor mapping - the device tree	383
Allocating and using GPIO	384
Putting it all together	386
The GPIO interface and the device tree	389
The legacy integer-based interface and device tree	390
GPIO mapping to IRQ in the device tree	393
GPIO and sysfs	394
Exporting a GPIO from kernel code	396
Summary	397

Chapter 15: GPIO Controller Drivers – gpio_chip — 399

Driver architecture and data structures	399
Pin controller guideline	404
Sysfs interface for GPIO controller	404
GPIO controllers and DT	405
Summary	405

Chapter 16: Advanced IRQ Management — 407

Multiplexing interrupts and interrupt controllers	410
Advanced peripheral IRQs management	419
Interrupt request and propagation	422
Chaining IRQ	423
Chained interrupts	423
Nested interrupts	424
Case study – GPIO and IRQ chip	424
Legacy GPIO and IRQ chip	425
New gpiolib irqchip API	427
Interrupt controller and DT	429
Summary	430

Chapter 17: Input Devices Drivers — 431
Input device structures — 431
Allocating and registering an input device — 434
Polled input device sub-class — 435
Generating and reporting an input event — 439
User space interface — 441
Putting it all together — 443
Driver examples — 445
Summary — 451

Chapter 18: RTC Drivers — 453
RTC framework data structures — 454
RTC API — 456
Reading and setting time — 457
Driver example — 460
Playing with alarms — 461
RTCs and user space — 464
The sysfs interface — 465
The hwclock utility — 466
Summary — 466

Chapter 19: PWM Drivers — 467
PWM controller driver — 469
Driver example — 471
PWM controller binding — 474
PWM consumer interface — 475
PWM clients binding — 477
Using PWMs with the sysfs interface — 479
Summary — 481

Chapter 20: Regulator Framework — 483
PMIC/producer driver interface — 484
Driver data structures — 484
Description structure — 485
Constraints structure — 486
init data structure — 487
Feeding init data into a board file — 488
Feeding init data into the DT — 489
Configuration structure — 491
Device operation structure — 491
Driver methods — 492
Probe function — 493
Remove function — 494

Case study: Intersil ISL6271A voltage regulator	494
Driver example	499
Regulators consumer interface	503
Regulator device requesting	504
Controlling the regulator device	505
Regulator output enable and disable	505
Voltage control and status	506
Current limit control and status	506
Operating mode control and status	507
Regulator binding	507
Summary	508

Chapter 21: Framebuffer Drivers — 509

Driver data structures	510
Device methods	514
Driver methods	516
Detailed fb_ops	518
Checking information	519
Set controller's parameters	520
Screen blanking	521
Accelerated methods	522
Putting it all together	523
Framebuffer from user space	523
Summary	526

Chapter 22: Network Interface Card Drivers — 527

Driver data structures	528
The socket buffer structure	528
Socket buffer allocation	530
Network interface structure	531
The device methods	533
Opening and closing	535
Packet handling	537
Packet reception	537
Packet transmission	540
Driver example	543
Status and control	546
The interrupt handler	546
Ethtool support	548
Driver methods	549
The probe function	550
Module unloading	552
Summary	552

Index

Preface

The Linux kernel is a complex, portable, modular, and widely used piece of software, running on around 80% of servers and embedded systems in more than half of the devices throughout the world. Device drivers play a critical role in the context of how well a Linux system performs. As Linux has turned out to be one of the most popular operating systems interest in developing personal device drivers is also increasing steadily.

A device driver is the link between the user space and devices, through the kernel.

This book will begins with two chapters that will help you understand the basics of drivers and prepare you for the long journey through the Linux kernel. This book will then cover driver development based on Linux subsystems such as memory management, PWM, RTC, IIO, GPIO, IRQ management. The book will also cover practical approach to direct memory access and network device drivers.

Source code in this book has been tested on both x86 PC and UDOO Quad from SECO, which is based on an ARM i.MX6 from NXP, with enough features and connections to allow us to cover all of tests discussed in the book. Some drivers are also provided for testing purposes for inexpensive components such as MCP23016 and 24LC512, which are I2C GPIO controller and eeprom memory respectively.

By the end of this book, you will be comfortable with the concept of device driver development and will be in a position to write any device driver from scratch using the latest kernel version (v4.13 at the time of writing).

What this book covers

Chapter 1, *Introduction to Kernel Development*, introduces the Linux kernel development process. The chapter will discuss the downloading, configuring, and compiling steps of a kernel, as well for x86 as for ARM-based systems

Chapter 2, *Device Driver Basis*, deals with Linux modularity by means of kernel modules, and describes their loading/unloading. It also describe a driver architecture and some basic concepts and some kernel best practices.

Chapter 3, *Kernel Facilities and Helper Functions*, walks through frequently used kernel functions and mechanisms, such as work queue, wait queue, mutexes, spinlock, and any other facilities that are useful for improved driver reliability.

Preface

Chapter 4, *Character Device Drivers*, focuses exporting a devices functionalities to the user space by means of character devices as well as supporting custom commands using the IOCTL interface.

Chapter 5, *Platform Device Drivers*, explains what a platform device is and introduces the concept of pseudo-platform bus, as well as the device and bus matching mechanism. This chapter describes platform driver architecture in a general manner, and how to handle platform data.

Chapter 6, *The Concept of Device Tree*, discusses the mechanism to feed device descriptions to the kernel. This chapter explains device addressing, resource handling, every data type supported in DT and their kernel APIs.

Chapter 7, *I2C Client Drivers*, dives into I2C device drivers architecture, the data structures and device addressing and accessing methods on the bus.

Chapter 8, *SPI Device Drivers*, describe SPI-based device driver architecture, as well as the data structures involved. The chapter discuss each device's access method and specificities, as well as traps one should avoid. SPI DT binding is discussed too.

Chapter 9, *Regmap API – A Register Map Abstraction*, provides an overview of the regmap API, and how it abstracts the underlying SPI and I2C transaction. This chapter describes the generic API, as well as the dedicated API.

Chapter 10, *IIO framework*, introduce the kernel data acquisition and measurement framework, to handle Digital to Analog Converters (DACs) and Analog to Digital Converters (ADCs). This walk through the IIO API, deals with triggered buffers and continuous data capture, and looks at single channel acquisition through the sysfs interface.

Chapter 11, *Kernel Memory Management*, first introduces the concept of virtual memory, in order to describe the whole kernel memory layout. This chapter walks through the kernel memory management subsystem, discussing memory allocation and mapping, their APIs and all devices involved in such mechanisms, as well as kernel caching mechanism.

Chapter 12, *DMA – Direct Memory Access*, introduce DMA and its new kernel API: the DMA Engine API. This chapter will talk about different DMA mappings and describes how to address cache coherency issues. In addition, the chapter summarize the whole concepts in use cases, based on i.MX6 SoC, from NXP.

Chapter 13, *Linux Device Model*, provides an overview of the heart of Linux, describing how objects are represented in the kernel, and how Linux is designed under the hood in a general manner, starting from kobject to devices, through buses, classes, and device drivers. This chapter also highlight sometime unknown side in user space, the kernel object hierarchy in sysfs.

Chapter 14, *Pin Control and GPIO Subsystem*, describes the kernel pincontrol API and GPIOLIB, which is the kernel API to handle GPIO. This chapter also discusses the old and deprecated integer-based GPIO interface, as well as the descriptor-based interface, which is the new one, and finally, the way they can be configured from within the DT.

Chapter 15, *GPIO Controller Drivers – gpio_chip*, necessary elements to write such device drivers. That says, its main data structure is struct gpio_chip. This structure is explained in detail in this chapter, along with a full and working driver provided in the source of the book.

Chapter 16, *Advanced IRQ Management*, demystifies the Linux IRQ core. This chapter walks through Linux IRQ management, starting from interrupt propagation over the system and moving to interrupt controller drivers, thus explaining the concept of IRQ multiplexing, using the Linux IRQ domain API

Chapter 17, *Input Devices Drivers*, provides a global view of input subsystems, dealing with both IRQ-based and polled input devices, and introducing both APIs. This chapter explains and shows how user space code deals with such devices.

Chapter 18, *RTC Drivers*, walks through and demystifies the RTC subsystem and its API. This chapter goes far enough and explains how to deal with alarms from within RTC drivers

Chapter 19, *PWM Drivers*, provides a full description of the PWM framework, talking about the controller side API as well the consumer side API. PWM management from the user space is discussed in the last section in this chapter.

Chapter 20, *Regulator Framework*, highlights how important power management is. The first part of the chapter deals with Power Management IC (PMIC) and explains its driver design and API. The second part focuses on the consumer side, talking about requesting and using regulators.

Chapter 21, *Framebuffer Drivers*, explains framebuffer concept and how it works. It also shows how to design framebuffer drivers, walks through its API, and discusses accelerated as well as non-accelerated methods. This chapter shows how drivers can expose framebuffer memory so that user space can write into, without worrying about underlying tasks.

Chapter 22, *Network Interface Card Drivers*, walk through the NIC driver's architecture and their data structures, thus showing you how to handle device configuration, data transfer, and socket buffers.

What you need for this book

This book assumes a medium level of understanding the Linux operating system, basic knowledge of C programming (at least pointer handling). That is all. If additional skill is required for a given chapter, links on document reference will be provided to readers to quickly learn these skills.

Linux kernel compiling is a quite long and heavy task. The minimum hardware or virtual requirements are as the follows:

- CPU: 4 cores
- Memory: 4 GB RAM
- Free disk space: 5 GB (large enough)

In this book, you will need the following software list:

- Linux operating system: preferably a Debian-based distribution, which is used for example in the book (Ubuntu 16.04)
- At least version 5 of both gcc and gcc-arm-linux (as used in the book)

Other necessary packages are described in dedicated chapter in the book. Internet connectivity is required for kernel sources downloading.

Who this book is for

To make usage of the content of this book, a basic prior knowledge of C programming and basics Linux commands is expected. This book covers Linux drivers development for widely used embedded devices, using the kernel version v4.1, and covers changes until the last version at the time of writing this book (v4.13). This book is essentially intended for embedded engineers, Linux system administrators, developer, and kernel hackers. Whether you are a software developer, a system architect, or maker willing to dive into Linux driver development, this book is for you.

Conventions

In this book, you will find a number of text styles that distinguish between different kinds of information. Here are some examples of these styles and an explanation of their meaning. Code words in text, database table names, folder names, filenames, file extensions, pathnames, dummy URLs, user input, and Twitter handles are shown as follows: "The .name field must be the same as the device's name you give when you register the device in the board specific file".

A block of code is set as follows:

```
#include <linux/of.h>
#include <linux/of_device.h>
```

Any command-line input or output is written as follows:

```
sudo apt-get update
sudo apt-get install linux-headers-$(uname -r)
```

New terms and **important words** are shown in bold.

Warnings or important notes appear like this.

Tips and tricks appear like this.

Reader feedback

Feedback from our readers is always welcome. Let us know what you think about this book-what you liked or disliked. Reader feedback is important for us as it helps us develop titles that you will really get the most out of. To send us general feedback, simply email feedback@packtpub.com, and mention the book's title in the subject of your message. If there is a topic that you have expertise in and you are interested in either writing or contributing to a book, see our author guide at www.packtpub.com/authors.

Customer support

Now that you are the proud owner of a Packt book, we have a number of things to help you to get the most from your purchase.

Downloading the example code

You can download the example code files for this book from your account at `http://www.packtpub.com`. If you purchased this book elsewhere, you can visit `http://www.packtpub.com/support` and register to have the files emailed directly to you. You can download the code files by following these steps:

1. Log in or register to our website using your email address and password.
2. Hover the mouse pointer on the **SUPPORT** tab at the top.
3. Click on **Code Downloads & Errata**.
4. Enter the name of the book in the **Search** box.
5. Select the book for which you're looking to download the code files.
6. Choose from the drop-down menu where you purchased this book from.
7. Click on **Code Download**.

Once the file is downloaded, please make sure that you unzip or extract the folder using the latest version of:

- WinRAR / 7-Zip for Windows
- Zipeg / iZip / UnRarX for Mac
- 7-Zip / PeaZip for Linux

The code bundle for the book is also hosted on GitHub at `https://github.com/PacktPublishing/Linux-Device-Drivers-Development`. We also have other code bundles from our rich catalog of books and videos available at `https://github.com/PacktPublishing/`. Check them out!

Downloading the color images of this book

We also provide you with a PDF file that has color images of the screenshots/diagrams used in this book. The color images will help you better understand the changes in the output. You can download this file from `https://www.packtpub.com/sites/default/files/downloads/LinuxDeviceDriversDevelopment_ColorImages.pdf`.

Errata

Although we have taken every care to ensure the accuracy of our content, mistakes do happen. If you find a mistake in one of our books-maybe a mistake in the text or the code- we would be grateful if you could report this to us. By doing so, you can save other readers from frustration and help us improve subsequent versions of this book. If you find any errata, please report them by visiting `http://www.packtpub.com/submit-errata`, selecting your book, clicking on the **Errata Submission Form** link, and entering the details of your errata. Once your errata are verified, your submission will be accepted and the errata will be uploaded to our website or added to any list of existing errata under the Errata section of that title. To view the previously submitted errata, go to `https://www.packtpub.com/books/content/support` and enter the name of the book in the search field. The required information will appear under the **Errata** section.

Piracy

Piracy of copyrighted material on the internet is an ongoing problem across all media. At Packt, we take the protection of our copyright and licenses very seriously. If you come across any illegal copies of our works in any form on the internet, please provide us with the location address or website name immediately so that we can pursue a remedy. Please contact us at `copyright@packtpub.com` with a link to the suspected pirated material. We appreciate your help in protecting our authors and our ability to bring you valuable content.

Questions

If you have a problem with any aspect of this book, you can contact us at `questions@packtpub.com`, and we will do our best to address the problem.

1
Introduction to Kernel Development

Linux started as a hobby project in 1991 for a Finnish student, Linus Torvalds. The project has gradually grown and still does, with roughly 1000 contributors around the world. Nowadays, Linux is a must, in embedded systems as well as on servers. A kernel is a center part of an operating system, and its development is not so obvious.

Linux offers many advantages over other operating systems:

- It is free of charge
- Well documented with a large community
- Portable across different platforms
- Provides access to the source code
- Lots of free open source software

This book tries to be as generic as possible. There is a special topic, device tree, which is not a full x86 feature yet. That topic will then be dedicated to ARM processors, and all those fully supporting the device tree. Why those architectures? Because they are most used on the desktop and servers (for x86) and on embedded systems (ARM).

This chapter deals among others with:

- Development environment setup
- Getting, configure, and build kernel sources
- Kernel source code organization
- Introduction to kernel coding style

Environment setup

Before one starts any development, you need to set an environment up. The environment dedicated to Linux development is quite simple, at least, on Debian based systems:

```
$ sudo apt-get update
$ sudo apt-get install gawk wget git diffstat unzip texinfo \
gcc-multilib build-essential chrpath socat libsdl1.2-dev \
xterm ncurses-dev lzop
```

There are parts of codes in this book that are compatible with ARM **system on chip** (**SoC**). One should install `gcc-arm` as well:

```
sudo apt-get install gcc-arm-linux-gnueabihf
```

I'm running Ubuntu 16.04, on an ASUS RoG, with an Intel core i7 (8 physical cores), 16 GB of RAM, 256 GB of SSD, and 1 TB of magnetic hard drive. My favorite editor is Vim, but you are free to use the one you are most comfortable with.

Getting the sources

In the early kernel days (until 2003), odd–even versioning styles were used; where odd numbers were stable, even numbers were unstable. When the 2.6 version was released, the versioning scheme switched to X.Y.Z, where:

- X: This was the actual kernel's version, also called major, it incremented when there were backwards-incompatible API changes
- Y: This was the minor revision, it incremented after adding a functionality in a backwards-compatible manner
- Z: This is also called PATCH, represented the version relative to bug fixes

This is called semantic versioning, and has been used until the 2.6.39 version; when Linus Torvalds decided to bump the version to 3.0, which also meant the end of semantic versioning in 2011, and then, an X.Y scheme was adopted.

When it came to the 3.20 version, Linus argued that he could no longer increase Y, and decided to switch to an arbitrary versioning scheme, incrementing X whenever Y got large enough that he ran out of fingers and toes to count it. This is the reason why the version has moved from 3.20 to 4.0 directly. Have a look at:
https://plus.google.com/+LinusTorvalds/posts/jmtzzLiiejc.

Now the kernel uses an arbitrary X.Y versioning scheme, which has nothing to do with semantic versioning.

Source organization

For the needs of this book, you must use Linus Torvald's Github repository.

```
git clone https://github.com/torvalds/linux
git checkout v4.1
ls
```

- `arch/`: The Linux kernel is a fast growing project that supports more and more architectures. That being said, the kernel wants to be as generic as possible. Architecture specific code is separated from the rest, and falls in this directory. This directory contains processor-specific subdirectories such as `alpha/`, `arm/`, `mips/`, `blackfin/`, and so on.
- `block/`: This directory contains codes for block storage devices, actually the scheduling algorithm.
- `crypto/`: This directory contains the cryptographic API and the encryption algorithms code.
- `Documentation/`: This should be your favorite directory. It contains the descriptions of APIs used for different kernel frameworks and subsystems. You should look here prior to asking any questions on forums.
- `drivers/`: This is the heaviest directory, continuously growing as device drivers get merged. It contains every device driver organized in various subdirectories.
- `fs/`: This directory contains the implementation of different filesystems that the kernel actually supports, such as NTFS, FAT, ETX{2,3,4}, sysfs, procfs, NFS, and so on.
- `include/`: This contains kernel header files.
- `init/`: This directory contains the initialization and start up code.
- `ipc/`: This contains implementation of the **Inter-Process Communication** (**IPC**) mechanisms, such as message queues, semaphores, and shared memory.
- `kernel/`: This directory contains architecture-independent portions of the base kernel.
- `lib/`: Library routines and some helper functions live here. They are: generic **kernel object** (**kobject**) handlers and **Cyclic Redundancy Code** (**CRC**) computation functions, and so on.

- `mm/`: This contains memory management code.
- `net/`: This contains networking (whatever network type it is) protocols code.
- `scripts/`: This contains scripts and tools used during the kernel development. There are other useful tools here.
- `security/`: This directory contains the security framework code.
- `sound/`: Audio subsystems codes fall here.
- `usr/`: This currently contains the initramfs implementation.

The kernel must remain portable. Any architecture-specific code should be located in the `arch` directory. Of course, the kernel code related to user space API does not change (system calls, `/proc`, `/sys`), as it would break the existing programs.

The book deals with version 4.1 of the kernel. Therefore, any changes made until v4.11 version are covered too, at least this can be said about the frameworks and subsystems.

Kernel configuration

The Linux kernel is a makefile-based project, with 1000s of options and drivers. To configure your kernel, either use `make menuconfig` for an ncurse-based interface or `make xconfig` for an X-based interface. Once chosen, options will be stored in a `.config` file, at the root of the source tree.

In most of the cases, there will be no need to start a configuration from scratch. There are default and useful configuration files available in each `arch` directory, which you can use as a start point:

```
ls arch/<you_arch>/configs/
```

For ARM-based CPUs, these configs files are located in `arch/arm/configs/`, and for an i.MX6 processor, the default file config is `arch/arm/configs/imx_v6_v7_defconfig`. Similarly for x86 processors, we find the files in `arch/x86/configs/`, with only two default configuration files, `i386_defconfig` and `x86_64_defconfig`, for 32 and 64 bits versions respectively. It is quite straightforward for an x86 system:

```
make x86_64_defconfig
make zImage -j16
make modules
makeINSTALL_MOD_PATH </where/to/install> modules_install
```

Given an i.MX6-based board, one can start with `ARCH=arm make imx_v6_v7_defconfig`, and then `ARCH=arm make menuconfig`. With the former command, you will store the default option in `.config` file, and with the latter, you can update add/remove options, depending on the needs.

One may run into a Qt4 error with `xconfig`. In such a case, one should just use the following command:

```
sudo apt-get install   qt4-dev-tools qt4-qmake
```

Build your kernel

Building the kernel requires you to specify the architecture for which it is built for, as well as the compiler. That says, it is not necessary for a native build.

```
ARCH=arm make imx_v6_v7_defconfig
ARCH=arm CROSS_COMPILE=arm-linux-gnueabihf- make zImage -j16
```

After that, one will see something like:

```
    [...]
    LZO     arch/arm/boot/compressed/piggy_data
    CC      arch/arm/boot/compressed/misc.o
    CC      arch/arm/boot/compressed/decompress.o
    CC      arch/arm/boot/compressed/string.o
    SHIPPED arch/arm/boot/compressed/hyp-stub.S
    SHIPPED arch/arm/boot/compressed/lib1funcs.S
    SHIPPED arch/arm/boot/compressed/ashldi3.S
    SHIPPED arch/arm/boot/compressed/bswapsdi2.S
    AS      arch/arm/boot/compressed/hyp-stub.o
    AS      arch/arm/boot/compressed/lib1funcs.o
    AS      arch/arm/boot/compressed/ashldi3.o
    AS      arch/arm/boot/compressed/bswapsdi2.o
    AS      arch/arm/boot/compressed/piggy.o
    LD      arch/arm/boot/compressed/vmlinux
    OBJCOPY arch/arm/boot/zImage
    Kernel: arch/arm/boot/zImage is ready
```

From the kernel build, the result will be a single binary image, located in `arch/arm/boot/`. Modules are built with the following command:

```
ARCH=arm CROSS_COMPILE=arm-linux-gnueabihf- make modules
```

You can install them using the following command:

```
ARCH=arm CROSS_COMPILE=arm-linux-gnueabihf- make modules_install
```

Introduction to Kernel Development

The `modules_install` target expects an environment variable, `INSTALL_MOD_PATH`, which specifies where you should install the modules. If not set, the modules will be installed at `/lib/modules/$(KERNELRELEASE)/kernel/`. This is discussed in `Chapter 2`, *Device Driver Basis*.

i.MX6 processors support device trees, which are files you use to describe the hardware (this is discussed in detail in `Chapter 6`, *The Concept of Device Tree*). However, to compile every `ARCH` device tree, you can run the following command:

```
ARCH=arm CROSS_COMPILE=arm-linux-gnueabihf- make dtbs
```

However, the `dtbs` option is not available on all platforms that support device tree. To build a standalone DTB, you should use:

```
ARCH=arm CROSS_COMPILE=arm-linux-gnueabihf- make imx6d-    sabrelite.dtb
```

Kernel habits

The kernel code tries to follow standard rules through its evolution. In this chapter, we will just be introduced to them. They are all discussed in a dedicated chapter, starting from `Chapter 3`, *Kernel Facilities and Helper Functions*, we get a better overview of the kernel development process and tips, till `Chapter 13`, *Linux Device Model*.

Coding style

Before going deep in this section, you should always refer to the kernel coding style manual, at `Documentation/CodingStyle` in the kernel source tree. This coding style is a set of rules you should respect, at least if you need to get its patches accepted by kernel developers. Some of these rules concern indentation, program flow, naming convention, and so on.

Most popular ones are:

- Always use tab indentation of 8 characters, and the line should be 80 columns long. If the indentation prevents you from writing your function, it is because this one has too many nesting levels. One can size the tabs and verify the line size using `scripts/cleanfile` script in from the kernel source:

    ```
    scripts/cleanfile my_module.c
    ```

- You can also indent the code correctly using the `indent` tool:

  ```
  sudo apt-get install indent
  scripts/Lindent my_module.c
  ```

- Every function/variable that is not exported should be declared as static.
- No spaces should be added around (inside) parenthesized expressions. *s = size of (struct file);* is accepted, whereas *s = size of(struct file);* is not.
- Using `typdefs` is forbidden.
- Always use `/* this */` comment style, not `// this`
    - BAD: `// do not use this please`
    - GOOD: `/* Kernel developers like this */`
- You should capitalise macros, but functional macros can be in lowercase.
- A comment should not replace a code that is not illegible. Prefer rewriting the code rather than adding a comment.

Kernel structures allocation/initialization

The kernel always offers two possible allocation mechanisms for its data structures and facilities.

Some of these structures are:

- Workqueue
- List
- Waitqueue
- Tasklet
- Timer
- Completion
- mutex
- spinlock

Dynamical initializers are all macros it means they are always capitalized: `INIT_LIST_HEAD()`, `DECLARE_WAIT_QUEUE_HEAD()`, `DECLARE_TASKLET( )`, and so on.

Introduction to Kernel Development

That being said, these are all discussed in `Chapter 3`, *Kernel Facilities and Helper Functions*. Therefore, data structures that represent framework devices are always allocated dynamically, each of which having its own allocation and deallocation API. These framework device types are:

- Network
- Input device
- Char device
- IIO device
- Class
- Framebuffer
- Regulator
- PWM device
- RTC

Scope of the static objects is visible in the whole driver, and by every device this driver manages. Dynamically allocated objects are visible only by the device that is actually using a given instance of the module.

Classes, objects, and OOP

The kernel implements OOP by means of a device and a class. Kernel subsystems are abstracted by means of classes. There are almost as many subsystems as there are directories under `/sys/class/`. The `struct kobject` structure is the center piece of this implementation. It even brings in a reference counter, so that the kernel may know how many users actually use the object. Every object has a parent, and has an entry in `sysfs` (if mounted).

Every device that falls into a given subsystem has a pointer to an **operations** (**ops**) structure, which exposes operations that can be executed on this device.

Summary

This chapter explained in a very short and simple manner how you should download the Linux source and process a first build. It also deals with some common concepts. That said, this chapter is quite brief and may not be enough, but never mind, it is just an introduction. That is why the next chapter gets more into the details of the kernel building process, how to actually compile a driver, either externally or as a part of the kernel, as well as some basics that one should learn before starting the long journey that kernel development represents.

2
Device Driver Basis

A driver is a piece of software whose aim is to control and manage a particular hardware device; hence the name device driver. From an operating system point of view, it can be either in the kernel space (running in privileged mode) or in the user space (with lower privilege). This book only deals with kernel space drivers, especially Linux kernel drivers. Our definition is a device driver exposes the functionality of the hardware to user programs.

This book's aim is not to teach you how to become a Linux guru—I'm not even one at all—but there are some concepts you should understand prior to writing a device driver. C programming skills are mandatory; you should be at least familiar with pointers. You should also be familiar with some of the manipulating functions. Some hardware skills are required too. So this chapter essentially discusses:

- Module building processes, as well as their loading and unloading
- Driver skeletons, and debugging message management
- Error handling in the driver

User space and kernel space

The concept of kernel space and user space is a bit abstract. It is all about memory and access rights. One may consider the kernel to be privileged, whereas the user apps are restricted. It is a feature of a modern CPU, allowing it to operate either in privileged or unprivileged mode. This concept will be clearer to you in `Chapter 11`, *Kernel Memory Management*.

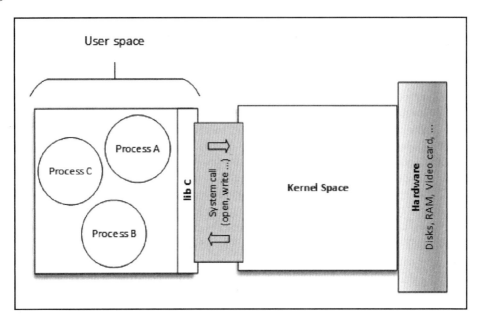

User space and kernel space

The preceding figure introduces the separation between kernel and user space, and highlights the fact that system calls represent the bridge between them (we discuss this later in this chapter). One can describe each space as following:

- **Kernel space:** This is a set of addresses where the kernel is hosted and where it runs. Kernel memory (or kernel space) is a memory range, owned by the kernel, protected by access flags, preventing any user apps from messing with the kernel (un)knowingly. On the other hand the kernel can access the whole system memory, since it runs with the higher priority on the system. In kernel mode, the CPU can access the whole memory (both kernel space and user space).

- **User space:** This is a set of addresses (locations) where normal programs (such as gedit and so on) are restricted to run. You may consider it as a sand-box or a jail, so that a user program can't mess with memory or any other resource owned by another program. In user mode, the CPU can only access memory tagged with user-space access rights. The only way for the user app to run into the kernel space is through system calls. Some of these are read, write, open, close, mmap, and so on. User-space code runs with lower priority. When a process performs a system call, a software interrupt is sent to the kernel, which turns on privileged mode so that the process can run in kernel space. When the system call returns, the kernel turns off the privileged mode and the process is jailed again.

The concept of modules

A module is to the Linux kernel what a plugin (add-on) is to user software (Firefox is an example). It dynamically extends the kernel functionalities without even the need to restart the computer. Most of the time, kernel modules are plug and play. Once inserted, they are ready to be used. In order to support modules, the kernel must have been built with the following option enabled:

```
CONFIG_MODULES=y
```

Module dependencies

In Linux, a module can provide functions or variables, exporting them using the EXPORT_SYMBOL macro, which makes them available for other modules. These are called symbols. A dependency of module B on module A is that module B is using one of the symbols exported by module A.

depmod utility

depmod is a tool that you run during the kernel build process to generate module dependency files. It does that by reading each module in /lib/modules/<kernel_release>/ to determine what symbols it should export and what symbols it needs. The result of that process is written to the file modules.dep, and its binary version modules.dep.bin. It is a kind of module indexing.

Device Driver Basis

Module loading and unloading

For a module to be operational, one should load it into the kernel, either by using `insmod` given the module path as argument, which is the preferred method during development, or by using `modprobe`, a clever command but that one prefered in production systems.

Manual loading

Manual loading needs the intervention of a user, which should have root access. The two classical methods to achieve this are described as follows:

modprobe and insmod

During development, one usually uses `insmod` in order to load a module and it should be given the path of the module to load:

```
insmod /path/to/mydrv.ko
```

It is low-level form of module loading, which forms the base of other module loading methods, and the one we will use in this book. On the other hand, there is `modprobe`, mostly used by sysadmin or in a production system. `modprobe` is a clever command that parses the file `modules.dep` in order to load dependencies first, prior to loading the given module. It automatically handles module dependencies, as a package manager does:

```
modprobe mydrv
```

Whether one can use `modprobe` or not depends on `depmod` being aware of module installation.

/etc/modules-load.d/<filename>.conf

If you want some module to be loaded at boot time, just create the file `/etc/modules-load.d/<filename>.conf`, and add the module's name that should be loaded, one per line. `<filename>` should be meaningful to you, and people usually use module: `/etc/modules-load.d/modules.conf`. You may create as many `.conf` files as you need:

An example of `/etc/modules-load.d/mymodules.conf` is as follows:

```
#this line is a comment
uio
iwlwifi
```

Auto-loading

The `depmod` utility doesn't only build `modules.dep` and `modules.dep.bin` files. It does more than that. When a kernel developer actually writes a driver, they know exactly what hardware the driver will support. They are then responsible for feeding the driver with the product and vendor IDs of all devices supported by the driver. `depmod` also processes module files in order to extract and gather that information, and generates a `modules.alias` file, located in `/lib/modules/<kernel_release>/modules.alias`, which will map devices to their drivers:

An excerpt of `modules.alias` is as follows:

```
alias usb:v0403pFF1Cd*dc*dsc*dp*ic*isc*ip*in* ftdi_sio
alias usb:v0403pFF18d*dc*dsc*dp*ic*isc*ip*in* ftdi_sio
alias usb:v0403pDAFFd*dc*dsc*dp*ic*isc*ip*in* ftdi_sio
alias usb:v0403pDAFEd*dc*dsc*dp*ic*isc*ip*in* ftdi_sio
alias usb:v0403pDAFDd*dc*dsc*dp*ic*isc*ip*in* ftdi_sio
alias usb:v0403pDAFCd*dc*dsc*dp*ic*isc*ip*in* ftdi_sio
alias usb:v0D8Cp0103d*dc*dsc*dp*ic*isc*ip*in* snd_usb_audio
alias usb:v*p*d*dc*dsc*dp*ic01isc03ip*in* snd_usb_audio
alias usb:v200Cp100Bd*dc*dsc*dp*ic*isc*ip*in* snd_usb_au
```

At this step, you'll need a user-space hot-plug agent (or device manager), usually `udev` (or `mdev`), that will register with the kernel in order to get notified when a new device appears.

The notification is done by the kernel, sending the device's description (pid, vid, class, device class, device subclass, interface, and all other information that may identify a device) to the hot-plug daemon, which in turn calls `modprobe` with this information. `modprobe` then parses the `modules.alias` file in order to match the driver associated with the device. Before loading the module, `modprobe` will look for its dependencies in `module.dep`. If it finds any, the dependencies will be loaded prior to the associated module loading; otherwise, the module is loaded directly.

Module unload

The usual command to unload a module is `rmmod`. One should prefer using this to unload a module loaded with `insmod` command. The command should be given the module name to unload as a parameter. Module unloading is a kernel feature that one can enable or disable, according to the value of the `CONFIG_MODULE_UNLOAD` config option. Without this option, one will not be able to unload any module. Let us enable module unloading support:

```
CONFIG_MODULE_UNLOAD=y
```

At runtime, the kernel will prevent from unloading modules that may break things, even if one asks it to do so. This is because the kernel keeps a reference count on module usage, so that it knows whether a module is actually in use or not. If the kernel believes it is unsafe to remove a module, it will not. Obviously, one can change this behavior:

```
MODULE_FORCE_UNLOAD=y
```

The preceding option should be set in the kernel config in order to force module unload:

```
rmmod -f mymodule
```

On the other hand, a higher level command to unload a module in a smart manner is `modeprobe -r`, which automatically unloads unused dependencies:

```
modeprobe -r mymodule
```

As you may have guessed, it is a really helpful option for developers. Finally, one can check whether a module is loaded or not with the following command:

```
lsmod
```

Driver skeletons

Let's consider the following `helloworld` module. It will be the basis for our work during the rest of this chapter:

helloworld.c

```
#include <linux/init.h>
#include <linux/module.h>
#include <linux/kernel.h>

static int __init helloworld_init(void) {
    pr_info("Hello world!\n");
    return 0;
}

static void __exit helloworld_exit(void) {
    pr_info("End of the world\n");
}

module_init(helloworld_init);
module_exit(helloworld_exit);
MODULE_AUTHOR("John Madieu <john.madieu@gmail.com>");
MODULE_LICENSE("GPL");
```

Module entry and exit point

Kernel drivers all have entry and exit points: the former corresponds to the function called when the module is loaded (`modprobe`, `insmod`) and the latter is the function executed at module unloading (at `rmmod or modprobe -r`).

We all remember the `main()` function, which is the entry point for every user-space program written in C/C++ that exits when that same function returns. With kernel modules, things are different. The entry point can have any name you want, and unlike a user-space program that exits when `main()` returns, the exit point is defined in another function. All you need to do is to inform the kernel which functions should be executed as an entry or exit point. The actual functions `hellowolrd_init` and `hellowolrd_exit` could be given any name. The only thing that is actually mandatory is to identify them as the corresponding loading and removing functions, giving them as parameters to the `module_init()` and `module_exit()` macros.

To sum up, `module_init()` is used to declare the function that should be called when the module is loaded (with `insmod` or `modprobe`). What is done in the initialization function will define the behavior of the module. `module_exit()` is used to declare the function that should be called when the module is unloaded (with `rmmod`).

Either the `init` function or the `exit` function is run once, right after the module is loaded or unloaded.

__init and __exit attributes

`__init` and `__exit` are actually kernel macros, defined in `include/linux/init.h`, shown as follows:

```
#define __init      __section(.init.text)
#define __exit      __section(.exit.text)
```

The `__init` keyword tells the linker to place the code in a dedicated section into the kernel object file. This section is known in advance to the kernel, and freed when the module is loaded and the init function finished. This applies only to built-in drivers, not to loadable modules. The kernel will run the init function of the driver for the first time during its boot sequence.

Device Driver Basis

Since the driver cannot be unloaded, its init function will not be called again until the next reboot. There is no need to keep references on its init function anymore. The same for the `__exit` keyword, whose corresponding code is omitted when the module is compiled statically into the kernel, or when module unloading support is not enabled, because in both cases, the exit function is never called. `__exit` has no effect on loadable modules.

Let's spend more time understanding how such attributes work. It is all about object files called **Executable and Linkable Format** (**ELF**). An ELF object file is made of various named sections. Some of these are mandatory and form the basis of the ELF standard, but one can make up any section one wants and have it used by special programs. This is what the kernel does. One can run `objdump -h module.ko` in order to print out different sections that constitute the given `module.ko` kernel module:

```
jma@jma:~/work/tutos/sources/helloworld$ objdump -h helloworld-params.ko

helloworld-params.ko:     file format elf64-x86-64

Sections:
Idx Name          Size      VMA               LMA               File off  Algn
  0 .note.gnu.build-id 00000024  0000000000000000  0000000000000000  00000040  2**2
                  CONTENTS, ALLOC, LOAD, READONLY, DATA
  1 .text         00000017  0000000000000000  0000000000000000  00000070  2**4
                  CONTENTS, ALLOC, LOAD, RELOC, READONLY, CODE
  2 .init.text    00000014  0000000000000000  0000000000000000  00000087  2**0
                  CONTENTS, ALLOC, LOAD, RELOC, READONLY, CODE
  3 .text.unlikely 00000000  0000000000000000  0000000000000000  0000009b  2**0
                  CONTENTS, ALLOC, LOAD, READONLY, CODE
  4 .rodata.str1.1 0000002a  0000000000000000  0000000000000000  0000009b  2**0
                  CONTENTS, ALLOC, LOAD, READONLY, DATA
  5 .modinfo      00000157  0000000000000000  0000000000000000  000000c8  2**3
                  CONTENTS, ALLOC, LOAD, READONLY, DATA
  6 __param       00000078  0000000000000000  0000000000000000  00000220  2**3
                  CONTENTS, ALLOC, LOAD, RELOC, READONLY, DATA
  7 .rodata       0000004c  0000000000000000  0000000000000000  000002a0  2**5
                  CONTENTS, ALLOC, LOAD, RELOC, READONLY, DATA
  8 __mcount_loc  00000008  0000000000000000  0000000000000000  000002f0  2**3
                  CONTENTS, ALLOC, LOAD, RELOC, READONLY, DATA
  9 __versions    00000180  0000000000000000  0000000000000000  00000300  2**5
                  CONTENTS, ALLOC, LOAD, READONLY, DATA
 10 .data         00000018  0000000000000000  0000000000000000  00000480  2**3
                  CONTENTS, ALLOC, LOAD, RELOC, DATA
 11 .gnu.linkonce.this_module 00000380  0000000000000000  0000000000000000  000004c0  2**6
                  CONTENTS, ALLOC, LOAD, RELOC, DATA, LINK_ONCE_DISCARD
 12 .bss          00000000  0000000000000000  0000000000000000  00000840  2**0
                  ALLOC
 13 .comment      0000006a  0000000000000000  0000000000000000  00000840  2**0
                  CONTENTS, READONLY
 14 .note.GNU-stack 00000000  0000000000000000  0000000000000000  000008aa  2**0
                  CONTENTS, READONLY
```

List of sections of helloworld-params.ko module

Only a few of the sections in the caption are standerd ELF sections:

- `.text`, also called code, which contains program code
- `.data`, which contains initialized data, and is also called data segment
- `.rodata`, for read-only data
- `.comment`
- Uninitialized data segment, also called **block started by symbol (bss)**

Other sections are added on demand for the kernel purpose. The most important for this chapter are **.modeinfo** sections, which store information about the modules, and **.init.text** sections, which store code prefixed with the __init macro.

The linker (`ld` on Linux systems), which is a part of binutils, is responsible for the placement of symbols (data, code, and so on) in the appropriate section in the generated binary in order to be processed by the loader when the program is executed. One may customize these sections, change their default location, or even add additional sections by providing a linker script, called a **linker definition file (LDF)** or **linker definition script (LDS)**. Now all you have to do is to inform the linker of the symbol placement through compiler directives. The GNU C compiler provides attributes for that purpose. In the case of the Linux kernel, there is a custom LDS file provided, located in `arch/<arch>/kernel/vmlinux.lds.S`. __init and __exit are then used to mark symbols to be placed onto dedicated sections mapped in kernel's LDS files.

In conclusion, __init and __exit are Linux directives (actually macros), which wrap the C compiler attribute used for symbol placement. They instruct the compiler to put the code they prefix respectively in `.init.text` and `.exit.text` sections, even though the kernel can access different object sections.

Module information

Even without having to read its code, one should be able to gather some information (for example, the author(s), parameter(s) description, the license) about a given module. A kernel module uses its `.modinfo` section to store informations about the module. Any MODULE_* macro will update the content of that section with the values passed as parameters. Some of these macros are MODULE_DESCRIPTION(), MODULE_AUTHOR(), and MODULE_LICENSE(). The real underlying macro provided by the kernel to add an entry in the module info section is MODULE_INFO(tag, info), which adds generic info of form tag = info. This means a driver author could add any free form info they want, such as:

```
MODULE_INFO(my_field_name, "What eeasy value");
```

Device Driver Basis

One can dump the content of the `.modeinfo` section of a kernel module using the `objdump -d -j .modinfo` command on the given module:

```
jma@jma:~/work/tutos/sources/helloworld$ objdump helloworld-params.ko -d  -j  .modinfo

helloworld-params.ko:     file format elf64-x86-64


Disassembly of section .modinfo:

0000000000000000 <__UNIQUE_ID_license7>:
   0:   6c 69 63 65 6e 73 65 3d 47 50 4c 00                  license=GPL.

000000000000000c <__UNIQUE_ID_author6>:
   c:   61 75 74 68 6f 72 3d 4a 6f 68 6e 20 4d 61 64 69      author=John Madi
  1c:   65 75 20 3c 6a 6f 68 6e 2e 6d 61 64 69 65 75 40      eu <john.madieu@
  2c:   66 6f 6f 62 61 72 2e 63 6f 6d 3e 00                  foobar.com>.

0000000000000038 <__UNIQUE_ID_myarr5>:
  38:   70 61 72 6d 3d 6d 79 61 72 72 3a 74 68 69 73 20      parm=myarr:this
  48:   69 73 20 6d 79 20 61 72 72 61 79 20 6f 66 20 69      is my array of i
  58:   6e 74 00                                             nt.

000000000000005b <__UNIQUE_ID_mystr4>:
  5b:   70 61 72 6d 3d 6d 79 73 74 72 3a 74 68 69 73 20      parm=mystr:this
  6b:   69 73 20 6d 79 20 63 68 61 72 20 70 6f 69 6e 74      is my char point
  7b:   65 72 20 76 61 72 69 61 62 6c 65 00                  er variable.

0000000000000087 <__UNIQUE_ID_myint3>:
  87:   70 61 72 6d 3d 6d 79 69 6e 74 3a 74 68 69 73 20      parm=myint:this
  97:   69 73 20 6d 79 20 69 6e 74 20 76 61 72 69 61 62      is my int variab
  a7:   6c 65 00                                             le.

00000000000000aa <__UNIQUE_ID_myarrtype2>:
  aa:   70 61 72 6d 74 79 70 65 3d 6d 79 61 72 72 3a 61      parmtype=myarr:a
  ba:   72 72 61 79 20 6f 66 20 69 6e 74 00                  rray of int.

00000000000000c6 <__UNIQUE_ID_mystrtype1>:
  c6:   70 61 72 6d 74 79 70 65 3d 6d 79 73 74 72 3a 63      parmtype=mystr:c
  d6:   68 61 72 70 00                                       harp.

00000000000000db <__UNIQUE_ID_myinttype0>:
  db:   70 61 72 6d 74 79 70 65 3d 6d 79 69 6e 74 3a 69      parmtype=myint:i
  eb:   6e 74 00 00 00                                       nt...

00000000000000f0 <__UNIQUE_ID_srcversion1>:
  f0:   73 72 63 76 65 72 73 69 6f 6e 3d 42 42 46 34 33      srcversion=BBF43
 100:   45 30 39 38 45 41 42 35 44 32 45 32 44 44 37 38      E098EAB5D2E2DD78
 110:   43 30 00 00 00 00 00 00                              C0......

0000000000000118 <__module_depends>:
 118:   64 65 70 65 6e 64 73 3d 00                           depends=.
```

Content of .modeinfo section of helloworld-params.ko module

The modinfo section can be seen as the data sheet of the module. The user-space tool that actually prints information in a stylized manner is `modinfo`:

```
jma@jma:~/work/tutos/sources/helloworld$ modinfo ./helloworld-params.ko
filename:       /home/jma/work/tutos/sources/helloworld/./helloworld-params.ko
license:        GPL
author:         John Madieu <john.madieu@foobar.com>
my_field_name:  What eeasy value
srcversion:     47B038B61944D8CD2E680DB
depends:
vermagic:       4.4.0-93-generic SMP mod_unload modversions
parm:           myint:this is my int variable (int)
parm:           mystr:this is my char pointer variable (charp)
parm:           myarr:this is my array of int (array of int)
```

modinfo output

Apart from the custom info one defines, there is standard info one should provide, and that the kernel provides macros for; these are license, module author, parameter description, module version, and module description.

Licensing

The license is defined in a given module by the `MODULE_LICENSE()` macro:

```
MODULE_LICENSE ("GPL");
```

The license will define how your source code should be shared (or not) with other developers. `MODULE_LICENSE()` tells the kernel what license our module is under. It has an effect on your module behavior, since a non GPL-compatible license will result in your module not being able to see/use services/functions exported by the kernel through the `EXPORT_SYMBOL_GPL()` macro, which shows the symbols to GPL-compatible modules only, which is the opposite of `EXPORT_SYMBOL()`, which exports functions for modules with any license. Loading a non GPL-compatible will also result in a tainted kernel; that means a non-open source or untrusted code has been loaded, and you will likely have no support from the community. Remember that the module without `MODULE_LICENSE()` is not considered open source and will taint the kernel too. The following is an excerpt of `include/linux/module.h`, describing the license supported by the kernel:

```
/*
 * The following license idents are currently accepted as indicating free
 * software modules
 *
 * "GPL"                        [GNU Public License v2 or later]
```

Device Driver Basis

```
 *   "GPL v2"                      [GNU Public License v2]
 *   "GPL and additional rights"   [GNU Public License v2 rights and more]
 *   "Dual BSD/GPL"                [GNU Public License v2
 *                                 or BSD license choice]
 *   "Dual MIT/GPL"                [GNU Public License v2
 *                                 or MIT license choice]
 *   "Dual MPL/GPL"                [GNU Public License v2
 *                                 or Mozilla license choice]
 *
 * The following other idents are available
 *
 *   "Proprietary"                 [Non free products]
 *
 * There are dual licensed components, but when running with Linux it is the
 * GPL that is relevant so this is a non issue. Similarly LGPL linked with GPL
 * is a GPL combined work.
 *
 * This exists for several reasons
 * 1.   So modinfo can show license info for users wanting to vet their setup
 *      is free
 * 2.   So the community can ignore bug reports including proprietary modules
 * 3.   So vendors can do likewise based on their own policies
 */
```

 It is mandatory for your module to be at least GPL-compatible in order for you to enjoy full kernel services.

Module author(s)

MODULE_AUTHOR() declares the module's author(s):

```
MODULE_AUTHOR("John Madieu <john.madieu@gmail.com>");
```

It is possible to have more than one author. In this case, each author must be declared with MODULE_AUTHOR():

```
MODULE_AUTHOR("John Madieu <john.madieu@gmail.com>");
MODULE_AUTHOR("Lorem Ipsum <l.ipsum@foobar.com>");
```

Module description

`MODULE_DESCRIPTION()` briefly describes what the module does:

```
MODULE_DESCRIPTION("Hello, world! Module");
```

Errors and message printing

Error codes are interpreted either by the kernel or by the user-space application (through the `errno` variable). Error handling is very important in software development, more than it is in kernel development. Fortunately, the kernel provides a couple of errors that cover almost every error you'll encounter, and sometimes you will need to print them out in order to help you debug.

Error handling

Return the wrong error code for a given error and it will result in either the kernel or user-space app producing unneeded behavior and making a wrong decision. To keep things clear, there are predefined errors in the kernel tree that cover almost every case you may face. Some of the errors (with their meaning) are defined in `include/uapi/asm-generic/errno-base.h`, and the rest of the list can be found in `include/uapi/asm-generic/errno.h`. The following is an excerpt of list of errors, from `include/uapi/asm-generic/errno-base.h`:

```
#define EPERM        1   /* Operation not permitted */
#define ENOENT       2    /* No such file or directory */
#define ESRCH        3   /* No such process */
#define EINTR        4   /* Interrupted system call */
#define EIO          5   /* I/O error */
#define ENXIO        6   /* No such device or address */
#define E2BIG        7   /* Argument list too long */
#define ENOEXEC      8    /* Exec format error */
#define EBADF        9   /* Bad file number */
#define ECHILD       10  /* No child processes */
#define EAGAIN       11  /* Try again */
#define ENOMEM       12  /* Out of memory */
#define EACCES       13  /* Permission denied */
#define EFAULT       14  /* Bad address */
#define ENOTBLK      15  /* Block device required */
#define EBUSY        16  /* Device or resource busy */
#define EEXIST       17   /* File exists */
#define EXDEV        18  /* Cross-device link */
```

Device Driver Basis

```
#define  ENODEV          19      /* No such device */
#define  ENOTDIR         20      /* Not a directory */
#define  EISDIR          21      /* Is a directory */
#define  EINVAL          22      /* Invalid argument */
#define  ENFILE          23      /* File table overflow */
#define  EMFILE          24      /* Too many open files */
#define  ENOTTY          25      /* Not a typewriter */
#define  ETXTBSY         26      /* Text file busy */
#define  EFBIG      27   /* File too large */
#define  ENOSPC          28      /* No space left on device */
#define  ESPIPE          29      /* Illegal seek */
#define  EROFS      30   /* Read-only file system */
#define  EMLINK          31      /* Too many links */
#define  EPIPE      32   /* Broken pipe */
#define  EDOM       33   /* Math argument out of domain of func */
#define  ERANGE          34      /* Math result not representable */
```

Most of time, the classical way to return an error is to do so in the form `return -ERROR`, especially when it comes to answering to system calls. For example, for an I/O error, the error code is `EIO` and one should `return -EIO`:

```
dev = init(&ptr);
if(!dev)
return -EIO
```

Errors sometimes cross the kernel space and propagate themselves to the user space. If the returned error is an answer to a system call (`open`, `read`, `ioctl`, `mmap`), the value will be automatically assigned to the user-space `errno` global variable, on which one can use `strerror(errno)` to translate the error into a readable string:

```
#include <errno.h>   /* to access errno global variable */
#include <string.h>
[...]
if(wite(fd, buf, 1) < 0) {
    printf("something gone wrong! %s\n", strerror(errno));
}
[...]
```

When you face an error, you must undo everything that has been set until the error occurs. The usual way to do this is to use the `goto` statement:

```
ptr = kmalloc(sizeof (device_t));
if(!ptr) {
        ret = -ENOMEM
        goto err_alloc;
}
dev = init(&ptr);
```

```
if(dev) {
        ret = -EIO
        goto err_init;
}
return 0;

err_init:
        free(ptr);
err_alloc:
        return ret;
```

The reason why one uses the `goto` statement is simple. When it comes to handling error, let's say at step 5, one has to clean previous operations (steps 4, 3, 2, 1). Instead of doing lot of nested checking operation shown as follows:

```
if (ops1() != ERR) {
    if (ops2() != ERR) {
        if ( ops3() != ERR) {
            if (ops4() != ERR) {
```

This may be confusing, and may lead to indentation issues. One prefers using the `goto` in order to have a straight control flow, shown as follows:

```
if (ops1() == ERR) // |
    goto error1;   // |
if (ops2() == ERR) // |
    goto error2;   // |
if (ops3() == ERR) // |
    goto error3;   // |
if (ops4() == ERR) // V
    goto error4;
error5:
[...]
error4:
[...]
error3:
[...]
error2:
[...]
error1:
[...]
```

This means, one should only use goto to move forward in a function.

Handling null pointer errors

When it comes to returning an error from functions that are supposed to return a pointer, functions often return the NULL pointer. It is a working but quite meaningless approach, since one does not exactly know why this null pointer is returned. For that purpose, the kernel provides three functions, ERR_PTR, IS_ERR, and PTR_ERR:

```
void *ERR_PTR(long error);
long IS_ERR(const void *ptr);
long PTR_ERR(const void *ptr);
```

The first actually returns the error value as a pointer. Given a function that is likely to return -ENOMEM after a failed memory allocation, we have to do something like return ERR_PTR(-ENOMEM);. The second is used to check whether the returned value is a pointer error or not, if (IS_ERR(foo)). The last returns the actual error code return PTR_ERR(foo);. The following is an example:

How to use ERR_PTR, IS_ERR, and PTR_ERR:

```
static struct iio_dev *indiodev_setup(){
    [...]
    struct iio_dev *indio_dev;
    indio_dev = devm_iio_device_alloc(&data->client->dev, sizeof(data));
    if (!indio_dev)
        return ERR_PTR(-ENOMEM);
    [...]
    return indio_dev;
}

static int foo_probe([...]){
    [...]
    struct iio_dev *my_indio_dev = indiodev_setup();
    if (IS_ERR(my_indio_dev))
        return PTR_ERR(data->acc_indio_dev);
    [...]
}
```

This is a plus on error handling, which is also an excerpt of the kernel coding style that says: If the name of a function is an action or an imperative command, the function should return an error-code integer. If the name is a predicate, the function should return a succeeded Boolean. For example, add work is a command, and the add_work() function returns 0 for success or -EBUSY for failure. In the same way, PCI device present is a predicate, and the pci_dev_present() function returns 1 if it succeeds in finding a matching device or 0 if it doesn't.

Message printing – printk()

The printk() is to the kernel what printf() is to the user-space. Lines written by printk() can be displayed through the dmesg command. Depending on how important the message you need to print is, you can choose between eight log-level messages, defined in include/linux/kern_levels.h, along with their meaning:

The following is the list of kernel log levels. Each of these levels correspond to a number in a string, whose priority is inverted proportional to the value of the number. For example, 0 is higher priority:

```
#define KERN_SOH        "\001"          /* ASCII Start Of Header */
#define KERN_SOH_ASCII      '\001'

#define KERN_EMERG   KERN_SOH "0"       /* system is unusable */
#define KERN_ALERT   KERN_SOH "1"       /* action must be taken immediately */
#define KERN_CRIT    KERN_SOH "2"       /* critical conditions */
#define KERN_ERR     KERN_SOH "3"       /* error conditions */
#define KERN_WARNING KERN_SOH "4"       /* warning conditions */
#define KERN_NOTICE  KERN_SOH "5"       /* normal but significant condition */
#define KERN_INFO    KERN_SOH "6"       /* informational */
#define KERN_DEBUG   KERN_SOH "7"       /* debug-level messages */
```

The following code shows how one can print a kernel message along with a log level:

```
printk(KERN_ERR "This is an error\n");
```

If you omit the debug level (`printk("This is an error\n")`), the kernel will provide one to the function, depending on the `CONFIG_DEFAULT_MESSAGE_LOGLEVEL` config option, which is the default kernel log level. One may actually use one of the following, much more meaningful macros, which are wrappers around those defined previously: `pr_emerg`, `pr_alert`, `pr_crit`, `pr_err`, `pr_warning`, `pr_notice`, `pr_info`, and `pr_debug`:

```
pr_err("This is the same error\n");
```

For new drivers, it is recommended to use these wrappers. The reality of `printk()` is that, whenever it is called, the kernel compares the message log level with the current console log level; if the former is higher (lower value) than the latter, the message will be immediately printed to the console. You can check your log level parameters with:

```
cat /proc/sys/kernel/printk
4 4 1 7
```

In this code, the first value is the current log level (4), and the second is the default one, according to the `CONFIG_DEFAULT_MESSAGE_LOGLEVEL` option. Other values are not relevant for the purpose of this chapter, so let us ignore these.

A list of kernel log levels is as follows:

```
/* integer equivalents of KERN_<LEVEL> */
#define LOGLEVEL_SCHED          -2      /* Deferred messages from sched code
                                 * are set to this special level */
#define LOGLEVEL_DEFAULT        -1      /* default (or last) loglevel */
#define LOGLEVEL_EMERG          0       /* system is unusable */
#define LOGLEVEL_ALERT          1       /* action must be taken immediately */
#define LOGLEVEL_CRIT           2       /* critical conditions */
#define LOGLEVEL_ERR            3       /* error conditions */
#define LOGLEVEL_WARNING        4       /* warning conditions */
#define LOGLEVEL_NOTICE         5       /* normal but significant condition */
#define LOGLEVEL_INFO           6       /* informational */
#define LOGLEVEL_DEBUG          7       /* debug-level messages */
```

The current log level can be changed with:

```
# echo <level> > /proc/sys/kernel/printk
```

Device Driver Basis

 `printk()` never blocks and is safe enough to be called even from atomic contexts. It tries to lock the console and print the message. If locking fails, the output will be written into a buffer and the function will return, never blocking. The current console holder will then be notified about new messages and will print them before releasing the console.

The kernel supports other debug methods too, either dynamically or by using `#define DEBUG` on top of the file. People interested in such debugging style can refer to kernel documentation in *Documentation/dynamic-debug-howto.txt* file.

Module parameters

As a user program does, a kernel module can accept arguments from the command line. This allows dynamically changing the behavior of the module according to given parameters, and can help the developer not having to indefinitely change/compile the module during a test/debug session. In order to set this up, one should first declare the variables that will hold the values of command line arguments, and use the `module_param()` macro on each of these. The macro is defined in `include/linux/moduleparam.h` (this should be included in the code too: `#include <linux/moduleparam.h>`) shown as follows:

```
module_param(name, type, perm);
```

This macro contains the following elements:

- `name`: The name of the variable used as the parameter
- `type`: The parameter's type (bool, charp, byte, short, ushort, int, uint, long, ulong), where `charp` stands for char pointer
- `perm`: This represents the `/sys/module/<module>/parameters/<param>` file permissions. Some of them are `S_IWUSR`, `S_IRUSR`, `S_IXUSR`, `S_IRGRP`, `S_WGRP`, and `S_IRUGO`, where:
    - `S_I` is just a prefix
    - `R`: read, `W`: write, `X`: execute
    - `USR`: user, `GRP`: group, `UGO`: user, group, others

[37]

One can eventually use a | (OR operation) to set multiple permissions. If perm is 0, the file parameter in sysfs will not be created. You should use only S_IRUGO read-only parameters, which I highly recommend; by making a | (OR) with other properties, you can obtain fine-grained properties.

When using module parameters, one should use MODULE_PARM_DESC in order to describe each of them. This macro will populate the module info section with each parameter's description. The following is a sample, from the helloworld-params.c source file provided with the code repository of the book:

```c
#include <linux/moduleparam.h>
[...]

static char *mystr = "hello";
static int myint = 1;
static int myarr[3] = {0, 1, 2};

module_param(myint, int, S_IRUGO);
module_param(mystr, charp, S_IRUGO);
module_param_array(myarr, int,NULL, S_IWUSR|S_IRUSR); /*  */

MODULE_PARM_DESC(myint,"this is my int variable");
MODULE_PARM_DESC(mystr,"this is my char pointer variable");
MODULE_PARM_DESC(myarr,"this is my array of int");

static int foo()
{
    pr_info("mystring is a string: %s\n", mystr);
    pr_info("Array elements: %d\t%d\t%d", myarr[0], myarr[1], myarr[2]);
    return myint;
}
```

To load the module and feed our parameter, we do the following:

```
# insmod hellomodule-params.ko mystring="packtpub" myint=15 myArray=1,2,3
```

One could have used `modinfo` prior to loading the module in order to display description of parameters supported by the module:

```
$ modinfo ./helloworld-params.ko
filename: /home/jma/work/tutos/sources/helloworld/./helloworld-params.ko
license: GPL
author: John Madieu <john.madieu@gmail.com>
srcversion: BBF43E098EAB5D2E2DD78C0
depends:
vermagic: 4.4.0-93-generic SMP mod_unload modversions
parm: myint:this is my int variable (int)
parm: mystr:this is my char pointer variable (charp)
parm: myarr:this is my array of int (array of int)
```

Building your first module

There are two places to build a module. It depends on whether you want people to enable the module by themselves or not using the kernel config interface.

The module's makefile

A makefile is a special file used to execute a set of actions, among which the most important is the compilation of programs. There is a dedicated tool to parse makefiles, called `make`. Prior to jumping to the description of the whole make file, let us introduce the `obj-<X>` kbuild variable.

In almost every kernel makefile, one will see at least one instance of an `obj<-X>` variable. This actually corresponds to the `obj-<X>` pattern, where <X> should be either y, m, left blank, or n. This is used by the kernel makefile from the head of the kernel build system in a general manner. These lines define the files to be built, any special compilation options, and any subdirectories to be entered recursively. A simple example is:

```
obj-y += mymodule.o
```

This tells kbuild that there is one object in the current directory named `mymodule.o`. `mymodule.o` will be built from `mymodule.c` or `mymodule.S`. How and if `mymodule.o` will be built or linked depends on the value of `<X>`:

- If `<X>` is set to `m`, the variable `obj-m` is used, and `mymodule.o` will be built as a module.
- If `<X>` is set to `y`, the variable `obj-y` is used, and `mymodule.o` will be built as part of the kernel. One then says foo is a built-in module.
- If `<X>` is set to `n`, the variable `obj-m` is used, and `mymodule.o` will not be built at all.

Therefore, the pattern `obj-$(CONFIG_XXX)` is often used, where `CONFIG_XXX` is a kernel config option, set or not during the kernel configuration process. An example is:

```
obj-$(CONFIG_MYMODULE) += mymodule.o
```

`$(CONFIG_MYMODULE)` evaluates to either `y` or `m` according to its value during the kernel configuration (remember `make menuconfig`). If `CONFIG_MYMODULE` is neither `y` nor `m`, then the file will not be compiled nor linked. `y` means built-in (it stands for yes in the kernel config process), and `m` stands for module. `$(CONFIG_MYMODULE)` pulls the right answer from the normal config process. This is explained in the next section.

The last use case is:

```
obj-<X> += somedir/
```

This means that kbuild should go into the directory named `somedir`; look for any makefile inside and process it in order to decide what objects should be built.

Back to the makefile, the following is the content makefile we will use to build each of the modules introduced in the book:

```
obj-m := helloworld.o

KERNELDIR ?= /lib/modules/$(shell uname -r)/build

all default: modules
install: modules_install

modules modules_install help clean:
$(MAKE) -C $(KERNELDIR) M=$(shell pwd) $@
```

- `obj-m := hellowolrd.o`: `obj-m` lists modules we want to build. For each `<filename>.o`, the build system will look for a `<filename>.c` to build. `obj-m` is used to build a module, whereas `obj-y` will result in a built-in object.
- `KERNELDIR := /lib/modules/$(shell uname -r)/build`: KERNELDIR is the location of the prebuilt kernel source. As we said earlier, we need a prebuilt kernel in order to build any module. If you have built your kernel from the source, one should set this variable with the absolute path of the built source directory. `-C` instructs to make utility to change into the specified directory prior to reading the makefiles or doing anything else.
- `M=$(shell pwd)`: This is relevant to the kernel build system. The kernel Makefile uses this variable to locate the directory of the external module to build. Your .c files should be placed.
- `all default: modules`: This line instructs the `make` utility to execute the `modules` target, whether `all` or `default` targets, which are classical targets when it comes to building user apps. In other words, `make default` or `make all` or simply `make` commands will be translated into `make modules`.
- `modules modules_install help clean::` This line represents the list target valid in this Makefile .
- `$(MAKE) -C $(KERNELDIR ) M=$(shell pwd) $@`: This is the rule to be executed for each target enumerated above. `$@` will be replaced with the name of the target that caused the rule to run. In other words, if one calls make modules, `$@` will be replaced with modules, and the rule will become: `$(MAKE) -C $(KERNELDIR ) M=$(shell pwd) module`.

In the kernel tree

Before you can build your driver in the kernel tree, you should first identify which directory in drivers should host your .c file. Given your file name `mychardev.c`, which contains the source code of your special character driver, it should be placed to the `drivers/char` directory in the kernel source. Every subdirectory in drivers has both `Makefile` and `Kconfig` files.

Device Driver Basis

Add the following content to the `Kconfig` of that directory:

```
config PACKT_MYCDEV
    tristate "Our packtpub special Character driver"
    default m
    help
      Say Y here if you want to support the /dev/mycdev device.
      The /dev/mycdev device is used to access packtpub.
```

In the makefile of that same directory, add:

```
obj-$(CONFIG_PACKT_MYCDEV)      += mychardev.o
```

Be careful when updating the `Makefile`; the `.o` file name must match the exact name of your `.c` file. If your source file is `foobar.c`, you must use `foobar.o` in the `Makefile`. In order to have your driver built as a module, add the following line in your board defconfig in the `arch/arm/configs` directory:

```
CONFIG_PACKT_MYCDEV=m
```

You may also run `make menuconfig` to select it from the UI, and run `make`, to build the kernel, then `make modules` to build modules (including yours). To make the driver be built in, just replace `m` with `y`:

```
CONFIG_PACKT_MYCDEV=m
```

Everything described here is what embedded board manufacturers do in order to provide a **Board Support Package (BSP)** with their board, with a kernel that already contains their custom drivers:

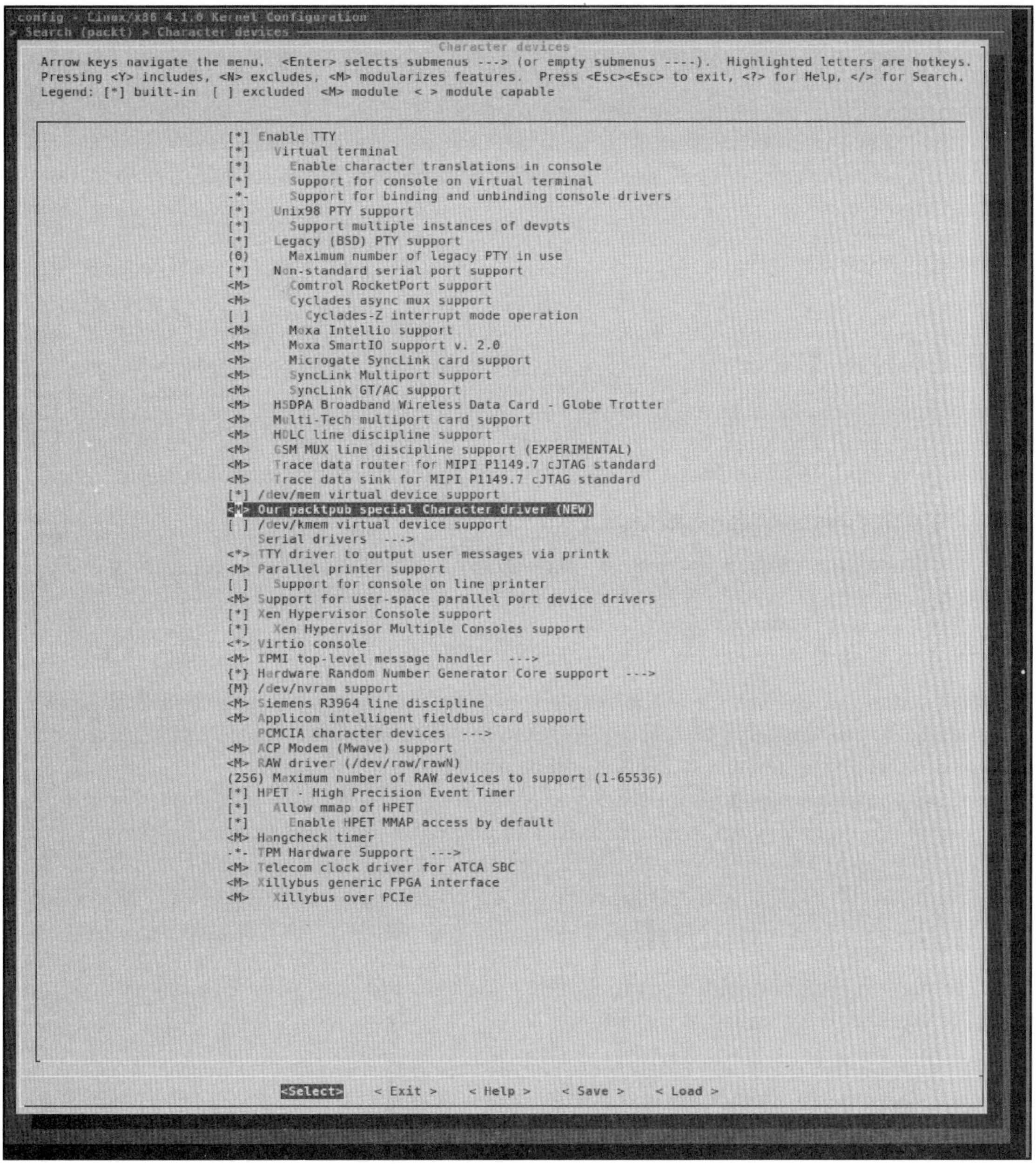

packt_dev module in kernel tree

Once configured, you can build the kernel with `make` and build modules with `make modules`.

Modules included in the kernel source tree are installed in `/lib/modules/$(KERNELRELEASE)/kernel/`. On your Linux system, it is `/lib/modules/$(uname -r)/kernel/`. Run the following command in order to install the modules:

```
make modules_install
```

Out of the tree

Before you can build an external module, you need to have a complete and precompiled kernel source-tree. The kernel source-tree version must be the same as the kernel you'll load and use your module with. There are two ways to obtain a prebuilt kernel version:

- Build it by yourself (discussed this earlier)
- Install the `linux-headers-*` package from your distribution repository

```
sudo apt-get update
sudo apt-get install linux-headers-$(uname -r)
```

This will install only headers, not the whole source tree. Headers will then be installed in `/usr/src/linux-headers-$(uname -r)`. On my computer, it is `/usr/src/linux-headers-4.4.0-79-generic/`. There will be a symlink, `/lib/modules/$(uname -r)/build`, pointing to the previously installed headers. It is the path you should specify as your kernel directory in your `Makefile`. It is all you have to do for a prebuilt kernel.

Building the module

Now, when you are done with your makefile, just change to your source directory and run the `make` command, or `make modules`:

```
jma@jma:~/work/tutos/sources/helloworld$ make
make -C /lib/modules/4.4.0-79-generic/build \
    M=/media/jma/DATA/work/tutos/sources/helloworld modules
make[1]: Entering directory '/usr/src/linux-headers-4.4.0-79-generic'
  CC [M] /media/jma/DATA/work/tutos/sources/helloworld/helloworld.o
  Building modules, stage 2.
  MODPOST 1 modules
```

```
            CC
/media/jma/DATA/work/tutos/sources/helloworld/helloworld.mod.o
         LD [M]
/media/jma/DATA/work/tutos/sources/helloworld/helloworld.ko
        make[1]: Leaving directory '/usr/src/linux-headers-4.4.0-79-
generic'
        jma@jma:~/work/tutos/sources/helloworld$ ls
        helloworld.c   helloworld.ko   helloworld.mod.c   helloworld.mod.o
helloworld.o   Makefile   modules.order   Module.symvers
        jma@jma:~/work/tutos/sources/helloworld$ sudo insmod   helloworld.ko
        jma@jma:~/work/tutos/sources/helloworld$ sudo rmmod helloworld
        jma@jma:~/work/tutos/sources/helloworld$ dmesg
        [...]
        [308342.285157] Hello world!
        [308372.084288] End of the world
```

The preceding example only dealt with native builds, compiling on an x86 machine for an x86 machine. What about cross-compilation ? This is the process by which one compiles on machine A, called host, a code that is intended to run on machine B, called target; host and target having different architectures. The classical use case is to build on an x86 machine a code that should run on an ARM architecture, which is exactly our situation.

When it comes to crosscompiling a kernel module, there are essentially two variables the kernel makefile needs to be aware of; these are: ARCH and CROSS_COMPILE, which respectively represent the target architecture and the compiler prefix name. So what change between native compilation and cross compilation of a kernel module is the make command. The following is the line to build for ARM:

```
make ARCH=arm CROSS_COMPILE=arm-none-linux-gnueabihf-
```

Summary

This chapter showed you the basics of driver development and explained the concept of module/built-in devices, as well as their loading and unloading. Even if you are not able to interact with the user space, you are ready to write a complete driver, print a formatted message, and understand the concept of init/exit. The next chapter will deal with character devices, with which you will be able to target enhanced features, write code accessible from the user space, and have a significant impact on the system.

3
Kernel Facilities and Helper Functions

The kernel is a standalone piece of software, as you'll see in this chapter, that does not make use of any C library. It implements any mechanism you may encounter in modern libraries, and even more, such as compression, string functions, and so on. We will walk step by step through the most important aspects of such capabilities.

In this chapter, we will cover the following topic:

- Introducing the kernel container data structure
- Dealing with the kernel sleeping mechanism
- Using timers
- Delving into the kernel locking mechanism (mutex, spnlock)
- Deferring work using a kernel dedicated API
- Using IRQs

Understanding container_of macro

When it comes to managing several data structures in the code, you'll almost always need to embed one structure into another and retrieve them at any moment without being asked questions about memory offset or boundaries. Let's say you have a `struct person`, as defined here:

```
struct person {
    int  age;
    char *name;
} p;
```

Kernel Facilities and Helper Functions

By only having a pointer on `age` or `name`, one can retrieve the whole structure wrapping (containing) that pointer. As the name says, `container_of` macro is used to find the container of the given field of a structure. The macro is defined in `include/linux/kernel.h` and looks like:

```
#define container_of(ptr, type, member) ({                      \
    const typeof(((type *)0)->member) * __mptr = (ptr);         \
    (type *)((char *)__mptr - offsetof(type, member)); })
```

Don't be afraid by the pointers; just see it as:

```
container_of(pointer, container_type, container_field);
```

Here are the elements of the preceding code fragment:

- `pointer`: This is the pointer to the field in the structure
- `container_type`: This is the type of structure wrapping (containing) the pointer
- `container_field`: This is the name of the field to which `pointer` points inside the structure

Let us consider the following container:

```
struct person {
    int   age;
    char *name;
};
```

Now let us consider one of its instance, along with a pointer to the `name` member:

```
struct person somebody;
[...]
char *the_name_ptr = somebody.name;
```

Along with a pointer to the `name` member (`the_name_ptr`), you can use the `container_of` macro in order to get a pointer to the whole structure (container) that wraps this member by using the following:

```
struct person *the_person;
the_person = container_of(the_name_ptr, struct person, name);
```

`container_of` takes the offset of `name` at the beginning of the struct into account to get the correct pointer location. If you subtract the offset of the field `name` from the pointer `the_name_ptr`, you will get the correct location. It is what the macro's last line does:

```
(type *)( (char *)__mptr - offsetof(type,member) );
```

Applying this to a real example, it gives the following:

```
struct family {
    struct person *father;
    struct person *mother;
    int number_of_suns;
    int salary;
} f;

/*
 * pointer to a field of the structure
 * (could be any member of any family)
 */
struct *person = family.father;
struct family *fam_ptr;

/* now let us retrieve back its family */
fam_ptr = container_of(person, struct family, father);
```

It's all you need to know about the `container_of` macro, and believe me, it is enough. In real drivers that we'll develop further in the book, it looks like the following:

```
struct mcp23016 {
    struct i2c_client *client;
    struct gpio_chip chip;
}

/* retrive the mcp23016 struct given a pointer 'chip' field */
static inline struct mcp23016 *to_mcp23016(struct gpio_chip *gc)
{
    return container_of(gc, struct mcp23016, chip);
}

static int mcp23016_probe(struct i2c_client *client,
              const struct i2c_device_id *id)
{
    struct mcp23016 *mcp;
    [...]
    mcp = devm_kzalloc(&client->dev, sizeof(*mcp), GFP_KERNEL);
    if (!mcp)
        return -ENOMEM;
    [...]
}
```

`controller_of` macro is mainly used in generic containers in the kernel. In some examples in this book (starting from Chapter 5, *Platform Device Drivers*), you will encounter the `container_of` macro.

Linked lists

Imagine you have a driver that manages more than one device, let's say five devices. You may need to keep a track of each of them in your driver. What you need here is a linked list. Two types of linked list actually exist:

- Simply linked list
- Doubly linked list

Therefore, kernel developers only implement circular doubly linked lists because this structure allows you to implement FIFO and LIFO, and kernel developers take care to maintain a minimal set of code. The header to be added in the code in order to support lists is <linux/list.h>. The data structure at the core of list implementation in the kernel is struct list_head structure, defined as the following:

```
struct list_head {
    struct list_head *next, *prev;
};
```

The struct list_head is used in both the head of the list and each node. In the world of the kernel, before a data structure can be represented as a linked list, that structure must embed a struct list_head field. For example, let's create a list of cars:

```
struct car {
    int door_number;
    char *color;
    char *model;
};
```

Before we can create a list for the car, we must change its structure in order to embed a struct list_head field. The structure becomes:

```
struct car {
    int door_number;
    char *color;
    char *model;
    struct list_head list; /* kernel's list structure */
};
```

First, we need to create a `struct list_head` variable that will always point to the head (first element) of our list. This instance of `list_head` is not associated to any car and is special:

```
static LIST_HEAD(carlist) ;
```

Now we can create cars and add them to our list—`carlist`:

```
#include <linux/list.h>

struct car *redcar = kmalloc(sizeof(*car), GFP_KERNEL);
struct car *bluecar = kmalloc(sizeof(*car), GFP_KERNEL);

/* Initialize each node's list entry */
INIT_LIST_HEAD(&bluecar->list);
INIT_LIST_HEAD(&redcar->list);

/* allocate memory for color and model field and fill every field */
 [...]
list_add(&redcar->list, &carlist) ;
list_add(&bluecar->list, &carlist) ;
```

It is as simple as that. Now, `carlist` contains two elements. Let us get deeper into the linked list API.

Creating and initializing the list

There are two ways to create and initialize the list:

Dynamic method

The dynamic method consists of a `struct list_head` and initializes it with the `INIT_LIST_HEAD` macro:

```
struct list_head mylist;
INIT_LIST_HEAD(&mylist);
```

The following is the expansion of `INIT_LIST_HEAD`:

```
static inline void INIT_LIST_HEAD(struct list_head *list)
{
    list->next = list;
    list->prev = list;
}
```

Static method

Static allocation is done through the `LIST_HEAD` macro:

```
LIST_HEAD(mylist)
```

`LIST_HEAD`s definition is defined as follows:

```
#define LIST_HEAD(name) \
    struct list_head name = LIST_HEAD_INIT(name)
```

The following is its expansion:

```
#define LIST_HEAD_INIT(name) { &(name), &(name) }
```

This assigns each pointer (`prev` and `next`) inside the `name` field to point to `name` itself (just like `INIT_LIST_HEAD` does).

Creating a list node

To create new nodes, just create our data struct instance, and initialize their embedded `list_head` field. Using the car example, it will give the following:

```
struct car *blackcar = kzalloc(sizeof(struct car), GFP_KERNEL);

/* non static initialization, since it is the embedded list field*/
INIT_LIST_HEAD(&blackcar->list);
```

As said earlier, use `INIT_LIST_HEAD`, which is a dynamically allocated list and usually part of another structure.

Adding a list node

The kernel provides `list_add` to add a new entry to the list, which is a wrapper around the internal function `__list_add`:

```
void list_add(struct list_head *new, struct list_head *head);
static inline void list_add(struct list_head *new, struct list_head *head)
{
    __list_add(new, head, head->next);
}
```

`__list_add` will take two known entries as a parameter, and inserts your elements between them. Its implementation in the kernel is quite easy:

```
static inline void __list_add(struct list_head *new,
                struct list_head *prev,
                struct list_head *next)
{
    next->prev = new;
    new->next = next;
    new->prev = prev;
    prev->next = new;
}
```

The following is an example of adding two cars in our list:

```
list_add(&redcar->list, &carlist);
list_add(&blue->list, &carlist);
```

This mode can be used to implement a stack. The other function to add an entry into the list is:

```
void list_add_tail(struct list_head *new, struct list_head *head);
```

This inserts the given new entry at the end of the list. Given our previous example, we can use the following:

```
list_add_tail(&redcar->list, &carlist);
list_add_tail(&blue->list, &carlist);
```

This mode can be used to implement a queue.

[53]

Deleting a node from the list

List handling is an easy task in kernel code. Deleting a node is straightforward:

```
void list_del(struct list_head *entry);
```

Following the preceding example, let us delete the red car:

```
list_del(&redcar->list);
```

list_del disconnects the prev and next pointers of the given entry, resulting in an entry removal. The memory allocated for the node is not freed yet; you need to do that manually with kfree.

Linked list traversal

We have the macro list_for_each_entry(pos, head, member) for list traversal.

- head is the list's head node.
- member is the name of the list struct list_head within our data struct (in our case, it is list).
- pos is used for iteration. It is a loop cursor (just like i in for(i=0; i<foo; i++)). head could be the head node of the linked list, or any entry, and we don't care since we are dealing with a doubly linked list:

```
struct car *acar; /* loop counter */
int blue_car_num = 0;

/* 'list' is the name of the list_head struct in our data structure */
list_for_each_entry(acar, carlist, list){
    if(acar->color == "blue")
        blue_car_num++;
}
```

Why do we need the name of the `list_head` type field in our data structure? Look at the `list_for_each_entry` definition:

```
#define list_for_each_entry(pos, head, member)         \
for (pos = list_entry((head)->next, typeof(*pos), member);   \
     &pos->member != (head);    \
     pos = list_entry(pos->member.next, typeof(*pos), member))

#define list_entry(ptr, type, member) \
    container_of(ptr, type, member)
```

Given this, we can understand that it is all about `container_of`'s power. Also bear in mind `list_for_each_entry_safe(pos, n, head, member)`.

Kernel sleeping mechanism

Sleeping is the mechanism by which a process relaxes a processor, with the possibility of handling another process. The reason why a processor can sleep could be for sensing data availability, or waiting for a resource to be free.

The kernel scheduler manages a list of tasks to run, known as a run queue. Sleeping processes are not scheduled anymore, since they are removed from that run queue. Unless its state changes (that is, it wakes up), a sleeping process will never be executed. You may relax a processor as soon as one is waiting for something (resource or anything else), and make sure a condition or someone else will wake it up. That said, the Linux kernel eases the implementation of the sleeping mechanism by providing a set of functions and data structures.

Wait queue

Wait queues are essentially used to process blocked I/O, to wait for particular conditions to be true, and to sense data or resource availability. To understand how it works, let's have a look at its structure in `include/linux/wait.h`:

```
struct __wait_queue {
    unsigned int flags;
#define WQ_FLAG_EXCLUSIVE 0x01
    void *private;
    wait_queue_func_t func;
    struct list_head task_list;
};
```

Kernel Facilities and Helper Functions

Let's pay attention to the `task_list` field. As you can see, it is a list. Every process you want to put to sleep is queued in that list (hence the name *wait queue*) and put into a sleep state until a condition becomes true. The wait queue can be seen as nothing but a simple list of processes and a lock.

The functions you will always face when dealing with wait queues are:

- Static declaration:

    ```
    DECLARE_WAIT_QUEUE_HEAD(name)
    ```

- Dynamic declaration:

    ```
    wait_queue_head_t my_wait_queue;
    init_waitqueue_head(&my_wait_queue);
    ```

- Blocking:

    ```
    /*
     * block the current task (process) in the wait queue if
     * CONDITION is false
     */
    int wait_event_interruptible(wait_queue_head_t q, CONDITION);
    ```

- Unblocking:

    ```
    /*
     * wake up one process sleeping in the wait queue if
     * CONDITION above has become true
     */
    void wake_up_interruptible(wait_queue_head_t *q);
    ```

`wait_event_interruptible` does not continuously poll, but simply evaluates the condition when it is called. If the condition is false, the process is put into a `TASK_INTERRUPTIBLE` state and removed from the run queue. The condition is then only rechecked each time you call `wake_up_interruptible` in the wait queue. If the condition is true when `wake_up_interruptible` runs, a process in the wait queue will be awakened, and its state set to `TASK_RUNNING`. Processes are awakened in the order they are put to sleep. To awaken all processes waiting in the queue, you should use `wake_up_interruptible_all`.

In fact, the main functions are wait_event, wake_up, and wake_up_all. They are used with processes in the queue in an exclusive (uninterruptible) wait, since they can't be interrupted by the signal. They should be used only for critical tasks. Interruptible functions are just optional (but recommended). Since they can be interrupted by signals, you should check their return value. A nonzero value means your sleep has been interrupted by some sort of signal, and the driver should return ERESTARTSYS.

If someone has called wake_up or wake_up_interruptible and the condition is still FALSE, then nothing will happen. Without wake_up (or wake_up_interuptible), process(es) will never be awakened. Here is an example of a wait queue:

```
#include <linux/module.h>
#include <linux/init.h>
#include <linux/sched.h>
#include <linux/time.h>
#include <linux/delay.h>
#include<linux/workqueue.h>

static DECLARE_WAIT_QUEUE_HEAD(my_wq);
static int condition = 0;

/* declare a work queue*/
static struct work_struct wrk;

static void work_handler(struct work_struct *work)
{
    printk("Waitqueue module handler %s\n", __FUNCTION__);
    msleep(5000);
    printk("Wake up the sleeping module\n");
    condition = 1;
    wake_up_interruptible(&my_wq);
}

static int __init my_init(void)
{
    printk("Wait queue example\n");

    INIT_WORK(&wrk, work_handler);
    schedule_work(&wrk);

    printk("Going to sleep %s\n", __FUNCTION__);
    wait_event_interruptible(my_wq, condition != 0);

    pr_info("woken up by the work job\n");
```

```
        return 0;
}

void my_exit(void)
{
    printk("waitqueue example cleanup\n");
}

module_init(my_init);
module_exit(my_exit);
MODULE_AUTHOR("John Madieu <john.madieu@foobar.com>");
MODULE_LICENSE("GPL");
```

In the preceding example, the current process (actually `insmod`) will be put into sleep in the wait queue for 5 seconds and woken up by the work handler. The `dmesg` output is as follows:

```
[342081.385491] Wait queue example
[342081.385505] Going to sleep my_init
[342081.385515] Waitqueue module handler work_handler
[342086.387017] Wake up the sleeping module
[342086.387096] woken up by the work job
[342092.912033] waitqueue example cleanup
```

Delay and timer management

Time is one of the most used resources, right after memory. It is used to do almost everything: defer work, sleep, scheduling, timeout, and many other tasks.

There are the two categories of time. The kernel uses absolute time to know what time it is, that is, the date and time of the day, whereas relative time is used by, for example, the kernel scheduler. For absolute time, there is a hardware chip called **real-time clock** (**RTC**). We will deal with such devices later in the book in `Chapter 18`, *RTC Drivers*. On the other side, to handle relative time, the kernel relies on a CPU feature (peripheral), called a timer, which, from the kernel's point of view, is called a *kernel timer*. Kernel timers are what we will talk about in this section.

Kernel timers are classified into two different parts:

- Standard timers, or system timers
- High-resolution timers

Standard timers

Standard timers are kernel timers operating on the granularity of jiffies.

Jiffies and HZ

A jiffy is a kernel unit of time declared in `<linux/jiffies.h>`. To understand jiffies, we need to introduce a new constant HZ, which is the number of times `jiffies` is incremented in one second. Each increment is called a *tick*. In other words, HZ represents the size of a jiffy. HZ depends on the hardware and on the kernel version, and also determines how frequently the clock interrupt fires. This is configurable on some architecture, fixed on other ones.

What it means is that `jiffies` is incremented HZ times every second. If HZ = 1,000, then it is incremented 1,000 times (that is, one tick every 1/1,000 seconds). Once defined, the **programmable interrupt timer** (**PIT**), which is a hardware component, is programmed with that value in order to increment jiffies when the PIT interrupt comes in.

Depending on the platform, jiffies can lead to overflow. On a 32-bit system, HZ = 1,000 will result in about 50 days duration only, whereas the duration is about 600 million years on a 64-bit system. By storing jiffies in a 64-bit variable, the problem is solved. A second variable has then been introduced and defined in `<linux/jiffies.h>`:

```
extern u64 jiffies_64;
```

In this manner on 32-bit systems, `jiffies` will point to low-order 32-bits, and `jiffies_64` will point to high-order bits. On 64-bit platforms, `jiffies = jiffies_64`.

Timers API

A timer is represented in the kernel as an instance of `timer_list`:

```
#include <linux/timer.h>

struct timer_list {
    struct list_head entry;
    unsigned long expires;
    struct tvec_t_base_s *base;
    void (*function)(unsigned long);
    unsigned long data;
};
```

`expires` is an absolute value in jiffies. `entry` is a doubly linked list, and `data` is optional, and passed to the callback function.

Timer setup initialization

The following are steps to initialize timers:

1. **Setting up the timer:** Set up the timer, feeding the user-defined callback and data:

    ```
    void setup_timer( struct timer_list *timer, \
             void (*function)(unsigned long), \
             unsigned long data);
    ```

    One can also use this:

    ```
    void init_timer(struct timer_list *timer);
    ```

    `setup_timer` is a wrapper around `init_timer`.

2. **Setting the expiration time:** When the timer is initialized, we need to set its expiration before the callback gets fired:

    ```
    int mod_timer( struct timer_list *timer, unsigned long expires);
    ```

3. **Releasing the timer:** When you are done with the timer, it needs to be released:

    ```
    void del_timer(struct timer_list *timer);
    int del_timer_sync(struct timer_list *timer);
    ```

    `del_timer` returns `void` whether it has deactivated a pending timer or not. Its return value is 0 on an inactive timer, or 1 on an active one. The last, `del_timer_sync`, waits for the handler to finish its execution, even those that may happen on another CPU. You should not hold a lock preventing the handler's completion, otherwise it will result in a dead lock. You should release the timer in the module cleanup routine. You can independently check whether the timer is running or not:

    ```
    int timer_pending( const struct timer_list *timer);
    ```

    This function checks whether there are any fired timer callbacks pending.

Standard timer example

```c
#include <linux/init.h>
#include <linux/kernel.h>
#include <linux/module.h>
#include <linux/timer.h>

static struct timer_list my_timer;

void my_timer_callback(unsigned long data)
{
    printk("%s called (%ld).\n", __FUNCTION__, jiffies);
}
static int __init my_init(void)
{
    int retval;
    printk("Timer module loaded\n");

    setup_timer(&my_timer, my_timer_callback, 0);
    printk("Setup timer to fire in 300ms (%ld)\n", jiffies);

    retval = mod_timer( &my_timer, jiffies + msecs_to_jiffies(300) );
    if (retval)
        printk("Timer firing failed\n");
    return 0;
}
static void my_exit(void)
{
    int retval;
    retval = del_timer(&my_timer);
    /* Is timer still active (1) or no (0) */
    if (retval)
        printk("The timer is still in use...\n");

    pr_info("Timer module unloaded\n");
}

module_init(my_init);
module_exit(my_exit);
MODULE_AUTHOR("John Madieu <john.madieu@gmail.com>");
MODULE_DESCRIPTION("Standard timer example");
MODULE_LICENSE("GPL");
```

High resolution timers (HRTs)

Standard timers are less accurate and do not suit real-time applications. High-resolution timers, introduced in kernel v2.6.16 (and enabled by the CONFIG_HIGH_RES_TIMERS option in the kernel configuration) have a resolution of microseconds (up to nanoseconds, depending on the platform), compared to milliseconds on standard timers. The standard timer depends on HZ (since they rely on jiffies), whereas HRT implementation is based on ktime.

Kernel and hardware must support an HRT before being used on your system. In other words, there must be an arch-dependent code implemented to access your hardware HRTs.

HRT API

The required headers are:

```
#include <linux/hrtimer.h>
```

An HRT is represented in the kernel as an instance of hrtimer:

```
struct hrtimer {
    struct timerqueue_node node;
    ktime_t _softexpires;
    enum hrtimer_restart (*function)(struct hrtimer *);
    struct hrtimer_clock_base *base;
    u8 state;
    u8 is_rel;
};
```

HRT setup initialization

1. **Initializing the hrtimer**: Before hrtimer initialization, you need to set up a ktime, which represents time duration. We will see how to achieve that in the following example:

   ```
   void hrtimer_init( struct hrtimer *time, clockid_t which_clock,
                      enum hrtimer_mode mode);
   ```

2. **Starting hrtimer**: hrtimer can be started as shown in the following example:

   ```
   int hrtimer_start( struct hrtimer *timer, ktime_t time,
                      const enum hrtimer_mode mode);
   ```

mode represents the expiry mode. It should be HRTIMER_MODE_ABS for an absolute time value, or HRTIMER_MODE_REL for a time value relative to now.

3. **hrtimer cancellation**: You can either cancel the timer or see whether it is possible to cancel it or not:

```
int hrtimer_cancel( struct hrtimer *timer);
int hrtimer_try_to_cancel(struct hrtimer *timer);
```

Both return 0 when the timer is not active and 1 when the timer is active. The difference between these two functions is that hrtimer_try_to_cancel fails if the timer is active or its callback is running, returning -1, whereas hrtimer_cancel will wait until the callback finishes.

We can independently check whether the hrtimer's callback is still running with the following:

```
int hrtimer_callback_running(struct hrtimer *timer);
```

Remember, hrtimer_try_to_cancel internally calls hrtimer_callback_running.

In order to prevent the timer from automatically restarting, the hrtimer callback function must return HRTIMER_NORESTART.

You can check whether HRTs are available on your system by doing the following:

- By looking in the kernel config file, which should contain something like CONFIG_HIGH_RES_TIMERS=y: zcat /proc/configs.gz | grep CONFIG_HIGH_RES_TIMERS.
- By looking at the cat /proc/timer_list or cat /proc/timer_list | grep resolution result. The .resolution entry must show 1 nsecs and the event_handler must show hrtimer_interrupts.
- By using the clock_getres system call.
- From within the kernel code, by using #ifdef CONFIG_HIGH_RES_TIMERS.

With HRTs enabled on your system, the accuracy of sleep and timer system calls do not depend on jiffies anymore, but they are still as accurate as HRTs are. It is the reason why some systems do not support nanosleep(), for example.

Dynamic tick/tickless kernel

With previous HZ options, the kernel is interrupted HZ times per second in order to reschedule tasks, even in an idle state. If HZ is set to 1,000, there will be 1,000 kernel interruptions per second, preventing the CPU from being idle for a long time, thus affecting CPU power consumption.

Now let's look at a kernel with no fixed or predefined ticks, where the ticks are disabled until some task needs to be performed. We call such a kernel a **tickless kernel**. In fact, tick activation is scheduled, based on the next action. The right name should be **dynamic tick kernel**. The kernel is responsible for task scheduling, and maintains a list of runnable tasks (the run queue) in the system. When there is no task to schedule, the scheduler switches to the idle thread, which enables dynamic tick by disabling the periodic tick until the next timer expires (a new task is queued for processing).

Under the hood, the kernel also maintains a list of the tasks timeouts (it then knows when and how long it has to sleep). In an idle state, if the next tick is further away than the lowest timeout in the tasks list timeout, the kernel programs the timer with that timeout value. When the timer expires, the kernel re-enables the periodic ticks back and invokes the scheduler, which then schedules the task associated with the timeout. This is how the tickless kernel removes the periodic tick and saves power when idle.

Delays and sleep in the kernel

Without going deep into the details, there are two types of delays, depending on the context your code runs in: atomic or nonatomic. The mandatory header to handle delays in the kernel is `#include <linux/delay>`.

Atomic context

Tasks in the atomic context (such as ISR) can't sleep, and can't be scheduled; it is the reason why busy-wait loops are used for delaying purposes in an atomic context. The kernel exposes the `Xdelay` family of functions that will spend time in a busy loop, long (based on jiffies) enough to achieve the desired delay:

- `ndelay(unsigned long nsecs)`
- `udelay(unsigned long usecs)`
- `mdelay(unsigned long msecs)`

You should always use `udelay()` since `ndelay()` precision depends on how accurate your hardware timer is (not always the case on an embedded SOC). Use of `mdelay()` is also discouraged.

Timer handlers (callbacks) are executed in an atomic context, meaning that sleeping is not allowed at all. By *sleeping*, I mean any function that may result in sending the caller to sleep, such as allocating memory, locking a mutex, an explicit call to `sleep()` function, and so on.

Nonatomic context

In a nonatomic context, the kernel provides the `sleep[_range]` family of functions and which function to use depends on how long you need to delay by:

- `udelay(unsigned long usecs)`: Busy-wait loop based. You should use this function if you need to sleep for a few μsecs (< ~10 us).
- `usleep_range(unsigned long min, unsigned long max)`: Relies on hrtimers, and it is recommended to let this sleep for few ~μsecs or small msecs (10 us - 20 ms), avoiding the busy-wait loop of `udelay()`.
- `msleep(unsigned long msecs)`: Backed by jiffies/legacy_timers. You should use this for larger, msecs sleep (10 ms+).

Sleep and delay topics are well explained in *Documentation/timers/timers-howto.txt* in the kernel source.

Kernel locking mechanism

Locking is a mechanism that helps shares resources between different threads or processes. A shared resource is a data or a device that can be accessed by at least two user, simultaneously or no. Locking mechanisms prevent abusive access, for example, a process writing data when another one is reading in the same place, or two processes accessing the same device (the same GPIO for example). The kernel provides several locking mechanisms. The most important are:

- Mutex
- Semaphore
- Spinlock

We will only learn about mutexes and spinlock, since they are widely used in device drivers.

Mutex

Mutual exclusion (**mutex**) is the de facto most used locking mechanism. To understand how it works, let's see what its structure looks like in `include/linux/mutex.h`:

```
struct mutex {
    /* 1: unlocked, 0: locked, negative: locked, possible waiters */
    atomic_t count;
    spinlock_t wait_lock;
    struct list_head wait_list;
    [...]
};
```

As we have seen in the section *wait queue*, there is also a `list` type field in the structure: `wait_list`. The principle of sleeping is the same.

Contenders are removed from the scheduler run queue and put onto the wait list (`wait_list`) in a sleep state. The kernel then schedules and executes other tasks. When the lock is released, a waiter in the wait queue is woken, moved off the `wait_list`, and scheduled back.

Mutex API

Using mutex requires only a few basic functions:

Declare

- Statically:

    ```
    DEFINE_MUTEX(my_mutex);
    ```

- Dynamically:

    ```
    struct mutex my_mutex;
    mutex_init(&my_mutex);
    ```

Acquire and release

- Lock:

```
void mutex_lock(struct mutex *lock);
int  mutex_lock_interruptible(struct mutex *lock);
int  mutex_lock_killable(struct mutex *lock);
```

- Unlock:

```
void mutex_unlock(struct mutex *lock);
```

Sometimes, you may only need to check whether a mutex is locked or not. For that purpose, you can use the `int mutex_is_locked(struct mutex *lock)` function.

```
int mutex_is_locked(struct mutex *lock);
```

What this function does is just check whether the mutex's owner is empty (`NULL`) or not. There is also `mutex_trylock`, that acquires the mutex if it is not already locked, and returns 1; otherwise, it returns 0:

```
int mutex_trylock(struct mutex *lock);
```

As with the wait queue's interruptible family function, `mutex_lock_interruptible()`, which is recommended, will result in the driver being able to be interrupted by any signal, whereas with `mutex_lock_killable()`, only signals killing the process can interrupt the driver.

You should be very careful with `mutex_lock()`, and use it when you can guarantee that the mutex will be released, whatever happens. In the user context, it is recommended you always use `mutex_lock_interruptible()` to acquire the mutex, since `mutex_lock()` will not return if a signal is received (even a *ctrl* + *c*).

Here is an example of a mutex implementation:

```
struct mutex my_mutex;
mutex_init(&my_mutex);

/* inside a work or a thread */
mutex_lock(&my_mutex);
access_shared_memory();
mutex_unlock(&my_mutex);
```

Kernel Facilities and Helper Functions

Please have a look at `include/linux/mutex.h` in the kernel source to see the strict rules you must respect with mutexes. Here are some of them:

- Only one task can hold the mutex at a time; this is actually not a rule, but a fact
- Multiple unlocks are not permitted
- They must be initialized through the API
- A task holding the mutex may not exit, since the mutex will remain locked, and possible contenders will wait (will sleep) forever
- Memory areas where held locks reside must not be freed
- Held mutexes must not be reinitialized
- Since they involve rescheduling, mutexes may not be used in atomic contexts, such as tasklets and timers

As with `wait_queue`, there is no polling mechanism with mutexes. Every time that `mutex_unlock` is called on a mutex, the kernel checks for waiters in `wait_list`. If any, one (and only one) of them is awakened and scheduled; they are woken in the same order in which they were put to sleep.

Spinlock

Like mutex, spinlock is a mutual exclusion mechanism; it only has two states:

- locked (aquired)
- unlocked (released)

Any thread that needs to acquire the spinlock will active loop until the lock is acquired, which breaks out of the loop. This is the point where mutex and spinlock differ. Since spinlock heavily consumes the CPU while looping, it should be used for very quick acquires, especially when time to hold the spinlock is less than time to reschedule. Spinlock should be released as soon as the critical task is done.

In order to avoid wasting CPU time by scheduling a thread that may probably spin, trying to acquire a lock held by another thread moved off the run queue, the kernel disables preemption whenever a code holding a spinlock is running. With preemption disabled, we prevent the spinlock holder from being moved off the run queue, which could lead waiting processes to spin for a long time and consume CPU.

As long as one holds a spinlock, other tasks may be spinning while waiting on it. By using spinlock, you asserts and guarantee that it will not be held for a long time. You can say it is better to spin in a loop, wasting CPU time, than the cost of sleeping your thread, context-shifting to another thread or process, and being woken up afterward. Spinning on a processor means no other task can run on that processor; it then makes no sense to use spinlock on a single core machine. In the best case, you will slow down the system; in the worst case, you will deadlock, as with mutexes. For this reason, the kernel just disables preemption in response to the `spin_lock(spinlock_t *lock)` function on single processor. On a single processor (core) system, you should use `spin_lock_irqsave()` and `spin_unlock_irqrestore()`, which will respectively disable the interrupts on the CPU, preventing interrupt concurrency.

Since you do not know in advance what system you will write the driver for, it is recommended you acquire a spinlock using `spin_lock_irqsave(spinlock_t *lock, unsigned long flags)`, which disables interrupts on the current processor (the processor where it is called) before taking the spinlock. `spin_lock_irqsave` internally calls `local_irq_save(flags);`, an architecture-dependent function to save the IRQ status, and `preempt_disable()` to disable preemption on the relevant CPU. You should then release the lock with `spin_unlock_irqrestore()`, which does the reverse operations that we previously enumerated. This is a code that does lock acquire and release. It is an IRQ handler, but let's just focus on the lock aspect. We will discuss more about IRQ handlers in the next section:

```
/* some where */
spinlock_t my_spinlock;
spin_lock_init(my_spinlock);

static irqreturn_t my_irq_handler(int irq, void *data)
{
    unsigned long status, flags;

    spin_lock_irqsave(&my_spinlock, flags);
    status = access_shared_resources();

    spin_unlock_irqrestore(&gpio->slock, flags);
    return IRQ_HANDLED;
}
```

Spinlock versus mutexes

Used for concurrency in the kernel, spinlocks and mutexes each have their own objectives:

- Mutexes protect the process's critical resource, whereas spinlock protects the IRQ handler's critical sections
- Mutexes put contenders to sleep until the lock is acquired, whereas spinlocks infinitely spin in a loop (consuming CPU) until the lock is acquired
- Because of the previous point, you can't hold spinlock for a long time, since waiters will waste CPU time waiting for the lock, whereas a mutex can be held as long as the resource needs to be protected, since contenders are put to sleep in a wait queue

When dealing with spinlocks, please keep in mind that preemption is disabled only for threads holding spinlocks, not for spinning waiters.

Work deferring mechanism

Deferring is a method by which you schedule a piece of work to be executed in the future. It's a way to report an action later. Obviously, the kernel provides facilities to implement such a mechanism; it allows you to defer functions, whatever their type, to be called and executed later. There are three of them in the kernel:

- **SoftIRQs**: Executed in an atomic context
- **Tasklets**: Executed in an atomic context
- **Workqueues**: Executed in a process context

Softirqs and ksoftirqd

Software IRQ (softirq), or software interrupt is a deferring mechanism used only for very fast processing, since it runs with a disabled scheduler (in an interrupt context). You'll rarely (almost never) want to deal with softirq directly. There are only networks and block device subsystems using softirq. Tasklets are an instantiation of softirqs, and will be sufficient in almost every case that you feel the need to use softirqs.

ksoftirqd

In most cases, softirqs are scheduled in hardware interrupts, which may arrive very quickly, faster than they can be serviced. They are then queued by the kernel in order to be processed later. **Ksoftirqds** are responsible for late execution (process context this time). A ksoftirqd is a per-CPU kernel thread raised to handle unserviced software interrupts:

In the preceding `top` sample from my personal computer, you can see `ksoftirqd/n` entries, where n is the CPU number that the ksoftirqd runs on. CPU-consuming ksoftirqd may indicate an overloaded system or a system under **interrupts storm**, which is never good. You can have a look at `kernel/softirq.c` to see how ksoftirqds are designed.

Tasklets

Tasklets are a bottom-half (we will see what this means later) mechanism built on top of softirqs. They are represented in the kernel as instances of struct `tasklet_struct`:

```
struct tasklet_struct
{
    struct tasklet_struct *next;
    unsigned long state;
    atomic_t count;
    void (*func)(unsigned long);
    unsigned long data;
};
```

Tasklets are not re-entrant by nature. A code is called reentrant if it can be interrupted anywhere in the middle of its execution, and then be safely called again. Tasklets are designed such that a tasklet can run on one and only one CPU simultaneously (even on an SMP system), which is the CPU it was scheduled on, but different tasklets may be run simultaneously on different CPUs. The tasklet API is quite basic and intuitive.

Declaring a tasklet

- Dynamically:

  ```
  void tasklet_init(struct tasklet_struct *t,
            void (*func)(unsigned long), unsigned long data);
  ```

- Statically:

  ```
  DECLARE_TASKLET( tasklet_example, tasklet_function, tasklet_data );
  DECLARE_TASKLET_DISABLED(name, func, data);
  ```

There is one difference between the two functions; the former creates a tasklet already enabled and ready to be scheduled without any other function call, done by setting the `count` field to 0, whereas the latter creates a tasklet disabled (done by setting `count` to 1), on which one has to call `tasklet_enable()` before the tasklet can be schedulable:

```
#define DECLARE_TASKLET(name, func, data) \
    struct tasklet_struct name = { NULL, 0, ATOMIC_INIT(0), func, data }

#define DECLARE_TASKLET_DISABLED(name, func, data) \
    struct tasklet_struct name = { NULL, 0, ATOMIC_INIT(1), func, data }
```

Globally, setting the `count` field to 0 means that the tasklet is disabled and cannot be executed, whereas a nonzero value means the opposite.

Enabling and disabling a tasklet

There is one function to enable a tasklet:

```
void tasklet_enable(struct tasklet_struct *);
```

`tasklet_enable` simply enables the tasklet. In older kernel versions, you may find void `tasklet_hi_enable(struct tasklet_struct *)` is used, but those two functions do exactly the same thing. To disable a tasklet, call:

```
void tasklet_disable(struct tasklet_struct *);
```

You can also call:

```
void tasklet_disable_nosync(struct tasklet_struct *);
```

`tasklet_disable` will disable the tasklet and return only when the tasklet has terminated its execution (if it was running), whereas `tasklet_disable_nosync` returns immediately, even if the termination has not occurred.

Tasklet scheduling

There are two scheduling functions for tasklet, depending on whether your tasklet has normal or higher priority:

```
void tasklet_schedule(struct tasklet_struct *t);
void tasklet_hi_schedule(struct tasklet_struct *t);
```

Kernel Facilities and Helper Functions

The kernel maintains normal priority and high priority tasklets in two different lists. `tasklet_schedule` adds the tasklet into the normal priority list, scheduling the associated softirq with a `TASKLET_SOFTIRQ` flag. With `tasklet_hi_schedule`, the tasklet is added into the high priority list, scheduling the associated softirq with a `HI_SOFTIRQ` flag. High priority tasklets are meant to be used for soft interrupt handlers with low latency requirements. There are some properties associated with tasklets you should know:

- Calling `tasklet_schedule` on a tasklet already scheduled, but whose execution has not started, will do nothing, resulting in the tasklet being executed only once.
- `tasklet_schedule` can be called in a tasklet, meaning that a tasklet can reschedule itself.
- High priority tasklets are always executed before normal ones. Abusive use of high priority tasks will increase the system latency. Only use them for really quick stuff.

You can stop a tasklet using the `tasklet_kill` function that will prevent the tasklet from running again or wait for its completion before killing it if the tasklet is currently scheduled to run:

```
void tasklet_kill(struct tasklet_struct *t);
```

Let us check. Look at the following example:

```
#include <linux/kernel.h>
#include <linux/module.h>
#include <linux/interrupt.h>    /* for tasklets API */

char tasklet_data[]="We use a string; but it could be pointer to a structure";

/* Tasklet handler, that just print the data */
void tasklet_work(unsigned long data)
{
    printk("%s\n", (char *)data);
}

DECLARE_TASKLET(my_tasklet, tasklet_function, (unsigned long) tasklet_data);

static int __init my_init(void)
{
    /*
     * Schedule the handler.
     * Tasklet arealso scheduled from interrupt handler
     */
```

```
        tasklet_schedule(&my_tasklet);
        return 0;
}

void my_exit(void)
{
    tasklet_kill(&my_tasklet);
}

module_init(my_init);
module_exit(my_exit);
MODULE_AUTHOR("John Madieu <john.madieu@gmail.com>");
MODULE_LICENSE("GPL");
```

Work queues

Added since Linux kernel 2.6, the most used and simple deferring mechanism is the work queue. It is the last one we will talk about in this chapter. As a deferring mechanism, it takes an opposite approach to the others we've seen, running only in a preemptible context. It is the only choice when you need to sleep in your bottom half (I will explain what a bottom half is later in the next section). By sleep, I mean process I/O data, hold mutexes, delay, and all the other tasks that may lead to sleep or move the task off the run queue.

Keep in mind that work queues are built on top of kernel threads, and this is the reason why I decided not to talk about the kernel thread as a deferring mechanism at all. However, there are two ways to deal with work queues in the kernel. First, there is a default shared work queue, handled by a set of kernel threads, each running on a CPU. Once you have work to schedule, you queue that work into the global work queue, which will be executed at the appropriate moment. The other method is to run the work queue in a dedicated kernel thread. It means whenever your work queue handler needs to be executed, your kernel thread is woken up to handle it, instead of one of the default predefined threads.

Structures and functions to call are different, depending on whether you chose a shared work queue or dedicated ones.

Kernel-global workqueue – the shared queue

Unless you have no choice, or you need critical performance, or you need to control everything from the work queue initialization to the work scheduling, and if you only submit tasks occasionally, you should use the shared work queue provided by the kernel. With that queue being shared over the system, you should be nice, and should not monopolize the queue for a long time.

Kernel Facilities and Helper Functions

Since the execution of the pending task on the queue is serialized on each CPU, you should not sleep for a long time because no other task on the queue will run until you wake up. You won't even know who you share the work queue with, so don't be surprised if your task takes longer to get the CPU. Work in the shared work queues is executed in a per-CPU thread called events/n, created by the kernel.

In this case, the work must also be initialized with the `INIT_WORK` macro. Since we are going to use the shared work queue, there is no need to create a work queue structure. We only need the `work_struct` structure that will be passed as an argument. There are three functions to schedule work on the shared work queue:

- The version that ties the work on the current CPU:

    ```
    int schedule_work(struct work_struct *work);
    ```

- The same but delayed function:

    ```
    static inline bool schedule_delayed_work(struct delayed_work *dwork,
                                    unsigned long delay)
    ```

- The function that actually schedules the work on a given CPU:

    ```
    int schedule_work_on(int cpu, struct work_struct *work);
    ```

- The same as shown previously, but with a delay:

    ```
    int scheduled_delayed_work_on(int cpu, struct delayed_work *dwork,
    unsigned long delay);
    ```

All of these functions schedule the work given as an argument on to the system's shared work queue `system_wq`, defined in `kernel/workqueue.c`:

```
struct workqueue_struct *system_wq __read_mostly;
EXPORT_SYMBOL(system_wq);
```

A work already submitted to the shared queue can be cancelled with the `cancel_delayed_work` function. You can flush the shared workqueue with:

```
void flush_scheduled_work(void);
```

Kernel Facilities and Helper Functions

Since the queue is shared over the system, one can't really know how long `flush_scheduled_work()` may last before it returns:

```
#include <linux/module.h>
#include <linux/init.h>
#include <linux/sched.h>      /* for sleep */
#include <linux/wait.h>       /* for wait queue */
#include <linux/time.h>
#include <linux/delay.h>
#include <linux/slab.h>          /* for kmalloc() */
#include <linux/workqueue.h>

//static DECLARE_WAIT_QUEUE_HEAD(my_wq);
static int sleep = 0;

struct work_data {
    struct work_struct my_work;
    wait_queue_head_t my_wq;
    int the_data;
};

static void work_handler(struct work_struct *work)
{
    struct work_data *my_data = container_of(work, \
                                struct work_data, my_work);
    printk("Work queue module handler: %s, data is %d\n", __FUNCTION__, my_data->the_data);
    msleep(2000);
    wake_up_interruptible(&my_data->my_wq);
    kfree(my_data);
}

static int __init my_init(void)
{
    struct work_data * my_data;

    my_data = kmalloc(sizeof(struct work_data), GFP_KERNEL);
    my_data->the_data = 34;

    INIT_WORK(&my_data->my_work, work_handler);
    init_waitqueue_head(&my_data->my_wq);

    schedule_work(&my_data->my_work);
    printk("I'm goint to sleep ...\n");
    wait_event_interruptible(my_data->my_wq, sleep != 0);
    printk("I am Waked up...\n");
    return 0;
}
```

Kernel Facilities and Helper Functions

```
static void __exit my_exit(void)
{
    printk("Work queue module exit: %s %d\n", __FUNCTION__, __LINE__);
}

module_init(my_init);
module_exit(my_exit);
MODULE_LICENSE("GPL");
MODULE_AUTHOR("John Madieu <john.madieu@gmail.com> ");
MODULE_DESCRIPTION("Shared workqueue");
```

In order to pass data to my work queue handler, you may have noticed that in both examples, I've embedded my `work_struct` structure inside my custom data structure, and used `container_of` to retrieve it. It is the common way to pass data to the work queue handler.

Dedicated work queue

Here, the work queue is represented as an instance of `struct workqueue_struct`. The work to be queued into the work queue is represented as an instance of `struct work_struct`. There are four steps involved prior to scheduling your work in your own kernel thread:

1. Declare/initialize a `struct workqueue_struct`.
2. Create your work function.
3. Create a `struct work_struct` so that your work function will be embedded into it.
4. Embed your work function in the `work_struct`.

Programming syntax

The following functions are defined in `include/linux/workqueue.h`:

- Declare work and work queue:

  ```
  struct workqueue_struct *myqueue;
  struct work_struct thework;
  ```

- Define the worker function (the handler):

  ```
  void dowork(void *data) {  /* Code goes here */ };
  ```

Kernel Facilities and Helper Functions

- Initialize our work queue and embed our work into:

  ```
  myqueue = create_singlethread_workqueue( "mywork" );
  INIT_WORK( &thework, dowork, <data-pointer> );
  ```

  We could have also created our work queues through a macro called `create_workqueue`. The difference between `create_workqueue` and `create_singlethread_workqueue` is that the former will create a work queue that in turn will create a separate kernel thread on each and every processor available.

- Scheduling work:

  ```
  queue_work(myqueue, &thework);
  ```

  Queue after the given delay to the given worker thread:

  ```
  queue_dalayed_work(myqueue, &thework, <delay>);
  ```

  These functions return `false` if the work was already on a queue and `true` if otherwise. `delay` represents the number of jiffies to wait before queueing. You may use the helper function `msecs_to_jiffies` in order to convert the standard ms delay into jiffies. For example, to queue a work after 5 ms, you can use `queue_delayed_work(myqueue, &thework, msecs_to_jiffies(5));`.

- Wait on all pending work on the given work queue:

  ```
  void flush_workqueue(struct workqueue_struct *wq)
  ```

  `flush_workqueue` sleeps until all queued work has finished their execution. New incoming (enqueued) work does not affect the sleep. One may typically use this in driver shutdown handlers.

- Cleanup:

  Use `cancel_work_sync()` or `cancel_delayed_work_sync` for synchronous cancellation, which will cancel the work if it is not already running, or block until the work has completed. The work will be cancelled even if it requeues itself. You must also ensure that the work queue on which the work was last queued can't be destroyed before the handler returns. These functions are to be used respectively for nondelayed or delayed work:

  ```
  int cancel_work_sync(struct work_struct *work);
  int cancel_delayed_work_sync(struct delayed_work *dwork);
  ```

Kernel Facilities and Helper Functions

Since Linux kernel v4.8, it is possible to use `cancel_work` or `cancel_delayed_work`, which are asynchronous forms of cancellation. One must check whether the function returns true or no, and makes sure the work does not requeue itself. You must then explicitly flush the work queue:

```
if ( !cancel_delayed_work( &thework) ){
flush_workqueue(myqueue);
destroy_workqueue(myqueue);
}
```

The other is a different version of the same method and will create only a single thread for all the processors. In case you need a delay before the work is enqueued, feel free to use the following work initialization macro:

```
INIT_DELAYED_WORK(_work, _func);
INIT_DELAYED_WORK_DEFERRABLE(_work, _func);
```

Using the preceding macros would imply that you should use the following functions to queue or schedule the work in the work queue:

```
int queue_delayed_work(struct workqueue_struct *wq,
        struct delayed_work *dwork, unsigned long delay)
```

`queue_work` ties the work to the current CPU. You can specify the CPU on which the handler should run using the `queue_work_on` function:

```
int queue_work_on(int cpu, struct workqueue_struct *wq,
        struct work_struct *work);
```

For delayed work, you can use:

```
int queue_delayed_work_on(int cpu, struct workqueue_struct *wq,
    struct delayed_work *dwork, unsigned long delay);
```

The following is an example of using dedicated work queue:

```
#include <linux/init.h>
#include <linux/module.h>
#include <linux/workqueue.h>    /* for work queue */
#include <linux/slab.h>         /* for kmalloc() */

struct workqueue_struct *wq;
struct work_data {
    struct work_struct my_work;
    int the_data;
};
static void work_handler(struct work_struct *work)
{
```

```c
        struct work_data * my_data = container_of(work,
                            struct work_data, my_work);
        printk("Work queue module handler: %s, data is %d\n",
            __FUNCTION__, my_data->the_data);
        kfree(my_data);
}

static int __init my_init(void)
{
    struct work_data * my_data;

    printk("Work queue module init: %s %d\n",
            __FUNCTION__, __LINE__);
    wq = create_singlethread_workqueue("my_single_thread");
    my_data = kmalloc(sizeof(struct work_data), GFP_KERNEL);

    my_data->the_data = 34;
    INIT_WORK(&my_data->my_work, work_handler);
    queue_work(wq, &my_data->my_work);
    return 0;
}

static void __exit my_exit(void)
{
    flush_workqueue(wq);
    destroy_workqueue(wq);
    printk("Work queue module exit: %s %d\n",
            __FUNCTION__, __LINE__);
}

module_init(my_init);
module_exit(my_exit);
MODULE_LICENSE("GPL");
MODULE_AUTHOR("John Madieu <john.madieu@gmail.com>");
```

Predefined (shared) workqueue and standard workqueue functions

The predefined work queue is defined in `kernel/workqueue.c` as follows:

```
struct workqueue_struct *system_wq __read_mostly;
```

It is nothing more than a standard work for which the kernel provides a custom API that simply wraps around the standard one.

Comparisons between kernel predefined work queue functions and standard work queue functions are mentioned as follows:

Predefined work queue function	Equivalent standard work queue function
`schedule_work(w)`	`queue_work(keventd_wq,w)`
`schedule_delayed_work(w,d)`	`queue_delayed_work(keventd_wq,w,d)` (on any CPU)
`schedule_delayed_work_on(cpu,w,d)`	`queue_delayed_work(keventd_wq,w,d)` (on a given CPU)
`flush_scheduled_work()`	`flush_workqueue(keventd_wq)`

Kernel threads

Work queues run on top of kernel threads. You already use kernel threads when you use work queues. It is the reason why I have decided not to talk about the kernel thread API.

Kernel interruption mechanism

An interrupt is the way a device halts the kernel, telling it that something interesting or important has happened. These are called IRQs on Linux systems. The main advantage interrupts offer is to avoid devices polling. It is up to the device to tell if there is a change in its state; it is not up to us to poll it.

In order to get notified when an interrupt occurs, you need to register to that IRQ, providing a function called interrupt handler that will be called every time that interrupt is raised.

Registering an interrupt handler

You can register a callback to be run when the interruption (or interrupt line) you are interested in gets fired. You can achieve that with the function `request_irq()`, declared in `<linux/interrupt.h>`:

```
int request_irq(unsigned int irq, irq_handler_t handler,
    unsigned long flags, const char *name, void *dev)
```

`request_irq()` may fail, and return 0 on success. Other elements of the preceding code are outlined in detail as follows:

- `flags`: These should be a bitmask of the masks defined in `<linux/interrupt.h>`. The most used are:
    - `IRQF_TIMER`: Informs the kernel that this handler is originated by a system timer interrupt.
    - `IRQF_SHARED`: Used for interrupt lines that can be shared by two or more devices. Each device sharing the same line must have this flag set. If omitted, only one handler can be registered for the specified IRQ line.
    - `IRQF_ONESHOT`: Used essentially in the threaded IRQ. It instructs the kernel not to re-enable the interrupt when the hardirq handler has finished. It will remain disabled until the threaded handler has been run.
    - In older kernel versions (until v2.6.35), there were `IRQF_DISABLED` flags, which asked the kernel to disable all interrupts when the handler is running. This flag is no longer used.
- `name`: This is used by the kernel to identify your driver in `/proc/interrupts` and `/proc/irq`.
- `dev`: Its primary goal is to pass as argument to the handler. This should be unique to each registered handler, since it is used to identify the device. It can be `NULL` for nonshared IRQs, but not for shared ones. The common way of using it is to provide a `device` structure, since it is both unique and may potentially be useful to the handler. That said, a pointer to any per-device data structure is sufficient:

```
struct my_data {
   struct input_dev *idev;
   struct i2c_client *client;
   char name[64];
   char phys[32];
};
static irqreturn_t my_irq_handler(int irq, void *dev_id)
{
   struct my_data *md = dev_id;
   unsigned char nextstate = read_state(lp);
   /* Check whether my device raised the irq or no */
   [...]
   return IRQ_HANDLED;
}
/* some where in the code, in the probe function */
int ret;
```

Kernel Facilities and Helper Functions

```
struct my_data *md;
md = kzalloc(sizeof(*md), GFP_KERNEL);
ret = request_irq(client->irq, my_irq_handler,
                  IRQF_TRIGGER_LOW | IRQF_ONESHOT,
                  DRV_NAME, md);
/* far in the release function */
free_irq(client->irq, md);
```

- handler: This is the callback function that will run when the interrupt is fired. An interrupt handler's structure looks like:

  ```
  static irqreturn_t my_irq_handler(int irq, void *dev)
  ```

- This contains the following code elements:
    - irq: The numeric value of the IRQ (the same used in request_irq).
    - dev: The same as used in request_irq.

Both parameters are given to your handler by the kernel. There are only two values the handler can return, depending on whether your device originated the IRQ or not:

- IRQ_NONE: Your device is not the originator of that interrupt (it especially happens on shared IRQ lines)
- IRQ_HANDLED: Your device caused the interrupt

Depending on the processing, one may use the IRQ_RETVAL(val) macro, which will return IRQ_HANDLED if the value is nonzero, or IRQ_NONE otherwise.

When writing the interrupt handler, you don't have to worry about reentrancy, since the IRQ line serviced is disabled on all processors by the kernel in order to avoid recursive interrupt.

The associated function to free the previously registered handler is:

```
void free_irq(unsigned int irq, void *dev)
```

If the specified IRQ is not shared, free_irq will not only remove the handler, but will also disable the line. If it is shared, only the handler identified through dev (which should be the same as that used in request_irq) is removed, but the interrupt line still remains, and will be disabled only when the last handler is removed. free_irq will block until any executing interrupts for the specified IRQ have completed. You must then avoid both request_irq and free_irq in the interrupt context.

Interrupt handler and lock

It goes without saying that you are in an atomic context and must only use spinlock for concurrency. Whenever there is global data accessible by both user code (the user task; that is, the system call) and interrupt code, this shared data should be protected by `spin_lock_irqsave()` in the user code. Let's see why we can't just use `spin_lock`. An interrupt handler will always have priority on the user task, even if that task is holding a spinlock. Simply disabling IRQ is not sufficient. An interrupt may happen on another CPU. It would be a disaster if a user task updating the data gets interrupted by an interrupt handler trying to access the same data. Using `spin_lock_irqsave()` will disable all interrupts on the local CPU, preventing the system call from being interrupted by any kind of interrupt:

```
ssize_t my_read(struct file *filp, char __user *buf, size_t count,
    loff_t *f_pos)
{
    unsigned long flags;
    /* some stuff */
    [...]
    unsigned long flags;
    spin_lock_irqsave(&my_lock, flags);
    data++;
    spin_unlock_irqrestore(&my_lock, flags)
    [...]
}

static irqreturn_t my_interrupt_handler(int irq, void *p)
{
    /*
     * preemption is disabled when running interrupt handler
     * also, the serviced irq line is disabled until the handler has completed
     * no need then to disable all other irq. We just use spin_lock and
     * spin_unlock
     */
    spin_lock(&my_lock);
    /* process data */
    [...]
    spin_unlock(&my_lock);
    return IRQ_HANDLED;
}
```

When sharing data between different interrupt handlers (that is, the same driver managing two or more devices, each having its own IRQ line), one should also protect that data with `spin_lock_irqsave()` in those handlers, in order to prevent the other IRQs from being triggered and uselessly spinning.

Concept of bottom halves

Bottom halves are mechanisms by which you split interrupt handlers into two part. This introduces another term, which is top half. Before discussing each of them, let us talk about their origin, and what problem they solve.

The problem – interrupt handler design limitations

Whether an interrupt handler holds a spinlock or not, preemption is disabled on the CPU running that handler. The more one wastes time in the handler, the less CPU is granted to the other task, which may considerably increase latency of other interrupts and so increase the latency of the whole system. The challenge is to acknowledge the device that raised the interrupt as quickly as possible in order to keep the system responsive.

On Linux systems (actually on all OS, by hardware design), any interrupt handler runs with its current interrupt line disabled on all processors, and sometimes you may need to disable all interrupts on the CPU actually running the handler, but you definitely don't want to miss an interrupt. To meet this need, the concept of *halves* has been introduced.

The solution – bottom halves

This idea consists of splitting the handler into two parts:

- The first part, called the top half or hard-IRQ, which is the registered function using `request_irq()` that will eventually mask/hide interrupts (on the current CPU, except the one being serviced since it is already disabled by the kernel before running the handler) depending on the needs, performs quick and fast operations (essentially time-sensitive tasks, read/write hardware registers, and fast processing of this data), schedules the second and next part, and then acknowledges the line. All interrupts that are disabled must have been re-enabled just before exiting the bottom half.
- The second part, called the bottom half, will process time-consuming stuff, and run with interrupt re-enabled. This way, you have the chance not to miss an interrupt.

Bottom halves are designed using a work-deferring mechanism, which we have seen previously. Depending on which one you choose, it may run in a (software) interrupt context, or in a process context. Bottom halves' mechanisms are:

- Softirqs
- Tasklets
- Workqueues
- Threaded IRQs

Softirqs and tasklets execute in a (software) interrupt context (meaning that preemption is disabled), Workqueues and threaded IRQs are executed in a process (or simply task) context, and can be preempted, but nothing prevents us from changing their real-time properties to fit your needs and change their preemption behavior (see CONFIG_PREEMPT or CONFIG_PREEMPT_VOLUNTARY. This also impacts the whole system). Bottom halves are not always possible. But when it is possible, it is certainly the best thing to do.

Tasklets as bottom halves

The tasklet deferring mechanism is most used in DMA, network, and block device drivers. Just try the following command in the kernel source:

```
grep -rn tasklet_schedule
```

Now let's see how to implement such a mechanism in our interrupt handler:

```
struct my_data {
    int my_int_var;
    struct tasklet_struct the_tasklet;
    int dma_request;
};

static void my_tasklet_work(unsigned long data)
{
    /* Do what ever you want here */
}

struct my_data *md = init_my_data;

/* somewhere in the probe or init function */
[...]
    tasklet_init(&md->the_tasklet, my_tasklet_work,
                 (unsigned long)md);
[...]
```

Kernel Facilities and Helper Functions

```
static irqreturn_t my_irq_handler(int irq, void *dev_id)
{
    struct my_data *md = dev_id;

    /* Let's schedule our tasklet */
    tasklet_schedule(&md.dma_tasklet);

    return IRQ_HANDLED;
}
```

In the preceding sample, our tasklet will execute the function `my_tasklet_work()`.

Workqueue as bottom halves

Let's just start with a sample:

```
static DECLARE_WAIT_QUEUE_HEAD(my_wq);   /* declare and init the wait queue */
static struct work_struct my_work;

/* some where in the probe function */
/*
 * work queue initialization. "work_handler" is the call back that will be
 * executed when our work is scheduled.
 */
INIT_WORK(my_work, work_handler);

static irqreturn_t my_interrupt_handler(int irq, void *dev_id)
{
    uint32_t val;
    struct my_data = dev_id;

    val = readl(my_data->reg_base + REG_OFFSET);
    if (val == 0xFFCD45EE)) {
        my_data->done = true;
        wake_up_interruptible(&my_wq);
    } else {
        schedule_work(&my_work);
    }

    return IRQ_HANDLED;
};
```

In the preceding sample, we used either a wait queue or a work queue in order to wake up a possibly sleeping process waiting for us, or schedule a work depending on the value of a register. We have no shared data or resource, so there is no need to disable all other IRQs (`spin_lock_irq_disable`).

Softirqs as bottom half

As said in the beginning of this chapter, we will not discuss softirq. Tasklets will be enough everywhere you feel the need to use softirqs. Anyway, let's talk about their defaults.

Softirqs run in a software interrupt context, with preemption disabled, holding the CPU until they complete. Softirq should be fast; otherwise they may slow the system down. When, for any reason, a softirq prevents the kernel from scheduling other tasks, any new incoming softirq will be handled by **ksoftirqd** threads, running in a process context.

Threaded IRQs

The main goal of threaded IRQs is reducing the time spent with interrupts disabled to a bare minimum. With threaded IRQs, the way you register an interrupt handler is a bit simplified. You does not even have to schedule the bottom half yourself. The core does that for us. The bottom half is then executed in a dedicated kernel thread. We do not use `request_irq()` anymore, but `request_threaded_irq()`:

```
int request_threaded_irq(unsigned int irq, irq_handler_t handler,\
                 irq_handler_t thread_fn, \
                 unsigned long irqflags, \
                 const char *devname, void *dev_id)
```

Kernel Facilities and Helper Functions

The `request_threaded_irq()` function accepts two functions in its parameters:

- **@handler function**: This is the same function as the one registered with `request_irq()`. It represents the top-half function, which runs in an atomic context (or hard-IRQ). If it can process the interrupt faster so that you can get rid of the bottom half at all, it should return `IRQ_HANDLED`. But, if the interrupt processing needs more than 100 µs, as discussed previously, you should use the bottom half. In this case, it should return `IRQ_WAKE_THREAD`, which will result in scheduling the `thread_fn` function that must have been provided.
- **@thread_fn function**: This represents the bottom half, as you would have scheduled in your top half. When the hard-IRQ handler (handler function) function returns `IRQ_WAKE_THREAD`, the kthread associated with this bottom half will be scheduled, invoking the `thread_fn` function when it comes to run the ktread. The `thread_fn` function must return `IRQ_HANDLED` when complete. After being executed, the kthread will not be rescheduled again until the IRQ is triggered again and the hard-IRQ returns `IRQ_WAKE_THREAD`.

Everywhere that you would have used the work queue to schedule the bottom half, threaded IRQs can be used. `handler` and `thread_fn` must be defined in order to have a proper threaded IRQ. A default hard-IRQ handler will be installed by the kernel if `handler` is `NULL` and `thread_fn != NULL` (see the following), which will simply return `IRQ_WAKE_THREAD` to schedule the bottom half. `handler` is always called in an interrupt context, whether it has been provided by yourself or by the kernel by default:

```
/*
 * Default primary interrupt handler for threaded interrupts. Is
 * assigned as primary handler when request_threaded_irq is called
 * with handler == NULL. Useful for oneshot interrupts.
 */
static irqreturn_t irq_default_primary_handler(int irq, void *dev_id)
{
    return IRQ_WAKE_THREAD;
}

request_threaded_irq(unsigned int irq, irq_handler_t handler,
                     irq_handler_t thread_fn, unsigned long irqflags,
                     const char *devname, void *dev_id)
{
    [...]
    if (!handler) {
        if (!thread_fn)
            return -EINVAL;
        handler = irq_default_primary_handler;
    }
```

Kernel Facilities and Helper Functions

```
        [...]
}
EXPORT_SYMBOL(request_threaded_irq);
```

With threaded IRQs, the handler definition does not change, but the way it is registered changes a little bit.

```
request_irq(unsigned int irq, irq_handler_t handler, \
            unsigned long flags, const char *name, void *dev)
{
    return request_threaded_irq(irq, handler, NULL, flags, \
                                name, dev);
}
```

Threaded bottom half

The simple following excerpt is a demonstration of how you can implement the threaded bottom half mechanism:

```
static irqreturn_t pcf8574_kp_irq_handler(int irq, void *dev_id)
{
    struct custom_data *lp = dev_id;
    unsigned char nextstate = read_state(lp);

    if (lp->laststate != nextstate) {
        int key_down = nextstate < ARRAY_SIZE(lp->btncode);
        unsigned short keycode = key_down ?
            p->btncode[nextstate] : lp->btncode[lp->laststate];

        input_report_key(lp->idev, keycode, key_down);
        input_sync(lp->idev);
        lp->laststate = nextstate;
    }
    return IRQ_HANDLED;
}

static int pcf8574_kp_probe(struct i2c_client *client, \
                            const struct i2c_device_id *id)
{
    struct custom_data *lp = init_custom_data();
    [...]
    /*
     * @handler is NULL and @thread_fn != NULL
     * the default primary handler is installed, which will
     * return IRQ_WAKE_THREAD, that will schedule the thread
     * asociated to the bottom half. the bottom half must then
```

Kernel Facilities and Helper Functions

```
         * return IRQ_HANDLED when finished
         */
        ret = request_threaded_irq(client->irq, NULL, \
                            pcf8574_kp_irq_handler, \
                            IRQF_TRIGGER_LOW | IRQF_ONESHOT, \
                            DRV_NAME, lp);
        if (ret) {
            dev_err(&client->dev, "IRQ %d is not free\n", \
                        client->irq);
            goto fail_free_device;
        }
        ret = input_register_device(idev);
        [...]
}
```

When an interrupt handler is executed, the serviced IRQ is always disabled on all CPUs, and re-enabled when the hard-IRQ (top-half) finishes. But if for any reason you need the IRQ line not to be re-enabled after the top half, and to remain disabled until the threaded handler has been run, you should request the threaded IRQ with the flag IRQF_ONESHOT enabled (by just doing an OR operation as shown previously). The IRQ line will then be re-enabled after the bottom half has finished.

Invoking user-space applications from the kernel

User-space applications are most of the time called from within the user space by other applications. Without going deep into the details, let's see an example:

```
#include <linux/init.h>
#include <linux/module.h>
#include <linux/workqueue.h>    /* for work queue */
#include <linux/kmod.h>

static struct delayed_work initiate_shutdown_work;
static void delayed_shutdown( void )
{
    char *cmd = "/sbin/shutdown";
    char *argv[] = {
            cmd,
            "-h",
            "now",
```

[92]

```
            NULL,
    };
    char *envp[] = {
            "HOME=/",
            "PATH=/sbin:/bin:/usr/sbin:/usr/bin",
            NULL,
    };

    call_usermodehelper(cmd, argv, envp, 0);
}
static int __init my_shutdown_init( void )
{
    schedule_delayed_work(&delayed_shutdown, msecs_to_jiffies(200));
    return 0;
}

static void __exit my_shutdown_exit( void )
{
  return;
}
module_init( my_shutdown_init );
module_exit( my_shutdown_exit );

MODULE_LICENSE("GPL");
MODULE_AUTHOR("John Madieu", <john.madieu@gmail.com>);
MODULE_DESCRIPTION("Simple module that trigger a delayed shut down");
```

In the preceding example, the API used (`call_usermodehelper`) is a part of the Usermode-helper API, with all functions defined in `kernel/kmod.c`. Its use is quite simple; just a look inside `kmod.c` will give you an idea. You may be wondering what this API was defined for. It is used by the kernel, for example, for module (un)loading and cgroups management.

Summary

In this chapter, we discussed about the fundamental elements to start driver development, presenting every mechanism frequently used in drivers. This chapter is very important, since it discusses topics other chapters in this book rely on. The next chapter for example, dealing with character devices, will use some of elements discussed in this chapter.

4
Character Device Drivers

Character devices transfer data to or from a user application by means of characters, in a stream manner (one character after another), like a serial port does. A character device driver exposes the properties and functionalities of a device by means of a special file in the /dev directory, which one can use to exchange data between the device and user application, and also allows you to control the real physical device. This is the basic concept of Linux that says *everything is a file*. A character device driver represents the most basic device driver in the kernel source. Character devices are represented in the kernel as instances of struct cdev, defined in include/linux/cdev.h:

```
struct cdev {
    struct kobject kobj;
    struct module *owner;
    const struct file_operations *ops;
    struct list_head list;
    dev_t dev;
    unsigned int count;
};
```

This chapter will walk through the specificities of character device drivers, explain how they create, identify, and register the devices with the system, and also give a better overview of the device file methods, which are methods by which the kernel exposes the device capabilities to user space, accessible by using file related system calls (read, write, select, open, close and so on), described in struct file_operations structures, which you have certainly heard of before.

The concept behind major and minor

Character devices are populated in the `/dev` directory. Do note that, they are not only files present in that directory. A character device file is recognizable to its type, which we can display thanks to the command `ls -l`. Major and minor identify and tie the devices with the drivers. Let us see how it works, by listing the content of the `/dev` directory (`ls -l /dev`):

```
[...]
drwxr-xr-x 2 root root    160 Mar 21 08:57 input
crw-r----- 1 root kmem  1,   2 Mar 21 08:57 kmem
lrwxrwxrwx 1 root root     28 Mar 21 08:57 log -> /run/systemd/journal/dev-log
crw-rw---- 1 root disk 10, 237 Mar 21 08:57 loop-control
brw-rw---- 1 root disk  7,   0 Mar 21 08:57 loop0
brw-rw---- 1 root disk  7,   1 Mar 21 08:57 loop1
brw-rw---- 1 root disk  7,   2 Mar 21 08:57 loop2
brw-rw---- 1 root disk  7,   3 Mar 21 08:57 loop3
```

Given the preceding excerpt, the first character of the first column identifies the file type. Possible values are:

- `c`: This is for character device files
- `b`: This is for block device file
- `l`: This is for symbolic link
- `d`: This is for directory
- `s`: This is for socket
- `p`: This is for named pipe

For `b` and `c` file types, the fifth and sixth columns right before the date respect the `<X, Y>` pattern. `X` represents the major, and `Y` is the minor. For example, the third line is `<1, 2>` and the last one is `<7, 3>`. That is one of the classical methods for identifying a character device file from user space, as well as its major and minor.

The kernel holds the numbers that identify a device in `dev_t` type variables, which are simply `u32` (32-bit unsigned long). The major is represented with only 12 bits, whereas the minor is coded on the 20 remaining bits.

As one can see in `include/linux/kdev_t.h`, given a `dev_t` type variable, one may need to extract the minor or the major. The kernel provides a macro for these purposes:

```
MAJOR(dev_t dev);
MINOR(dev_t dev);
```

On the other hand, you may have a minor and a major, and need to build a `dev_t`. The macro you should use is `MKDEV(int major, int minor);`:

```
#define MINORBITS    20
#define MINORMASK    ((1U << MINORBITS) - 1)
#define MAJOR(dev)   ((unsigned int) ((dev) >> MINORBITS))
#define MINOR(dev)   ((unsigned int) ((dev) & MINORMASK))
#define MKDEV(ma,mi) (((ma) << MINORBITS) | (mi))
```

The device is registered with a major number that identifies the device, and a minor, which one may use as an array index to a local list of devices, since one instance of the same driver may handle several devices while different drivers may handle different devices of the same type.

Device number allocation and freeing

Device numbers identify device files across the system. That means, there are two ways to allocate these device numbers (actually major and minor):

- **Statically**: Guessing a major not yet used by another driver using the `register_chrdev_region()` function. One should avoid using this as much as possible. Its prototype looks this:

    ```
    int register_chrdev_region(dev_t first, unsigned int count, \
                               char *name);
    ```

    This method returns 0 on success, or a negative error code on failure. `first` is made of the major number that we need along with the first minor of the desired range. One should use `MKDEV(ma,mi)`. `count` is the number of consecutive device numbers required, and `name` should be the name of the associated device or driver.

- **Dynamically**: Letting the kernel do the job for us, using the `alloc_chrdev_region()` function. This is the recommended way to obtain a valid device number. Its prototype is as follows:

  ```
  int alloc_chrdev_region(dev_t *dev, unsigned int firstminor, \
                          unsigned int count, char *name);
  ```

  This method returns 0 on success, or a negative error code on failure. `dev` is the ony output parameter. It represents the first number the kernel assigned. `firstminor` is the first of the requested range of minor numbers, `count` the number of minors one requires, and `name` should be the name of the associated device or driver.

The difference between the two is that with the former, one should know in advance what number we need. This is registration: one tells the kernel what device numbers we want. This may be used for pedagogic purposes, and works as long as the only user of the driver is you. When it comes to loading the driver on another machine, there is no guarantee the chosen number is free on that machine, and this will lead to conflicts and trouble. The second method is cleaner and much safer, since the kernel is responsible for guessing the right numbers for us. We do not even have to care about what the behavior would be on loading the module on to another machine, since the kernel will adapt accordingly.

Anyway, the preceding functions are generally not called directly from the driver, but masked by the framework on which the driver relies (IIO framework, input framework, RTC, and so on), by means of dedicated API. These frameworks are all discussed in further chapters in the book.

Introduction to device file operations

Operations that one can perform on files depend on the drivers that manage those files. Such operations are defined in the kernel as instances of `struct file_operations`. `struct file_operations` exposes a set of callbacks that will handle any user-space system call on a file. For example, if one wants users to be able to perform a `write` on the file representing our device, one must implement the callback corresponding to that `write` function and add it into the `struct file_operations` that will be tied to your device. Let's fill in a file operations structure:

```
struct file_operations {
    struct module *owner;
    loff_t (*llseek) (struct file *, loff_t, int);
    ssize_t (*read) (struct file *, char __user *, size_t, loff_t *);
    ssize_t (*write) (struct file *, const char __user *, size_t, loff_t
```

Character Device Drivers

```
*);
    unsigned int (*poll) (struct file *, struct poll_table_struct *);
    int (*mmap) (struct file *, struct vm_area_struct *);
    int (*open) (struct inode *, struct file *);
    long (*unlocked_ioctl) (struct file *, unsigned int, unsigned long);
    int (*release) (struct inode *, struct file *);
    int (*fsync) (struct file *, loff_t, loff_t, int datasync);
    int (*fasync) (int, struct file *, int);
    int (*lock) (struct file *, int, struct file_lock *);
    int (*flock) (struct file *, int, struct file_lock *);
    [...]
};
```

The preceding excerpt only lists important methods of the structure, especially the ones that are relevant for the needs of this book. One can find the full description in include/linux/fs.h in kernel sources. Each of these callbacks is linked with a system call, and none of them is mandatory. When a user code calls a files-related system call on a given file, the kernel looks for the driver responsible for that file (especially the one that created the file), locates its struct file_operations structure, and checks whether the method that matches the system call is defined or not. If yes, it simply runs it. If not, it returns an error code that varies depending on the system call. For example, an undefined (*mmap) method will return -ENODEV to user, whereas an undefined (*write) method will return -EINVAL.

File representation in the kernel

The kernel describes files as instances of struct inode (not struct file) structure, defined in include/linux/fs.h:

```
struct inode {
    [...]
    struct pipe_inode_info *i_pipe;    /* Set and used if this is a
*linux kernel pipe */
    struct block_device *i_bdev;   /* Set and used if this is a
* a block device */
    struct cdev         *i_cdev;   /* Set and used if this is a
* character device */
    [...]
}
```

Character Device Drivers

The `struct inode` is a filesystem data structure holding information, which is only relevant to the OS, about a file (whatever its type, character, block, pipe, and so on) or directory (yes!! from a kernel point of view, a directory is a file that on entry points to other files) on disk.

The `struct file` structure (also defined in `include/linux/fs.h`) is actually a higher level of file description that represents an open file in the kernel and which relies on the lower `struct inode` data structure:

```
struct file {
    [...]
    struct path f_path;                 /* Path to the file */
    struct inode *f_inode;              /* inode associated to this file */
    const struct file_operations *f_op; /* operations that can be
          * performed on this file
          */
    loff_t f_pos;                       /* Position of the cursor in
   * this file */
    /* needed for tty driver, and maybe others */
    void *private_data;      /* private data that driver can set
                              * in order to share some data between file
                              * operations. This can point to any data
                              * structure.
     */
    [...]
}
```

The difference between `struct inode` and `struct file` is that an inode doesn't track the current position within the file or the current mode. It only contains stuff that helps the OS find the contents of the underlying file structure (pipe, directory, regular disk file, block/character device file, and so on). On the other hand, the `struct file` is used as a generic structure (it actually holds a pointer to a `struct inode` structure) that represents and open file and provides a set of functions related to methods one can perform on the underlying file structure. Such methods are: `open`, `write`, `seek`, `read`, `select`, and so on. All this reinforces the philosophy of UNIX systems that says *everything is file*.

In other words, a `struct inode` represents a file in the kernel, and a `struct file` describes it when it is actually open. There may be different file descriptors that represent the same file opened several times, but these will point to the same inode.

Allocating and registering a character device

Character devices are represented in the kernel as instances of `struct cdev`. When writing a character device driver, your goal is to finally create and register an instance of that structure associated with a `struct file_operations`, exposing a set of operations (functions) the user-space can perform on the device. To reach that goal, there are some steps we must go through, which are as follows:

1. Reserve a major and a range of minors with `alloc_chrdev_region()`.
2. Create a class for your devices with `class_create()`, visible in /sys/class/.
3. Set up a `struct file_operation` (to be given to `cdev_init`), and for each device one needs to create, call `cdev_init()` and `cdev_add()` to register the device.
4. Then `create a device_create()` for each device, with a proper name. It will result in your device being created in the /dev directory:

```
#define EEP_NBANK 8
#define EEP_DEVICE_NAME "eep-mem"
#define EEP_CLASS "eep-class"

struct class *eep_class;
struct cdev eep_cdev[EEP_NBANK];
dev_t dev_num;

static int __init my_init(void)
{
    int i;
    dev_t curr_dev;

    /* Request the kernel for EEP_NBANK devices */
    alloc_chrdev_region(&dev_num, 0, EEP_NBANK, EEP_DEVICE_NAME);

    /* Let's create our device's class, visible in /sys/class */
    eep_class = class_create(THIS_MODULE, EEP_CLASS);

    /* Each eeprom bank represented as a char device (cdev)   */
    for (i = 0; i < EEP_NBANK; i++) {

        /* Tie file_operations to the cdev */
        cdev_init(&my_cdev[i], &eep_fops);
        eep_cdev[i].owner = THIS_MODULE;

        /* Device number to use to add cdev to the core */
        curr_dev = MKDEV(MAJOR(dev_num), MINOR(dev_num) + i);
```

```
            /* Now make the device live for the users to access */
            cdev_add(&eep_cdev[i], curr_dev, 1);

            /* create a device node each device /dev/eep-mem0, /dev/eep-mem1,
             * With our class used here, devices can also be viewed under
             * /sys/class/eep-class.
             */
            device_create(eep_class,
                        NULL,      /* no parent device */
                        curr_dev,
                        NULL,      /* no additional data */
                        EEP_DEVICE_NAME "%d", i); /* eep-mem[0-7] */
    }
    return 0;
}
```

Writing file operations

After introducing the preceding file operations, it is time to implement them in order to enhance the driver capabilities and expose the device's methods to the user space (by means of system calls or course). Each of these methods has its particularities, which we will highlight in this section.

Exchanging data between kernel space and user space

This section does not describe any driver file operation but instead, introduces some kernel facilities that one may use to write these driver methods. The driver's write() method consists of reading data from user space to kernel space, and then processing that data from the kernel. Such processing could be something like *pushing* the data to the device, for example. On the other hand, the driver's read() method consists of copying data from the kernel to the user space. Both of these methods introduces new elements we need to discuss prior to jumping to their respective steps. The first one is __user. __user is a cookie used by sparse (a semantic checker used by the kernel to find possible coding faults) to let the developer know he is actually about to use an untrusted pointer (or a pointer that may be invalid in the current virtual address mapping) improperly and that he should not dereference but instead, use dedicated kernel functions to access the memory to which this pointer points.

This allows us to introduce different kernel functions needed to access such memory, either to read or write. These are `copy_from_user()` and `copy_from_user()` respectively to copy a buffer from user space to kernel space, and vice versa, to copy a buffer from kernel to user space:

```
unsigned long copy_from_user(void *to, const void __user *from,
                             unsigned long n)
unsigned long copy_to_user(void __user *to, const void *from,
                           unsigned long n)
```

In both cases, pointers prefixed with `__user` point to user space (untrusted) memory. `n` represents the number of bytes to copy. `from` represents the source address, and `to` is the destination address. Each of these returns the number of bytes that could not be copied. On success, the return value should be 0.

Please do note that with `copy_to_user()`, if some data could not be copied, the function will pad the copied data to the requested size using zero bytes.

A single value copy

When it comes to copying single and simple variables like `char` and `int` but not larger data types like structures nor arrays, the kernel offers dedicated macros in order to quickly perform the desired operation. These macros are `put_user(x, ptr)` and `get_used(x, ptr)`, which are explained as follows:

- `put_user(x, ptr);`: This macro copies a variable from kernel space to user space. `x` represents value to copy to user space, and `ptr` is the destination address in user space. The macro returns 0 on success, or `-EFAULT` on error. `x` must be assignable to the result of dereferencing `ptr`. In other words, they must have (or point to) the same type.
- `get_user(x, ptr);`: This macro copies a variable from user space to kernel space, and returns 0 on success or `-EFAULT` on error. Please do note that `x` is set to 0 on error. `x` represents the kernel variable to store the result, and `ptr` is the source address in user space. The result of dereferencing `ptr` must be assignable to `x` without a cast. Guess what it means.

The open method

open is the method called every time someone opens your device's file. Device opening will always success in case where this method is not defined. One usually uses this method to perform device and data structure initialization, and return a negative error code if something goes wrong, or 0. The prototype of open method is defined as follows:

```
int (*open)(struct inode *inode, struct file *filp);
```

Per-device data

For each open performed on your character device, the callback function will be given a struct inode as parameter, which is the kernel lower-level representation of the file. That struct inode structure has a field named i_cdev that points to the cdev we have allocated in the init function. By embedding the struct cdev in our device-specific data as in struct pcf2127 in the following example, we will be able to get a pointer on that specific data using the container_of macro. Here is an open method sample.

The following is our data structure:

```
struct pcf2127 {
    struct cdev cdev;
    unsigned char *sram_data;
    struct i2c_client *client;
    int sram_size;
    [...]
};
```

Given this data structure, the open method would look like this:

```
static unsigned int sram_major = 0;
static struct class *sram_class = NULL;

static int sram_open(struct inode *inode, struct file *filp)
{
    unsigned int maj = imajor(inode);
    unsigned int min = iminor(inode);

    struct pcf2127 *pcf = NULL;
    pcf = container_of(inode->i_cdev, struct pcf2127, cdev);
    pcf->sram_size = SRAM_SIZE;

    if (maj != sram_major || min < 0 ){
        pr_err ("device not found\n");
        return -ENODEV; /* No such device */
```

Character Device Drivers

```
    }

    /* prepare the buffer if the device is opened for the first time */
    if (pcf->sram_data == NULL) {
            pcf->sram_data = kzalloc(pcf->sram_size, GFP_KERNEL);
            if (pcf->sram_data == NULL) {
                    pr_err("Open: memory allocation failed\n");
                    return -ENOMEM;
            }
    }
    filp->private_data = pcf;
    return 0;
}
```

The release method

The `release` method is called when the device gets closed, the reverse of the `open` method. You must then undo everything you have done in the open task. What you have to do is roughly:

1. Free any private memory allocated during the `open()` step.
2. Shut down the device (if supported) and discard every buffer on the last closing (if the device supports multi opening, or if the driver can handle more than one device at a time).

The following is an excerpt of a `release` function:

```
static int sram_release(struct inode *inode, struct file *filp)
{
    struct pcf2127 *pcf = NULL;
    pcf = container_of(inode->i_cdev, struct pcf2127, cdev);

    mutex_lock(&device_list_lock);
    filp->private_data = NULL;

    /* last close? */
    pcf2127->users--;
    if (!pcf2127->users) {
            kfree(tx_buffer);
            kfree(rx_buffer);
            tx_buffer = NULL;
            rx_buffer = NULL;

            [...]
```

Character Device Drivers

```
            if (any_global_struct)
                    kfree(any_global_struct);
    }
    mutex_unlock(&device_list_lock);

    return 0;
}
```

The write method

The `write()` method is used to send data to the device; whenever a user app calls the `write` function on the device's file, the kernel implementation is called. Its prototype is as follows:

```
ssize_t(*write)(struct file *filp, const char __user *buf, size_t count,
loff_t *pos);
```

- The return value is the number of bytes (size) written
- `*buf` represents the data buffer coming from the user space
- `count` is the size of the requested transfer
- `*pos` indicates the start position from which data should be written in the file

Steps to write

The following steps do not describe any standard nor universal method to implement the driver's `write()` method. They are just an overview of what kind of operations one can perform in this method.

1. Check for bad or invalid requests coming from the user space. This step is relevant only if the device exposes its memory (eeprom, I/O memory, and so on), which may have size limitations:

    ```
    /* if trying to Write beyond the end of the file, return error.
     * "filesize" here corresponds to the size of the device memory (if
    any)
     */
    if ( *pos >= filesize ) return -EINVAL;
    ```

Character Device Drivers

2. Adjust count for the remaining bytes in order to not go beyond the file size. This step is not mandatory neither, and is relevant in the same condition as step 1:

   ```
   /* filesize coerresponds to the size of device memory */
   if (*pos + count > filesize)
       count = filesize - *pos;
   ```

3. Find the location from which you will start to write. This step is relevant only if the device has a memory in which the write() method is supposed to write given data. As steps 2 and 3, this step is not mandatory:

   ```
   /* convert pos into valid address */
   void *from = pos_to_address( *pos );
   ```

4. Copy data from the user space and write it into the appropriate kernel space:

   ```
   if (copy_from_user(dev->buffer, buf, count) != 0){
       retval = -EFAULT;
       goto out;
   }
   /* now move data from dev->buffer to physical device */
   ```

5. Write to the physical device and return an error on failure:

   ```
   write_error = device_write(dev->buffer, count);
   if ( write_error )
       return -EFAULT;
   ```

6. Increase the current position of the cursor in the file, according to the number of bytes written. Finally, return the number of bytes copied:

   ```
   *pos += count;
   Return count;
   ```

The following is an example of the write method. Once again, this is aimed to give an overview:

```
ssize_t
eeprom_write(struct file *filp, const char __user *buf, size_t count,
    loff_t *f_pos)
{
    struct eeprom_dev *eep = filp->private_data;
    ssize_t retval = 0;

    /* step (1) */
    if (*f_pos >= eep->part_size)
        /* Writing beyond the end of a partition is not allowed. */
```

```
            return -EINVAL;

    /* step (2) */
    if (*pos + count > eep->part_size)
        count = eep->part_size - *pos;

    /* step (3) */
    int part_origin = PART_SIZE * eep->part_index;
    int register_address = part_origin + *pos;

    /* step(4) */
    /* Copy data from user space to kernel space */
    if (copy_from_user(eep->data, buf, count) != 0)
        return -EFAULT;
        /* step (5) */
    /* perform the write to the device */
    if (write_to_device(register_address, buff, count) < 0){
        pr_err("ee24lc512: i2c_transfer failed\n");
        return -EFAULT;
     }

    /* step (6) */
    *f_pos += count;
    return count;
}
```

The read method

The prototype of the read() method is given as follows:

```
ssize_t (*read) (struct file *filp, char __user *buf, size_t count, loff_t *pos);
```

The return value is the size read. The rest of the method's elements are described here:

- *buf is the buffer we receive from the user space
- count is the size of the requested transfer (size of the user buffer)
- *pos indicates the start position from which data should be read in the file

Steps to read

1. Prevent from reading beyond the file size, and return end-of-file:

    ```
    if (*pos >= filesize)
        return 0; /* 0 means EOF */
    ```

2. The number of bytes read can't go beyond the file size. Adjust `count` appropriately:

    ```
    if (*pos + count > filesize)
        count = filesize - (*pos);
    ```

3. Find the location from which you will start the read:

    ```
    void *from = pos_to_address (*pos); /* convert pos into valid address */
    ```

4. Copy the data into the user-space buffer and return an error on failure:

    ```
    sent = copy_to_user(buf, from, count);
    if (sent)
        return -EFAULT;
    ```

5. Advance the file's current position according to the number of bytes read, and return the number of bytes copied:

    ```
    *pos += count;
    Return count;
    ```

The following is an example of a driver `read()` file operation, which is intended to give an overview of what can be done there:

```
ssize_t  eep_read(struct file *filp, char __user *buf, size_t count, loff_t *f_pos)
{
    struct eeprom_dev *eep = filp->private_data;

    if (*f_pos >= EEP_SIZE) /* EOF */
        return 0;

    if (*f_pos + count > EEP_SIZE)
        count = EEP_SIZE - *f_pos;

    /* Find location of next data bytes */
    int part_origin       =  PART_SIZE * eep->part_index;
    int eep_reg_addr_start  =  part_origin + *pos;
```

```
    /* perform the read from the device */
    if (read_from_device(eep_reg_addr_start, buff, count) < 0){
        pr_err("ee24lc512: i2c_transfer failed\n");
        return -EFAULT;
    }

    /* copy from kernel to user space */
    if(copy_to_user(buf, dev->data, count) != 0)
        return -EIO;

    *f_pos += count;
    return count;
}
```

The llseek method

The `llseek` function is called when one moves the cursor position within a file. The entry point of this method in user space is `lseek()`. One can refer to the man-page in order to print the full description of either method from user space: `man llseek` and `man lseek`. Its prototype looks as follows:

```
loff_t(*llseek) (structfile *filp, loff_t offset, int whence);
```

- The return value is the new position in the file
- `loff_t` is an offset, relative to the current file position, which defines how much it will be changed
- `whence` defines where to seek from. Possible values are:
    - SEEK_SET: This puts the cursor into a position relative to the beginning of the file
    - SEEK_CUR: This puts the cursor into a position relative to the current file position
    - SEEK_END: This adjusts the cursor to a position relative to end-of-file

Steps to llseek

1. Use the `switch` statement to check every possible `whence` case, since they are limited, and adjust `newpos` accordingly:

   ```
   switch( whence ){
       case SEEK_SET:/* relative from the beginning of file */
           newpos = offset; /* offset become the new position */
           break;
       case SEEK_CUR: /* relative to current file position */
           newpos = file->f_pos + offset; /* just add offset to the current position */
           break;
       case SEEK_END: /* relative to end of file */
           newpos = filesize + offset;
           break;
       default:
           return -EINVAL;
   }
   ```

2. Check whether `newpos` is valid:

   ```
   if ( newpos < 0 )
       return -EINVAL;
   ```

3. Update `f_pos` with the new position:

   ```
   filp->f_pos = newpos;
   ```

4. Return the new file-pointer position:

   ```
   return newpos;
   ```

The following is an example of a user program that successively reads and seeks into a file. The underlying driver will then execute the `llseek()` file operation entry:

```
#include <unistd.h>
#include <fcntl.h>
#include <sys/types.h>
#include <stdio.h>

#define CHAR_DEVICE "toto.txt"

int main(int argc, char **argv)
{
    int fd= 0;
    char buf[20];
```

```c
    if ((fd = open(CHAR_DEVICE, O_RDONLY)) < -1)
        return 1;

    /* Read 20 bytes */
    if (read(fd, buf, 20) != 20)
        return 1;
    printf("%s\n", buf);

    /* Move the cursor to 10 time, relative to its actual position */
    if (lseek(fd, 10, SEEK_CUR) < 0)
        return 1;
    if (read(fd, buf, 20) != 20)
        return 1;
    printf("%s\n",buf);

    /* Move the cursor ten time, relative from the beginig of the file */
    if (lseek(fd, 7, SEEK_SET) < 0)
        return 1;
    if (read(fd, buf, 20) != 20)
        return 1;
    printf("%s\n",buf);

    close(fd);
    return 0;
}
```

The code produces the following output:

```
jma@jma:~/work/tutos/sources$ cat toto.txt
Lorem ipsum dolor sit amet, consectetur adipiscing elit, sed do eiusmod
tempor incididunt ut labore et dolore magna aliqua.
jma@jma:~/work/tutos/sources$ ./seek
Lorem ipsum dolor si
nsectetur adipiscing
psum dolor sit amet,
jma@jma:~/work/tutos/sources$
```

The poll method

If one needs to implement a passive wait (not wasting CPU cycles while sensing the character device), one must implement the poll() function, which will be called whenever a user-space program performs a select() or poll() system calls on the file associated with the device:

```
unsigned int (*poll) (struct file *, struct poll_table_struct *);
```

Character Device Drivers

The kernel function at the heart of this method is `poll_wait()`, defined in `<linux/poll.h>`, which is the header one should include in driver code:

```
void poll_wait(struct file * filp, wait_queue_head_t * wait_address,
poll_table *p)
```

`poll_wait()` adds the device associated with a `struct file` structure (given as first parameter) to a list of those that can wake up processes (which have been put to sleep in the `struct wait_queue_head_t` structure given as second parameter), according to events registered into the `struct poll_table` structure given as third parameter. A user process can run `poll()`, `select()`, or `epoll()` system calls to add a set of files to a list on which it needs to wait, in order to be aware of the associated (if any) devices readiness. The kernel will then call the `poll` entry of the driver associated with each device file. The `poll` method of each driver should then call `poll_wait()` in order to register events for which the process needs to be notified with the kernel, put that process to sleep until one of these events occurs, and register the driver as one of those that can wake the process up. The usual way is to use a wait queue per event type (one for readability, another one for writability, and eventually one for exception if needed), according to events supported by the `select()` (or `poll()`) system call.

The return value of the `(*poll)` file operation must have `POLLIN | POLLRDNORM` set if there is data to read (at the moment, select or poll is called), `POLLOUT | POLLWRNORM` if the device is writable (at the moment, select or poll is called here as well), and 0 if there is no new data and the device is not yet writable. In the following example, we assume the device supports both blocking read and write. Of course one may implement only one of these. If the driver does not define this method, the device will be considered as always readable and writable, so that `poll()` or `select()` system calls return immediately.

Steps to poll

When one implements the `poll` function, either the `read` or `write` method may change:

1. Declare a wait queue for each event type (read, write, exception) one needs to implement passive wait, to put tasks in when there is no data to read, or when the device is not writable yet:

    ```
    static DECLARE_WAIT_QUEUE_HEAD(my_wq);
    static DECLARE_WAIT_QUEUE_HEAD(my_rq);
    ```

2. Implement the `poll` function like this:

   ```
   #include <linux/poll.h>
   static unsigned int eep_poll(struct file *file, poll_table *wait)
   {
       unsigned int reval_mask = 0;
       poll_wait(file, &my_wq, wait);
       poll_wait(file, &my_rq, wait);

       if (new-data-is-ready)
           reval_mask |= (POLLIN | POLLRDNORM);
       if (ready_to_be_written)
           reval_mask |= (POLLOUT | POLLWRNORM);
       return reval_mask;
   }
   ```

3. Notify the wait queue when there is new data or when the device is writable:

   ```
   wake_up_interruptible(&my_rq); /* Ready to read */
   wake_up_interruptible(&my_wq); /* Ready to be written to */
   ```

One can notify readable events either from within the driver's `write()` method, meaning that the written data can be read back, or from within an IRQ handler, meaning that an external device sent some data which can be read back. On the other hand, one can notify writable events either from within the driver's `read()` method, meaning that the buffer is empty and can be filled again, or from within an IRQ handler, meaning that the device has completed a data-send operation, and is ready to accept data again.

When using a sleepy input/output operation (blocked I/O), either the `read` or `write` method may change. The wait queue used in the poll must be used in read too. When the user needs to read, if there is data, that data will be sent immediately to the process and you must update the wait queue condition (set to `false`); if there is no data, the process is put to sleep in the wait queue.

If the `write` method is supposed to feed data, then in the `write` callback, you must fill the data buffer and update the wait queue condition (set to `true`), and wake up the reader (see the section *wait queue*). If it is an IRQ instead, these operations must be performed in their handler.

The following is an excerpt of a code that `select()` on a given char device in order to sense data availability:

```c
#include <unistd.h>
#include <fcntl.h>
#include <stdio.h>
#include <stdlib.h>
#include <sys/select.h>

#define NUMBER_OF_BYTE 100
#define CHAR_DEVICE "/dev/packt_char"

char data[NUMBER_OF_BYTE];

int main(int argc, char **argv)
{
    int fd, retval;
    ssize_t read_count;
    fd_set readfds;

    fd = open(CHAR_DEVICE, O_RDONLY);
    if(fd < 0)
        /* Print a message and exit*/
        [...]

    while(1){
        FD_ZERO(&readfds);
        FD_SET(fd, &readfds);

        /*
         * One needs to be notified of "read" events only, without timeout.
         * This call will put the process to sleep until it is notified the
         * event for which it registered itself
         */
        ret = select(fd + 1, &readfds, NULL, NULL, NULL);

        /* From this line, the process has been notified already */
        if (ret == -1) {
            fprintf(stderr, "select call on %s: an error ocurred",
CHAR_DEVICE);
            break;
        }
        /*
         * file descriptor is now ready.
         * This step assume we are interested in one file only.
         */
        if (FD_ISSET(fd, &readfds)) {
            read_count = read(fd, data, NUMBER_OF_BYTE);
```

```
            if (read_count < 0 )
                /* An error occured. Handle this */
                [...]

            if (read_count != NUMBER_OF_BYTE)
                /* We have read less than need bytes */
                [...] /* handle this */
            else
            /* Now we can process data we have read */
            [...]
        }
    }
    close(fd);
    return EXIT_SUCCESS;
}
```

The ioctl method

A typical Linux system contains around 350 system calls (syscalls), but only a few of them are linked with file operations. Sometimes devices may need to implement specific commands that are not provided by system calls, and especially the ones associated with files and thus device files. In this case, the solution is to use **input/output control(ioctl)**, which is a method by which one extends a list of syscalls (actually commands) associated with a device.. One can use it to send special commands to devices (reset, shutdown, configure, and so on). If the driver does not define this method, the kernel will return -ENOTTY error to any ioctl() system call.

In order to be valid and safe, an ioctl command needs to be identified by a number which should be unique to the system. The unicity of ioctl numbers across the system will prevent it from sending the right command to the wrong device, or passing the wrong argument to the right command (given a duplicated ioctl number). Linux provides four helper macros to create an ioctl identifier, depending on whether there is data transfer or not and on the direction of the transfer. Their respective prototypes are:

```
_IO(MAGIC, SEQ_NO)
_IOW(MAGIC, SEQ_NO, TYPE)
_IOR(MAGIC, SEQ_NO, TYPE)
_IORW(MAGIC, SEQ_NO, TYPE)
```

Their descriptions are as follows:

- `_IO`: The `ioctl` does not need data transfer
- `_IOW`: The `ioctl` needs write parameters (`copy_from_user` or `get_user`)
- `_IOR`: The `ioctl` needs read parameters (`copy_to_user` or `put_user`)
- `_IOWR`: The `ioctl` needs both write and read parameters

What their parameters mean (in the order they are passed) is described here:

1. A number coded on 8 bits (0 to 255), called magic number.
2. A sequence number or command ID, also on 8 bits.
3. A data type, if any, that will inform the kernel about the size to be copied.

It is well documented in *Documentation/ioctl/ioctl-decoding.txt* in the kernel source, and existing `ioctl` are listed in *Documentation/ioctl/ioctl-number.txt*, a good place to start when you need to create an `ioctl` command.

Generating ioctl numbers (command)

One should generate their own ioctl number in a dedicated header file. It is not mandatory, but it is recommended, since this header should be available in user space too. In other words, one should duplicate the ioctl header file so that there is one in the kernel and one in the user space, which one can include in user apps. Let's now generate ioctl numbers in a real example:

eep_ioctl.h:

```
#ifndef PACKT_IOCTL_H
#define PACKT_IOCTL_H
/*
 * We need to choose a magic number for our driver, and sequential numbers
 * for each command:
 */
#define EEP_MAGIC 'E'
#define ERASE_SEQ_NO 0x01
#define RENAME_SEQ_NO 0x02
#define ClEAR_BYTE_SEQ_NO 0x03
#define GET_SIZE 0x04

/*
 * Partition name must be 32 byte max
 */
#define MAX_PART_NAME 32
```

Character Device Drivers

```
/*
 * Now let's define our ioctl numbers:
 */
#define EEP_ERASE _IO(EEP_MAGIC, ERASE_SEQ_NO)
#define EEP_RENAME_PART _IOW(EEP_MAGIC, RENAME_SEQ_NO, unsigned long)
#define EEP_GET_SIZE _IOR(EEP_MAGIC, GET_SIZE, int *)
#endif
```

Steps for ioctl

First, let us have a look at its prototype. It look likes as follows:

```
long ioctl(struct file *f, unsigned int cmd, unsigned long arg);
```

There is only one step: use a switch ... case statement and return an -ENOTTY error when an undefined ioctl command is called. One can find more information at http://man7.org/linux/man-pages/man2/ioctl.2.html:

```
/*
 * User space code also need to include the header file in which ioctls
 * defined are defined. This is eep_ioctl.h in our case.
 */
#include "eep_ioctl.h"
static long eep_ioctl(struct file *f, unsigned int cmd, unsigned long arg)
{
    int part;
    char *buf = NULL;
    int size = 1300;

    switch(cmd){
        case EEP_ERASE:
            erase_eepreom();
            break;
        case EEP_RENAME_PART:
            buf = kmalloc(MAX_PART_NAME, GFP_KERNEL);
            copy_from_user(buf, (char *)arg, MAX_PART_NAME);
            rename_part(buf);
            break;
        case EEP_GET_SIZE:
            copy_to_user((int*)arg, &size, sizeof(int));
            break;
        default:
            return -ENOTTY;
    }
    return 0;
}
```

If you think your ioctl command will need more than one argument, you should gather those arguments in a structure and just pass a pointer from the structure to ioctl.

Now, from the user space, you must use the same ioctl header as in the driver's code:

my_main.c

```c
#include <stdio.h>
#include <stdlib.h>
#include <fcntl.h>
#include <unistd.h>
#include "eep_ioctl.h"  /* our ioctl header file */

int main()
{
    int size = 0;
    int fd;
    char *new_name = "lorem_ipsum"; /* must not be longer than MAX_PART_NAME */

    fd = open("/dev/eep-mem1", O_RDWR);
    if (fd == -1){
        printf("Error while opening the eeprom\n");
        return -1;
    }

    ioctl(fd, EEP_ERASE);   /* ioctl call to erase partition */
    ioctl(fd, EEP_GET_SIZE, &size); /* ioctl call to get partition size */
    ioctl(fd, EEP_RENAME_PART, new_name);  /* ioctl call to rename partition */

    close(fd);
    return 0;
}
```

Filling the file_operations structure

When writing kernel modules, it is better to use designated initializers when it comes to statically initialize structures with their parameters. It consists of naming the member one needs to assign a value to. The form is `.member-name` to designate what member should be initialized. This allows, among other things, initializing the members in an undefined order, or leaving unchanged the fields that we do not want to modify.

Once we have defined our functions, we just have to fill the structure as follows:

```
static const struct file_operations eep_fops = {
    .owner =    THIS_MODULE,
    .read =     eep_read,
    .write =    eep_write,
    .open =     eep_open,
    .release =  eep_release,
    .llseek =   eep_llseek,
    .poll =     eep_poll,
    .unlocked_ioctl = eep_ioctl,
};
```

Let us remember, the structure is given as a parameter to `cdev_init` in the `init` method.

Summary

In this chapter, we have demystified character devices and we have seen how to let users interact with our driver through device files. We learned how to expose file operations to the user space and control their behavior from within the kernel. We went so far that you are even able to implement multi-device support. The next chapter is a bit hardware oriented since it deals with platform drivers which expose hardware device capabilities to the user space. The power of character drivers combined with platform drivers is just amazing. See you in the next chapter.

5
Platform Device Drivers

We all know about plug and play devices. They are handled by the kernel as soon as they are plugged in. These may be USB or PCI express, or any other auto-discovered devices. Therefore, other device types also exist, which are not hot-pluggable, and which the kernel needs to know about prior to being managed. There are I2C, UART, SPI, and other devices not wired to enumeration-capable buses.

There are real physical buses you may already know: USB, I2S, I2C, UART, SPI, PCI, SATA, and so on. Such buses are hardware devices named controllers. Since they are a part of SoC, they can't be removed, are non-discoverable, and are also called platform devices.

People often say platform devices are on-chip devices (embedded in the SoC). In practice, it is partially true, since they are hard-wired into the chip and can't be removed. But devices connected to I2C or SPI are not on-chip, and are platform devices too, because they are not discoverable. Similarly, there may be on-chip PCI or USB devices, but they are not platform devices, because they are discoverable.

From an SoC point of view, those devices (buses) are connected internally through dedicated buses, and are most of the time proprietary and specific to the manufacturer. From the kernel point of view, these are root devices, and connected to nothing. That is where the *pseudo platform bus* comes in. The pseudo platform bus, also called platform bus is a kernel virtual bus for devices that do not seat on a physical bus known to the kernel. In this chapter, platform devices refer to devices that rely on the pseudo platform bus.

Dealing with platform devices essentially requires two steps:

- Register a platform driver (with a unique name) that will manage your devices
- Register your platform device with the same name as the driver, and their resources, in order to let the kernel know that your device is there

That being said, in this chapter, we will discuss the following:

- Platform devices along with their driver
- Devices and driver-matching mechanisms in the kernel
- Registering platform drivers with devices, as well as platform data

Platform drivers

Before going further, please pay attention to the following warning. Not all platform devices are handled by platform drivers (or should I say pseudo platform drivers). Platform drivers are dedicated to devices not based on conventional buses. I2C devices or SPI devices are platform devices, but respectively rely on I2C or SPI buses, not on the platform bus. Everything needs to be done manually with the platform driver. The platform driver must implement a `probe` function, called by the kernel when the module is inserted or when a device claims it. When developing platform drivers, the main structure one has to fill is `struct platform_driver`, and registering your driver with the platform bus core with dedicated functions shown as follows:

```
static struct platform_driver mypdrv = {
    .probe      = my_pdrv_probe,
    .remove     = my_pdrv_remove,
    .driver     = {
        .name   = "my_platform_driver",
        .owner  = THIS_MODULE,
    },
};
```

Let us see what the meaning is of each element that composes the structure, and what they are used for:

- `probe()`: This is the function that gets called when a device claims your driver after a match occurs. Later, we will see how `probe` is called by the core. Its declaration is as follows:

    ```
    static int my_pdrv_probe(struct platform_device *pdev)
    ```

- `remove()`: This is called to get rid of the driver when it is not needed anymore by devices, and its declaration looks like this:

    ```
    static int my_pdrv_remove(struct platform_device *pdev)
    ```

- `struct device_driver`: This describes the driver itself, providing a name, owner, and some field, which we will see later.

Registering a platform driver with the kernel is as simple as calling `platform_driver_register()` or `platform_driver_probe()` in the `init` function (when the module is loaded). The difference between those functions is that:

- `platform_driver_register()` registers and puts the driver into a list of drivers maintained by the kernel, so that its `probe()` function can be called on demand whenever a new match occurs. To prevent your driver from being inserted and registered in that list, just use the `next` function.
- With `platform_driver_probe()`, the kernel immediately runs the match loop, checks if there is a platform device with the matching name, and then calls the driver's `probe()` if a match occurred, meaning that the device is present. If not, the driver is ignored. This method prevents the deferred probe, since it does not register the driver on the system. Here, the `probe` function is placed in an `__init` section, which is freed when the kernel boot has completed, thus preventing the deferred probe and reducing the driver's memory footprint. Use this method if you are 100% sure the device is present in the system:

```
ret = platform_driver_probe(&mypdrv, my_pdrv_probe);
```

The following is a simple platform driver that registers itself with the kernel:

```
#include <linux/module.h>
#include <linux/kernel.h>
#include <linux/init.h>
#include <linux/platform_device.h>

static int my_pdrv_probe (struct platform_device *pdev){
    pr_info("Hello! device probed!\n");
    return 0;
}

static void my_pdrv_remove(struct platform_device *pdev){
    pr_info("good bye reader!\n");
}

static struct platform_driver mypdrv = {
    .probe      = my_pdrv_probe,
    .remove     = my_pdrv_remove,
    .driver = {
            .name  = KBUILD_MODNAME,
            .owner = THIS_MODULE,
    },
```

Platform Device Drivers

```
};

static int __init my_drv_init(void)
{
    pr_info("Hello Guy\n");

    /* Registering with Kernel */
    platform_driver_register(&mypdrv);
    return 0;
}
static void __exit my_pdrv_remove (void)
{
    Pr_info("Good bye Guy\n");

    /* Unregistering from Kernel */
    platform_driver_unregister(&my_driver);
}

module_init(my_drv_init);
module_exit(my_pdrv_remove);

MODULE_LICENSE("GPL");
MODULE_AUTHOR("John Madieu");
MODULE_DESCRIPTION("My platform Hello World module");
```

Our module does nothing else in the `init/exit` function but register/unregister with the platform bus core. This is the case with most drivers. In this case, we can get rid of `module_init` and `module_exit`, and use the `module_platform_driver` macro.

The `module_platform_driver` macro looks like as follows:

```
/*
 * module_platform_driver() - Helper macro for drivers that don't
 * do anything special in module init/exit. This eliminates a lot
 * of boilerplate.  Each module may only use this macro once, and
 * calling it replaces module_init() and module_exit()
 */
#define module_platform_driver(__platform_driver) \
module_driver(__platform_driver, platform_driver_register, \
platform_driver_unregister)
```

Platform Device Drivers

This macro will be responsible for registering our module with the platform driver core. No need for `module_init` and `module_exit` macros, nor `init` and `exit` function anymore. It does not mean that those functions are not called anymore, just that we can forgot writing them by ourselves.

The `probe` function is not a substitute to `init` function. The `probe` function is called every time when a given device matches with the driver, whereas the `init` function runs only once, when the module gets loaded.

```
[...]
static int my_driver_probe (struct platform_device *pdev){
    [...]
}

static void my_driver_remove(struct platform_device *pdev){
    [...]
}

static struct platform_drivermy_driver = {
    [...]
};
module_platform_driver(my_driver);
```

There are specific macros for each bus that one needs to register the driver with. The following list is not exhaustive:

- `module_platform_driver(struct platform_driver)` for platform drivers, dedicated to devices that do not sit on conventional physical buses (we just used it above)
- `module_spi_driver(struct spi_driver)` for SPI drivers
- `module_i2c_driver(struct i2c_driver)` for I2C drivers
- `module_pci_driver(struct pci_driver)` for PCI drivers
- `module_usb_driver(struct usb_driver)` for USB drivers
- `module_mdio_driver(struct mdio_driver)` for mdio
- [...]

If you don't know which bus your driver needs to sit on, then it is a platform driver, and you should use `platform_driver_register` or `platform_driver_probe` to register the driver.

[125]

Platform devices

Actually, we should have said pseudo platform device, since this section concerns devices that sit on pseudo platform buses. When you are done with the driver, you will have to feed the kernel with devices needing that driver. A platform device is represented in the kernel as an instance of `struct platform_device`, and looks as follows:

```
struct platform_device {
    const char *name;
    u32 id;
    struct device dev;
    u32 num_resources;
    struct resource *resource;
};
```

When it comes to the platform driver, before driver and device match, the `name` field of both `struct platform_device` and `static struct platform_driver.driver.name` must be the same. The `num_resources` and `struct resource *resource` field will be covered in the next section. Just remember that, since `resource` is an array, `num_resources` must contain the size of that array.

Resources and platform data

At the opposite end to hot-pluggable devices, the kernel has no idea of what devices are present on your system, what they are capable of, or what they need in order to work properly. There is no auto-negotiation process, so any information provided to the kernel would be welcome. There are two methods to inform the kernel about the resources (irq, dma, memory region, I/O ports, buses) and data (any custom and private data structure you may want to pass to the driver) that the device needs which are discussed as follows:

Device provisioning - the old and depreciated way

This method is to be used with the kernel version that does not support device tree. With this method, drivers remain generic, and devices are registered in board-related source files.

Resources

Resources represent all the elements that characterize the device from the hardware point of view, and which the device needs, in order to be set up and work properly. There are only six types of resources in the kernel, all listed in include/linux/ioport.h, and used as flags to describe the resource's type:

```
#define IORESOURCE_IO   0x00000100   /* PCI/ISA I/O ports */
#define IORESOURCE_MEM  0x00000200   /* Memory regions */
#define IORESOURCE_REG  0x00000300   /* Register offsets */
#define IORESOURCE_IRQ  0x00000400   /* IRQ line */
#define IORESOURCE_DMA  0x00000800   /* DMA channels */
#define IORESOURCE_BUS  0x00001000   /* Bus */
```

A resource is represented in the kernel as an instance of struct resource:

```
struct resource {
        resource_size_t start;
        resource_size_t end;
        const char *name;
        unsigned long flags;
};
```

Let us explain the meaning of each element in the structure:

- start/end: This represents where the resource begins/ends. For I/O or memory regions, it represents where they begin/end. For IRQ lines, buses or DMA channels, start/end must have the same value.
- flags: This is a mask that characterizes the type of resource, for example IORESOURCE_BUS.
- name: This identifies or describes the resource.

Once one has provided the resources, one needs to extract them back in the driver in order to work with them. The probe function is a good place to extract them. Before one goes further, let's remember the declaration of the probe function for a platform device driver:

```
int probe(struct platform_device *pdev);
```

pdev is automatically filled by the kernel, with the data and resource we registered earlier. Let's see how to pick them.

Platform Device Drivers

The `struct resource` embedded in `struct platform_device` can be retrieved with the `platform_get_resource()` function. The following is the prototype of `platform_get_resource`:

```
struct resource *platform_get_resource(structplatform_device *dev,
                unsigned int type, unsigned int num);
```

The first parameter is an instance of the platform device itself. The second parameter tells what kind of resource we need. For memory, it should be `IORESOURCE_MEM`. Again, please have a look at `include/linux/ioport.h` for more details. `num` parameter is an index that says which resource type is desired. Zero indicates the first one, and so on.

If the resource is an IRQ, we must use `int platform_get_irq(struct platform_device * pdev, unsigned intnum)`, where `pdev` is the platform device, and `num` is the IRQ index within the resource (in case there is more than one). The whole `probe` function which we can use to extract the platform data which we registered for our device can look as follows:

```
static int my_driver_probe(struct platform_device *pdev)
{
struct my_gpios *my_gpio_pdata =
                (struct my_gpios*)dev_get_platdata(&pdev->dev);

    int rgpio = my_gpio_pdata->reset_gpio;
    int lgpio = my_gpio_pdata->led_gpio;

    struct resource *res1, *res2;
    void *reg1, *reg2;
    int irqnum;

    res1 = platform_get_resource(pdev, IORESSOURCE_MEM, 0);
    if((!res1)){
        pr_err(" First Resource not available");
        return -1;
    }
    res2 = platform_get_resource(pdev, IORESSOURCE_MEM, 1);
    if((!res2)){
        pr_err(" Second Resource not available");
        return -1;
    }

    /* extract the irq */
    irqnum = platform_get_irq(pdev, 0);
    Pr_info("\n IRQ number of Device: %d\n", irqnum);

    /*
```

```
 * At this step, we can use gpio_request, on gpio,
 * request_irq on irqnum and ioremap() on reg1 and reg2.
 * ioremap() is discussed in chapter 11, Kernel Memory Management
 */
[...]
return 0;
}
```

Platform data

Any other data whose type is not a part of the resource types enumerated in the preceding section fall here (for example, GPIO). Whatever their type is, the struct platform_device contains a struct device field, which in turn contains a struct platform_data field. Usually, one should embed that data in a structure and pass it to the platform_device.device.platform_data field. Let's say, for example, that you declare a platform device that needs two gpios number as platform data, one irq number, and two memory regions as resource. The following example shows how to register platform data along with the device. Here, we use platform_device_register(struct platform_device *pdev) function, which one uses to register a platform device with the platform core:

```
/*
 * Other data than irq or memory must be embedded in a structure
 * and passed to "platform_device.device.platform_data"
 */
struct my_gpios {
    int reset_gpio;
    int led_gpio;
};

/*our platform data*/
static struct my_gpiosneeded_gpios = {
    .reset_gpio = 47,
    .led_gpio   = 41,
};

/* Our resource array */
static struct resource needed_resources[] = {
    [0] = { /* The first memory region */
        .start = JZ4740_UDC_BASE_ADDR,
        .end   = JZ4740_UDC_BASE_ADDR + 0x10000 - 1,
        .flags = IORESOURCE_MEM,
        .name  = "mem1",
    },
    [1] = {
        .start = JZ4740_UDC_BASE_ADDR2,
```

Platform Device Drivers

```
                .end   = JZ4740_UDC_BASE_ADDR2 + 0x10000 -1,
                .flags = IORESOURCE_MEM,
                .name  = "mem2",
        },
        [2] = {
                .start = JZ4740_IRQ_UDC,
                .end   = JZ4740_IRQ_UDC,
                .flags = IORESOURCE_IRQ,
                .name  = "mc",
        },
};

static struct platform_devicemy_device = {
    .name = "my-platform-device",
    .id   = 0,
    .dev  = {
        .platform_data      = &needed_gpios,
    },
    .resource               = needed_resources,
    .num_resources = ARRY_SIZE(needed_resources),
};
platform_device_register(&my_device);
```

In the preceding example, we have used `IORESOURCE_IRQ` and `IORESOURCE_MEM` in order to inform the kernel about what kind of resource we provided. To see all other flag types, have a look at `include/linux/ioport.h` in the kernel tree.

In order to retrieve the platform data we registered earlier, we could have just used `pdev->dev.platform_data` (remember the `struct platform_device` structure), but it is recommended to use the kernel-provided function (which does the same thing, admittedly):

```
void *dev_get_platdata(const struct device *dev)
struct my_gpios *picked_gpios = dev_get_platdata(&pdev->dev);
```

Where to declare platform devices?

Devices are registered along with their resources and data. In this old and depreciated method, they are declared a separate module, or in the board `init` file in the `arch/<arch>/mach-xxx/yyyy.c`, which is `arch/arm/mach-imx/mach-imx6q.c` in our case, since we use a UDOO quad based on an i.MX6Q from NXP. The function `platform_device_register()` lets you do that:

```
static struct platform_device my_device = {
        .name                   = "my_drv_name",
        .id                     = 0,
        .dev.platform_data      = &my_device_pdata,
        .resource               = jz4740_udc_resources,
        .num_resources          = ARRY_SIZE(jz4740_udc_resources),
};
platform_device_register(&my_device);
```

The name of the device is very important, and is used by the kernel to match the driver with the same name.

Device provisioning - the new and recommended way

In the first method, any modification will necessitate rebuilding the whole kernel. If the kernel had to include any application/board-specific configurations, its size would incredibly increase. In order to keep things simple, and separate devices declarations (since they are not really part of the kernel) from the kernel source, a new concept has been introduced: *the device tree*. The main goal of DTS is to remove very specific and never-tested code from kernel. With the device tree, platform data and resources are homogenous. The device tree is a hardware description file and has a format similar to a tree structure, where every device is represented with a node, and any data or resource or configuration data is represented as the node's property. This way, you only need to recompile the device tree when you make some modifications. The device tree forms the subject of the next chapter, and we will see how to introduce it to the platform device.

Devices, drivers, and bus matching

Before any match can occur, Linux calls the `platform_match(struct device *dev, struct device_driver *drv)`. Platform devices are matched with their drivers by means of strings. According to the Linux device model, the bus element is the most important part. Each bus maintains a list of drivers and devices that are registered with it. The bus driver is responsible for devices and drivers matching. Any time one connects a new device or adds a new driver to a bus, that bus starts the matching loop.

Now, suppose that you register a new I2C device using functions provided by the I2C core (discussed in next chapter). The kernel will trigger the I2C bus matching loop, by calling the I2C core match function registered with the I2C bus driver, to check if there is already a registered driver that matches with your device. If there is no match, nothing will happen. If a match occurs, the kernel will notify (by means of a communication mechanism called netlink socket) the device manager (udev/mdev), which will load (if not loaded yet) the driver your device matched with. Once the driver loads, its `probe()` function will immediately be executed. Not only does I2C work like that, but every bus has its own matching mechanism that is roughly the same. A bus matching loop is triggered at each device or driver registration.

We can sum up what we have said in the preceding section in the following figure:

List of I2C drivers	List of USB drivers			List of Platform drivers
Driver foo	Driver bar			our gpio_reset driver
driver XXX	driver YYY			our led-gpio driver
...	...			...
...	...			...
driver B	driver B			Our gpio_keyboard matrix
driver A	driver A	...	...	our other driver foobar
I2C bus driver	USB bus driver	...	...	Platform Bus
Physical BUS (SPI, I2C, PCI, SATA, MDIO, USB ...)				

Every registered drivers and devices sit on a bus. This makes a tree. USB buses may be children of PCI buses, whereas MDIO buses are generally children of other devices, and so on. Thus, our preceding figure changes as follows:

List of I2C drivers	List of USB drivers			List of Platform drivers
				Our gpio_keyboard matrix
				our other driver foobar
Driver foo	Driver bar			our gpio_reset driver
driver XXX	driver YYY			our led-gpio driver
...	...			SPI Bus driver
...	...			UART Bus driver
				...
driver B	driver B			...
driver A	driver A	...	...	PCIe Bus driver
I2C bus driver	USB bus driver	...	...	USB Bus driver
Pseudo Platform Bus				
Physical BUS (SPI, I2C, PCI, SATA, MDIO, USB ...)				

When you register a driver with the `platform_driver_probe()` function, the kernel walks through the table of registered platform devices and looks for a match. If any, it calls the matched driver's `probe` function with the platform data.

How can platform devices and platform drivers match?

So far, we have only discussed how to fill different structures of both devices and drivers.But now we will see how they are registered with the kernel, and how Linux knows which devices are handled by which driver. The answer is MODULE_DEVICE_TABLE. This macro lets a driver expose its ID table, which describes which devices it can support. In the meantime, if the driver can be compiled as a module, the `driver.name` field should match the module name. If it does not match, the module won't be automatically loaded, unless we have used the MODULE_ALIAS macro to add another name for the module. At compilation time, that information is extracted from all the drivers in order to build a device table. When the kernel has to find the driver for a device (when a matching needs to be performed), the device table is walked through by the kernel. If an entry is found matching the `compatible` (for device tree), `device/vendor id` or `name` (for device ID table or name) of the added device, then the module providing that match is loaded (running the module's `init` function), and the `probe` function is called. The MODULE_DEVICE_TABLE macro is defined in `linux/module.h`:

```
#define MODULE_DEVICE_TABLE(type, name)
```

The following is the description of each parameter given to this macro:

- `type`: This can be either `i2c`, `spi`, `acpi`, `of`, `platform`, `usb`, `pci` or any other bus which you may find in `include/linux/mod_devicetable.h`. It depends on the bus our device sits on, or on the matching mechanism we want to use.
- `name`: This is a pointer on a `XXX_device_id` array, used for device matching. If we were talking about I2C devices, the structure would be `i2c_device_id`. For SPI device, it should be `spi_device_id`, and so on. For the device tree **Open Firmware** (**OF**) matching mechanism, we must use `of_device_id`.

For new non-discoverable platform device drivers, it is recommended not to use platform data anymore, but to use device tree capabilities instead, with OF matching mechanism. Please do note that the two methods are not mutually exclusive, thus one can mix these together.

Let's get deeper into the details for matching mechanisms, except for the OF style match which we will discuss in `Chapter 6`, *The Concept of Device Tree*.

Kernel devices and drivers-matching function

The function responsible for platform devices and driver-matching functions in kernel is defined in /drivers/base/platform.c as follows:

```
static int platform_match(struct device *dev, struct device_driver *drv)
{
    struct platform_device *pdev = to_platform_device(dev);
    struct platform_driver *pdrv = to_platform_driver(drv);

    /* When driver_override is set, only bind to the matching driver */
    if (pdev->driver_override)
        return !strcmp(pdev->driver_override, drv->name);

    /* Attempt an OF style match first */
    if (of_driver_match_device(dev, drv))
        return 1;

    /* Then try ACPI style match */
    if (acpi_driver_match_device(dev, drv))
        return 1;

    /* Then try to match against the id table */
    if (pdrv->id_table)
        return platform_match_id(pdrv->id_table, pdev) != NULL;

    /* fall-back to driver name match */
    return (strcmp(pdev->name, drv->name) == 0);
}
```

We can enumerate four matching mechanisms. They are all based on the string compare. If we have a look at platform_match_id, we'll understand how things work underneath:

```
static const struct platform_device_id *platform_match_id(
                    const struct platform_device_id *id,
                    struct platform_device *pdev)
{
    while (id->name[0]) {
        if (strcmp(pdev->name, id->name) == 0) {
            pdev->id_entry = id;
            return id;
        }
        id++;
    }
    return NULL;
}
```

Now let's have a look at the `struct device_driver` structure we discussed in Chapter 4, *Character Device Drivers:*

```
struct device_driver {
        const char *name;
        [...]
        const struct of_device_id      *of_match_table;
        const struct acpi_device_id    *acpi_match_table;
};
```

I intentionally removed fields that we are not interested in. `struct device_driver` forms the basis of every device driver. Whether it is an I2C, SPI, TTY, or other device driver, they all embed a `struct device_driver` element.

OF style and ACPI match

OF style is explained in Chapter 6, *The Concept of Device Tree*. The second mechanism is an ACPI table-based matching. We'll not discuss it at all in this book, but for your information, it uses struct `acpi_device_id`.

ID table matching

This match style has been for a long time, and is based on the `struct device_id` structure. All device id structures are defined in `include/linux/mod_devicetable.h`. To find the right structure name, you need to prefix `device_id` with the bus name whom your device driver seats on. Examples are: `struct i2c_device_id` for I2C, `struct platform_device_id` for platform devices (that don't sit on a real physical bus), `spi_device_id` for SPI devices, `usb_device_id` for USB, and so on. The typical structure of a `device_id table` for a platform device is as follows:

```
struct platform_device_id {
   char name[PLATFORM_NAME_SIZE];
   kernel_ulong_t driver_data;
};
```

Anyway, if an ID table is registered, it will be walked through whenever the kernel has run the match function to find a driver for an unknown or new platform device. If there is a match, the `probe` function of the matched driver will be invoked, and given as a parameter a `struct platform_device`, which will hold a pointer to the matching ID table entry that originated the match. The `.driver_data` element is an `unsigned long`, which is sometimes casted into pointer addresses in order to point to anything, just like in the serial-imx driver. The following is an example with `platform_device_id` in `drivers/tty/serial/imx.c`:

```
static const struct platform_device_id imx_uart_devtype[] = {
        {
                .name = "imx1-uart",
                .driver_data = (kernel_ulong_t) &imx_uart_devdata[IMX1_UART],
        }, {
                .name = "imx21-uart",
                .driver_data = (kernel_ulong_t)
&imx_uart_devdata[IMX21_UART],
        }, {
                .name = "imx6q-uart",
                .driver_data = (kernel_ulong_t)
&imx_uart_devdata[IMX6Q_UART],
        }, {
                /* sentinel */
        }
};
```

The `.name` field must be the same as the device's name you give when you register the device in the board specific file. The function responsible for this match style is `platform_match_id`. If you look at its definition in `drivers/base/platform.c`, you'll see:

```
static const struct platform_device_id *platform_match_id(
        const struct platform_device_id *id,
        struct platform_device *pdev)
{
    while (id->name[0]) {
        if (strcmp(pdev->name, id->name) == 0) {
            pdev->id_entry = id;
            return id;
        }
        id++;
    }
    return NULL;
}
```

Platform Device Drivers

In the following example, which is an excerpt from `drivers/tty/serial/imx.c` in kernel sources, one can see how the platform data is converted back into the original data structure, just by casting. That is how people sometimes pass any data structure as platform data:

```
static void serial_imx_probe_pdata(struct imx_port *sport,
        struct platform_device *pdev)
{
    struct imxuart_platform_data *pdata = dev_get_platdata(&pdev->dev);

    sport->port.line = pdev->id;
    sport->devdata = (structimx_uart_data *) pdev->id_entry->driver_data;

    if (!pdata)
        return;
    [...]
}
```

`pdev->id_entry` is a `struct platform_device_id`, which is a pointer to the matching ID table entry made available by the kernel, and whose `driver_data` element is casted back to a pointer on the data structure.

Per device-specific data on ID table matching

In the previous section, we have used `platform_device_id.platform_data` as a pointer. Your driver may need to support more than one device type. In this situation, you will need specific device data for each device type you support. You should then use the device id as an index to an array that contains every possible device data, and not as a pointer address anymore. The following are detailed steps in an example:

1. We define an enumeration, depending on the device type that we need to support in our driver:

```
enum abx80x_chip {
    AB0801,
    AB0803,
    AB0804,
    AB0805,
    AB1801,
    AB1803,
    AB1804,
    AB1805,
    ABX80X
};
```

2. We define the specific data-type structure:

   ```
   struct abx80x_cap {
       u16 pn;
   boolhas_tc;
   };
   ```

3. We fill an array with default values, and depending on the index in `device_id`, we can pick the right data:

   ```
   static struct abx80x_cap abx80x_caps[] = {
       [AB0801] = {.pn = 0x0801},
       [AB0803] = {.pn = 0x0803},
       [AB0804] = {.pn = 0x0804, .has_tc = true},
       [AB0805] = {.pn = 0x0805, .has_tc = true},
       [AB1801] = {.pn = 0x1801},
       [AB1803] = {.pn = 0x1803},
       [AB1804] = {.pn = 0x1804, .has_tc = true},
       [AB1805] = {.pn = 0x1805, .has_tc = true},
       [ABX80X] = {.pn = 0}
   };
   ```

4. We define our `platform_device_id` with a specific index:

   ```
   static const struct i2c_device_id abx80x_id[] = {
       { "abx80x", ABX80X },
       { "ab0801", AB0801 },
       { "ab0803", AB0803 },
       { "ab0804", AB0804 },
       { "ab0805", AB0805 },
       { "ab1801", AB1801 },
       { "ab1803", AB1803 },
       { "ab1804", AB1804 },
       { "ab1805", AB1805 },
       { "rv1805", AB1805 },
       { }
   };
   ```

5. Here we just have to do the stuff in the `probe` function:

   ```
   static int rs5c372_probe(struct i2c_client *client,
   const struct i2c_device_id *id)
   {
       [...]

       /* We pick the index corresponding to our device */
   int index = id->driver_data;
   ```

```
            /*
             * And then, we can access the per device data
             * since it is stored in abx80x_caps[index]
             */
        }
```

Name matching - platform device name matching

Now-a-days most platform drivers do not provide any table at all; they simply fill the name of the driver itself in the driver's name field. But the matching works because, if you look at the `platform_match` function, you will see that at the end the match falls back to name matching, comparing the driver's name and the device's name. Some older drivers still use that matching mechanism. The following is name matching from `sound/soc/fsl/imx-ssi.c`:

```
        static struct platform_driver imx_ssi_driver = {
            .probe = imx_ssi_probe,
            .remove = imx_ssi_remove,

            /* As you can see here, only the 'name' field is filled */
            .driver = {
                    .name = "imx-ssi",
            },
        };

        module_platform_driver(imx_ssi_driver);
```

To add a device that matches this driver, one must call `platform_device_register` or `platform_add_devices`, with the same name `imx-ssi`, in the board-specific file (usually in `arch/<your_arch>/mach-*/board-*.c`). For our quad core i.MX6-based UDOO, it is `arch/arm/mach-imx/mach-imx6q.c`.

Summary

The kernel pseudo platform bus has no secrets for you anymore. With bus matching mechanisms, you are able to understand how, when, and why your driver has been loaded, as well as which device it was for.We can implement any `probe` function, based on the matching mechanism we want. Since the main purpose of a driver is to handle a device, we are now able to populate devices in the system (the old and depreciated way). To finish in beauty, the next chapter will exclusively deal with the device tree, which is the new mechanism used to populate devices, along with their configurations, on the system.

6
The Concept of Device Tree

The **Device Tree** (**DT**) is an easy to read hardware description file, with JSON like formatting style, which is a simple tree structure where devices are represented by nodes with their properties. Properties can be either empty (just the key, to describe boolean values), or key-value pairs where the value can contain an arbitrary byte stream. This chapter is a simple introduction to DT. Every kernel subsystem or framework has its own DT binding. We will talk about those specific bindings when we deal with concerned topics. The DT originated from OF, which is a standard endorsed by computer companies, and whose main purpose is defining interfaces for computer firmware systems. That said, one can find more on DT specification at http://www.devicetree.org/. Therefore, this chapter will cover the basics of DT, such as:

- Naming convention, as well as aliases and labeling
- Describing data types and their APIs
- Managing addressing schemes and accessing the device resources
- Implementing OF match style and providing application-specific data

Device tree mechanism

DT is enabled in the kernel by setting the option CONFIG_OF to Y. In order to pull the DT API from within your driver, you must add the following headers:

```
#include <linux/of.h>
#include <linux/of_device.h>
```

DT supports a few data types. Let us have a look at them with a sample node description:

```
/* This is a comment */
// This is another comment
node_label: nodename@reg{
```

The Concept of Device Tree

```
        string-property = "a string";
        string-list = "red fish", "blue fish";
        one-int-property = <197>; /* One cell in this property */
        int-list-property = <0xbeef 123 0xabcd4>; /*each number (cell) is a
                                                  *32 bit integer(uint32).
                                                  *There are 3 cells in
                                                  */this property
    mixed-list-property = "a string", <0xadbcd45>, <35>, [0x01 0x23 0x45]
    byte-array-property = [0x01 0x23 0x45 0x67];
    boolean-property;
};
```

The following are some definitions of data types used in device trees:

- Text strings are represented with double quotes. One can use commas to create a list of the strings.
- Cells are 32-bit unsigned integers delimited by angle brackets.
- Boolean data is nothing but an empty property. The true or false value depends on the property being there or not.

Naming convention

Every node must have a name in the form `<name>[@<address>]`, where `<name>` is a string that can be up to 31 characters in length, and `[@<address>]` is optional, depending on whether the node represents an addressable device or not. `<address>` should be the primary address used to access the device. An example of device naming is as follows:

```
expander@20 {
    compatible = "microchip,mcp23017";
    reg = <20>;
    [...]
};
```

Or

```
i2c@021a0000 {
    compatible = "fsl,imx6q-i2c", "fsl,imx21-i2c";
    reg = <0x021a0000 0x4000>;
    [...]
};
```

On the other hand, the `label` is optional. Labeling a node is useful only if the node is intended to be referenced from a property of another node. One can see a label as a pointer to node, as explained in the next section.

Aliases, labels, and phandle

It is very important to understand how these three elements work. They are frequently used in the DT. Let us take the following DT to explain how they work:

```
aliases {
    ethernet0 = &fec;
    gpio0 = &gpio1;
    gpio1 = &gpio2;
    mmc0 = &usdhc1;
    [...]
};
gpio1: gpio@0209c000 {
    compatible = "fsl,imx6q-gpio", "fsl,imx35-gpio";
    [...]
};
node_label: nodename@reg {
    [...];
    gpios = <&gpio1 7 GPIO_ACTIVE_HIGH>;
};
```

A label is nothing but a way to tag a node, to let the node be identified by a unique name. In the real world, that name is converted into a unique 32-bit value by the DT compiler. In the preceding example, `gpio1` and `node_label` are both labels. Labels can then be used to refer to a node, since a label is unique to a node.

A **pointer handle (phandle)** is a 32-bit value associated with a node that is used to uniquely identify that node so that the node can be referenced from a property in another node. Labels are used to have a pointer to the node. By using `<&mylabel>`, you point to the node whose label is `mylabel`.

> The use of `&` is just like in the C programming language; to obtain the address of an element.

In the preceding example, `&gpio1` is converted to the phandle so that it refers to `gpio1` node. The same goes for the following example:

```
thename@address {
    property = <&mylabel>;
};

mylabel: thename@adresss {
    [...]
}
```

The Concept of Device Tree

In order not to walk through the whole tree to look for a node, the concept of aliases has been introduced. In the DT, the `aliases` node can be seen like a quick lookup table, an index of another node. One can use the function `find_node_by_alias()` to find a node given its alias. The aliases are not used directly in the DT source, but are instead deferenced by the Linux kernel.

DT compiler

The DT comes in two forms: the textual form, which represents the sources also known as `DTS`, and the binary blob form, which represents the compiled DT, also known as `DTB`. Source files have the `.dts` extension. Actually, there are also `.dtsi` text files, which represent SoC level definitions, whereas `.dts` files represent board level definitions. One can see `.dtsi` as header files, that should be included in `.dts` one, which are source files, not the reverse, a bit like including header files (`.h`) in the source file (`.c`). On the other hand, binary files use the `.dtb` extension.

There is actually a third form, which is the runtime representation of the DT in `/proc/device-tree`.

As its name says, the tool used to compile the device tree is called the **device tree compiler** (**dtc**). From the root kernel source, one can compile either a standalone specific DT or all DTs for the specific architecture.

Let us compile all DT (`.dts`) files for arm SoC's:

> **ARCH=arm CROSS_COMPILE=arm-linux-gnueabihf- make dtbs**

For a standalone DT:

> **ARCH=arm CROSS_COMPILE=arm-linux-gnueabihf- make imx6dl-sabrelite.dtb**

In the preceding example, the name of the source file is `imx6dl-sabrelite.dts`.

Given a compiled device tree (`.dtb`) file, you can do the reverse operation and extract the source (`.dts`) file:

```
dtc -I dtb -O dtsarch/arm/boot/dts imx6dl-sabrelite.dtb
>path/to/my_devicetree.dts
```

For the purpose of debugging, it could be useful to expose the DT to the user space. The `CONFIG_PROC_DEVICETREE` configuration variable will do that for you. You can then explore and walk through the DT in `/proc/device-tree`.

[144]

Representing and addressing devices

Each device is given at least one node in the DT. Some properties are common to many device types, especially devices sitting on a bus known to the kernel (SPI, I2C, platform, MDIO, and so on). These properties are `reg`, `#address-cells`, and `#size-cells`. The purpose of these properties is device addressing on the bus they sit on. That said, the main addressing property is `reg`, which is a generic property and whose meaning depends on the bus the device sits on. The # (sharp) that prefixes `size-cell` and `address-cell` can be translated into `length`.

Each addressable device gets a `reg` property that is a list of tuples in the form `reg = <address0size0 [address1size1] [address2size2] ... >`, where each tuple represents an address range used by the device. `#size-cells` indicates how many 32 bit cells are used to represent size, and may be 0 if size is not relevant. On the other hand, `#address-cells` indicates how many 32 bit cells are used to represent address. In other word, the address element of each tuple is interpreted according to `#address-cell`; same for the size element, which is interpreted according to `#size-cell`.

Actually, addressable devices inherit from `#size-cell` and `#address-cell` of their parent, which is the node that represents the bus controller. The presence of `#size-cell` and `#address-cell` in a given device does not affect the device itself, but its children. In other words, before interpreting the `reg` property of a given node, one must know the parent node's `#address-cells` and `#size-cells` values. The parent node is free to define whatever addressing scheme is suitable for device sub-nodes (children).

SPI and I2C addressing

SPI and I2C devices both belong to non-memory mapped devices, because their addresses are not accessible to the CPU. Instead, the parent device's driver (which is the bus controller driver) would perform indirect access on behalf of the CPU. Each I2C/SPI device is always represented as a sub-node of the I2C/SPI bus node the device seats on. For nonmemory mapped device, the `#size-cells` property is 0, and the size element in addressing the tuple is empty. It means the `reg` property for this kind of device is always on cell:

```
&i2c3 {
    [...]
    status = "okay";

    temperature-sensor@49 {
        compatible = "national,lm73";
        reg = <0x49>;
```

The Concept of Device Tree

```
        };

        pcf8523: rtc@68 {
            compatible = "nxp,pcf8523";
            reg = <0x68>;
        };
    };

&ecspi1 {
fsl,spi-num-chipselects = <3>;
cs-gpios = <&gpio5 17 0>, <&gpio5 17 0>, <&gpio5 17 0>;
status = "okay";
[...]

    ad7606r8_0: ad7606r8@1 {
        compatible = "ad7606-8";
        reg = <1>;
        spi-max-frequency = <1000000>;
        interrupt-parent = <&gpio4>;
        interrupts = <30 0x0>;
        convst-gpio = <&gpio6 18 0>;
    };
};
```

If one looks at a SoC level file at `arch/arm/boot/dts/imx6qdl.dtsi`, one will notice that `#size-cells` and `#address-cells` are respectively set to 0 for the former, and 1 for the last, in both `i2c` and `spi` nodes, which are respectively parents of I2C and SPI devices enumerated in the preceding section. This helps us to understand their `reg` property, which is only one cell for the address value, and none for the size value.

I2C device's `reg` property is used to specify the device's address on the bus. For SPI devices, `reg` represents the index of the chip-select line assigned to the device among the list of chips-select the controller node has. For example, for the ad7606r8 ADC, the chip-select index is 1, which corresponds to `<&gpio5 17 0>` in `cs-gpios`, which is the list of chip-select of the controller node.

You may ask why I used the I2C/SPI node's phandle: the answer is because I2C/SPI devices should be declared at board level file (`.dts`), whereas I2C/SPI buses controller are declared at SoC level file (`.dtsi`).

Platform device addressing

This section address simple memory-mapped devices whose memory is accessible to the CPU. Here, the reg property still defines the device's address, which is a list of memory regions on which you can access the device. Each region is represented with a tuple of cells, where the first cell is the base address of the memory region, and the second tuple is the size of the region. It then has the form reg = <base0 length0 [base1 length1] [address2 length2] ... >. Each tuple represents an address range used by the device.

In the real world, one should not interpret the reg property without knowing the value of two other properties, #size-cells and #address-cells. #size-cells tell us how large the length field is in each child reg tuple. The same for #address-cell, which tell us how many cells we must use to specify an address.

This kind of device should be declared within a node with a special value compatible = "simple-bus", meaning a simple memory-mapped bus with no specific handling nor driver:

```
soc {
    #address-cells = <1>;
    #size-cells = <1>;
    compatible = "simple-bus";
    aips-bus@02000000 { /* AIPS1 */
        compatible = "fsl,aips-bus", "simple-bus";
        #address-cells = <1>;
        #size-cells = <1>;
        reg = <0x02000000 0x100000>;
        [...];

    spba-bus@02000000 {
        compatible = "fsl,spba-bus", "simple-bus";
        #address-cells = <1>;
        #size-cells = <1>;
        reg = <0x02000000 0x40000>;
        [...]
        ecspi1: ecspi@02008000 {
            #address-cells = <1>;
            #size-cells = <0>;
            compatible = "fsl,imx6q-ecspi", "fsl,imx51-ecspi";
            reg = <0x02008000 0x4000>;
            [...]
        };
        i2c1: i2c@021a0000 {
            #address-cells = <1>;
            #size-cells = <0>;
            compatible = "fsl,imx6q-i2c", "fsl,imx21-i2c";
```

```
                    reg = <0x021a0000 0x4000>;
                    [...]
                };
            };
        };
```

In the preceding example, child nodes whose parent has `simple-bus` in the compatible property will be registered as platform devices. One can also see how I2C and SPI bus controllers change the addressing scheme of their children by setting `#size-cells = <0>;` because it is not relevant to them. A well-known place to look for any binding information is in the kernel device tree's documentation: *Documentation/devicetree/bindings/*.

Handling resources

The main purpose of a driver is to handle and manage devices, and most of the time, expose their functionalities to the user-space. The objective here is to gather the device's configuration parameters, and especially resources (memory region, interrupt line, DMA channel, clocks, and so on).

The following is the device node with which we will work during this section. It is the i.MX6 UART device's node, defined in `arch/arm/boot/dts/imx6qdl.dtsi`:

```
uart1: serial@02020000 {
        compatible = "fsl,imx6q-uart", "fsl,imx21-uart";
reg = <0x02020000 0x4000>;
        interrupts = <0 26 IRQ_TYPE_LEVEL_HIGH>;
        clocks = <&clks IMX6QDL_CLK_UART_IPG>,
<&clks IMX6QDL_CLK_UART_SERIAL>;
        clock-names = "ipg", "per";
dmas = <&sdma 25 4 0>, <&sdma 26 4 0>;
dma-names = "rx", "tx";
        status = "disabled";
    };
```

Concept of named resources

When the driver expect a list of resources of a certain type, one has no guarantee the list is ordered in a manner the driver expects, since the guy who writes the board level device tree is usually not the one that wrote the driver. A driver may expect, for example, its device node with 2 IRQs lines, one for the Tx event at index 0, the other for Rx at index 1. What happens if the order is not respected? The driver will have an unwanted behavior. To avoid such mismatches, the concept of named resources (clock, irq, dma, reg) has been introduced. It consists of defining our resource list, and naming them, so that whatever their indexes are, a given name will always match the resource.

The corresponding properties to name the resources are as follows:

- reg-names: This is for a list of memory regions in reg property
- clock-names: This is to name clocks in the clocks property
- interrupt-names: This give a name to each interrupt in the interrupts property
- dma-names: This is for the dma property

Now let us create a fake device node entry to explain that:

```
fake_device {
    compatible = "packt,fake-device";
    reg = <0x4a064000 0x800>, <0x4a064800 0x200>, <0x4a064c00 0x200>;
    reg-names = "config", "ohci", "ehci";
    interrupts = <0 66 IRQ_TYPE_LEVEL_HIGH>, <0 67 IRQ_TYPE_LEVEL_HIGH>;
    interrupt-names = "ohci", "ehci";
    clocks = <&clks IMX6QDL_CLK_UART_IPG>, <&clks IMX6QDL_CLK_UART_SERIAL>;
    clock-names = "ipg", "per";
    dmas = <&sdma 25 4 0>, <&sdma 26 4 0>;
    dma-names = "rx", "tx";
};
```

The code in the driver to extract each named resource is as follows:

```
struct resource *res1, *res2;
res1 = platform_get_resource_byname(pdev, IORESOURCE_MEM, "ohci");
res2 = platform_get_resource_byname(pdev, IORESOURCE_MEM, "config");

struct dma_chan  *dma_chan_rx, *dma_chan_tx;
dma_chan_rx = dma_request_slave_channel(&pdev->dev, "rx");
dma_chan_tx = dma_request_slave_channel(&pdev->dev, "tx");

inttxirq, rxirq;
txirq = platform_get_irq_byname(pdev, "ohci");
```

The Concept of Device Tree

```
rxirq = platform_get_irq_byname(pdev, "ehci");

structclk *clck_per, *clk_ipg;
clk_ipg = devm_clk_get(&pdev->dev, "ipg");
clk_ipg = devm_clk_get(&pdev->dev, "pre");
```

This way, you are sure to map the right name to the right resource, without needing to play with the index anymore.

Accessing registers

Here, the driver will take ownership of the memory region and map it into the virtual address space. We will discuss more about this in Chapter 11, *Kernel Memory Management*.

```
struct resource *res;
void __iomem *base;

res = platform_get_resource(pdev, IORESOURCE_MEM, 0);
/*
 * Here one can request and map the memory region
 * using request_mem_region(res->start, resource_size(res), pdev->name)
 * and ioremap(iores->start, resource_size(iores)
 *
 * These function are discussed in chapter 11, Kernel Memory Management.
 */
base = devm_ioremap_resource(&pdev->dev, res);
if (IS_ERR(base))
    return PTR_ERR(base);
```

platform_get_resource() will set the start and end fields of struct res according to the memory region present in the first (index 0) reg assignment. Please remember the last argument of platform_get_resource() represents the resource index. In the preceding sample, 0 indexes the first value of that resource type, just in case the device is assigned more than one memory region in the DT node. In our example, it's reg = <0x02020000 0x4000>, meaning that the allocated region starts at physical address 0x02020000 and has the size of 0x4000 bytes. platform_get_resource() will then set res.start = 0x02020000 and res.end = 0x02023fff.

Handling interrupts

The interrupt interface is actually divided into two parts; the consumer side and the controller side. Four properties are used to describe interrupt connections in the DT:

The controller is the device that exposes IRQ lines to the consumer. In controller side, on has the following properties:

- `interrupt-controller`: An empty (Boolean) property that one should define in order to mark the device as being an interrupt controller
- `#interrupt-cells`: This is a property of interrupt controllers. It states how many cells are used to specify an interrupt for that interrupt controller

The consumer is the device that generate the IRQ. Consumer binding expects the following properties:

- `interrupt-parent`: For the device node that generates interrupt, it is a property that contains a pointer `phandle` to the interrupt controller node to which the device is attached. If omitted, the device inherits that property from its parent node.
- `interrupts`: It is the interrupt specifier.

Interrupt binding and interrupt specifiers are tied to the interrupt controller device. The number of cells used to define an interrupt input depends on the interrupt controller, which is the only one deciding, by mean of its `#interrupt-cells` property. In the case of i.MX6, the interrupt controller is a **Global Interrupt Controller** (**GIC**). Its binding is well explained in *Documentation/devicetree/bindings/arm/gic.txt*.

The interrupt handler

This consist of fetching the IRQ number from the DT, and mapping it into Linux IRQ, thus registering a function callback for it. The driver code to do this is quite simple:

```
int irq = platform_get_irq(pdev, 0);
ret = request_irq(irq, imx_rxint, 0, dev_name(&pdev->dev), sport);
```

The `platform_get_irq()` call will return the `irq` number; this number is usable by `devm_request_irq()` (`irq` is then visible in `/proc/interrupts`). The second argument, `0`, says that we need the first interrupt specified in the device node. If, there is more than one interrupt, we can change this index according to the interrupt we need, or just use the named resource.

In our preceding example, the device node contains an interrupt specifier, which looks like as follows:

```
interrupts = <0 66 IRQ_TYPE_LEVEL_HIGH>;
```

- According to ARM GIC, the first cell informs us about interrupt type:
    - 0: **Shared peripheral interrupt(SPI)**, for interrupts signal shared among cores, which can be routed by the GIC to any core
    - 1: **Private peripheral interrupt** (**PPI**), for interrupt signal private to an individual core

The documentation can be found at:http://infocenter.arm.com/help/index.jsp?topic=/com.arm.doc.ddi0407e/CCHDBEBE.html.

- The second cell holds the interrupt number. This number depends on whether the interrupt line is a PPI or SPI.
- The third cell, `IRQ_TYPE_LEVEL_HIGH` in our case, represents sense level. All of the available sense levels are defined in `include/linux/irq.h`.

Interrupt controller code

The `interrupt-controller` property is used to declare a device as an interrupt controller. The `#interrupt-cells` property defines how many cells must be used to define a single interrupt line. We will discuss this in detail in `Chapter 16`, *Advanced IRQ Management*.

Extract application-specific data

Application-specific data is data beyond the common properties (neither resources nor GPIOs, regulator, and so on). Those are arbitrary properties and child nodes that can be assigned to a device. Such property names are usually prefixed with manufacture codes. These can be any string, Boolean, or integer values, along with their API defined in `drivers/of/base.c` in the Linux sources. The following examples we discuss are not exhaustive. Let us now reuse the node defined earlier in this chapter:

```
node_label: nodename@reg{
  string-property = ""a string"";
  string-list = ""red fish"", ""blue fish"";
  one-int-property = <197>; /* One cell in this property */
  int-list-property = <0xbeef 123 0xabcd4>;/* each number (cell) is 32
a                                           * bit integer(uint32). There
                                            * are 3 cells in this property
                                            */
  mixed-list-property = "a string", <0xadbcd45>, <35>, [0x01 0x23 0x45]
  byte-array-property = [0x01 0x23 0x45 0x67];
  one-cell-property = <197>;
  boolean-property;
};
```

Text string

The following is one `string` property:

```
string-property = "a string";
```

Back in the driver, one should use `of_property_read_string()` to read a string value. Its prototype is defined as follows:

```
int of_property_read_string(const struct device_node *np, const
                    char *propname, const char **out_string)
```

The following code shows how you can use it:

```
const char *my_string = NULL;
of_property_read_string(pdev->dev.of_node, "string-property", &my_string);
```

Cells and unsigned 32-bit integers

The following are our `int` properties:

```
one-int-property = <197>;
int-list-property = <1350000 0x54dae47 1250000 1200000>;
```

One should use `of_property_read_u32()` to read a cell value. Its prototype is defined as follows:

```
int of_property_read_u32_index(const struct device_node *np,
                 const char *propname, u32 index, u32 *out_value)
```

Back in the driver,

```
unsigned int number;
of_property_read_u32(pdev->dev.of_node, "one-cell-property", &number);
```

One can use `of_property_read_u32_array` to read a list of cells. Its prototype is as follows:

```
int of_property_read_u32_array(const struct device_node *np,
                 const char *propname, u32 *out_values, size_tsz);
```

Here, `sz` is the number of array elements to read. Have a look at `drivers/of/base.c` to see how to interpret its return value:

```
unsigned int cells_array[4];
if (of_property_read_u32_array(pdev->dev.of_node, "int-list-property",
cells_array, 4)) {
    dev_err(&pdev->dev, "list of cells not specified\n");
    return -EINVAL;
}
```

Boolean

One should use `of_property_read_bool()` to read the Boolean property whose name is given in the second argument of the function:

```
bool my_bool = of_property_read_bool(pdev->dev.of_node, "boolean-property");
If(my_bool){
    /* boolean is true */
} else
    /* Bolean is false */
}
```

Extract and parse sub-nodes

You are allowed to add any sub-node in your device node. Given a node representing a flash memory device, partitions can be represented as sub-nodes. For a device that handles a set of input and output GPIO, each set can be represented as a sub-node. The sample node is as follows:

```
eeprom: ee24lc512@55 {
        compatible = "microchip,24xx512";
reg = <0x55>;

        partition1 {
            read-only;
            part-name = "private";
            offset = <0>;
            size = <1024>;
        };

        partition2 {
            part-name = "data";
            offset = <1024>;
            size = <64512>;
        };
    };
```

One can use `for_each_child_of_node()` to walk through sub-nodes of the given node:

```
struct device_node *np = pdev->dev.of_node;
struct device_node *sub_np;
for_each_child_of_node(np, sub_np) {
        /* sub_np will point successively to each sub-node */
        [...]
int size;
        of_property_read_u32(client->dev.of_node,
"size", &size);
        ...
}
```

Platform drivers and DT

Platform drivers also work with DT. That being said, it is the recommended way to deal with platform devices nowadays, and there is no need to touch board files anymore, or even to recompile the kernel when a device's property changes. If you remember, in the previous chapter we discussed OF match style, which is a matching mechanism based on the DT. Let us see in the following section how it works:

OF match style

OF match style is the first matching mechanism performed by the platform core in order to match devices with their drivers. It uses the device tree's `compatible` property to match the device entry in `of_match_table`, which is a field of the `struct driver` substructure. Each device node has a `compatible` property, which is a string, or a list of strings. Any platform driver that declares one of the strings listed in the `compatible` property will trigger a match and will see its `probe` function executed.

A DT match entry is described in the kernel as an instance of the `struct of_device_id` structure, which is defined in `linux/mod_devicetable.h` and looks like:

```
// we are only interested in the two last elements of the structure
struct of_device_id {
    [...]
    char    compatible[128];
    const void *data;
};
```

The following is the meaning of each element of the structure:

- `char compatible[128]`: This is the string used to match the device node's compatible property in the DT. They must be identical before a match occurs.
- `const void *data`: This can point to any structure, which can be used as per-device type configuration data.

Since the `of_match_table` is a pointer, you can pass an array of the `struct of_device_id` to make your driver compatible with more than one device:

```
static const struct of_device_id imx_uart_dt_ids[] = {
    { .compatible = "fsl,imx6q-uart", },
    { .compatible = "fsl,imx1-uart", },
    { .compatible = "fsl,imx21-uart", },
    { /* sentinel */ }
};
```

Once you have filled your array of ids, it must be passed to the `of_match_table` field of your platform driver, in the driver substructure:

```
static struct platform_driver serial_imx_driver = {
    [...]
    .driver     = {
        .name   = "imx-uart",
        .of_match_table = imx_uart_dt_ids,
        [...]
    },
};
```

At this step, only your driver is aware of your `of_device_id` array. To get the kernel informed too (so that it can store your IDs in the device list maintained by the platform core), your array has to be registered with MODULE_DEVICE_TABLE, as described in Chapter 5, *Platform Device Drivers*:

```
MODULE_DEVICE_TABLE(of, imx_uart_dt_ids);
```

That is all! Our driver is DT-compatible. Back in our DT, let's declare a device compatible with our driver:

```
uart1: serial@02020000 {
    compatible = "fsl,imx6q-uart", "fsl,imx21-uart";
    reg = <0x02020000 0x4000>;
    interrupts = <0 26 IRQ_TYPE_LEVEL_HIGH>;
    [...]
};
```

Two compatible strings are provided here. If the first one does not match any driver, the core will perform the match with the second.

When a match occurs, the `probe` function of your driver is called, with a `struct platform_device` structure as the parameter, which contains a `struct device dev` field, in which there is a field `struct device_node *of_node` that corresponds to the node associated to our device, so that one can use it to extract the device settings:

```
static int serial_imx_probe(struct platform_device *pdev)
{
    [...]
struct device_node *np;
np = pdev->dev.of_node;

    if (of_get_property(np, "fsl,dte-mode", NULL))
        sport->dte_mode = 1;
        [...]
}
```

One can check if the DT node is set to know whether the driver has been loaded in response to an `of_match`, or instantiated from within the board's `init` file. You should then use the `of_match_device` function, in order to pick the `struct *of_device_id` entry that originated the match, which may contain the specific data you have passed:

```
static int my_probe(struct platform_device *pdev)
{
struct device_node *np = pdev->dev.of_node;
const struct of_device_id *match;

    match = of_match_device(imx_uart_dt_ids, &pdev->dev);
    if (match) {
        /* Devicetree, extract the data */
        my_data = match->data
    } else {
        /* Board init file */
        my_data = dev_get_platdata(&pdev->dev);
    }
    [...]
}
```

Dealing with non-device tree platforms

DT support is enabled in the kernel with the CONFIG_OF option. One would probably want to avoid using the DT API when its support is not enabled in the kernel. The way one can achieve that is to check whether CONFIG_OF is set or not. People used to do something like as follows:

```
#ifdef CONFIG_OF
    static const struct of_device_id imx_uart_dt_ids[] = {
        { .compatible = "fsl,imx6q-uart", },
        { .compatible = "fsl,imx1-uart", },
        { .compatible = "fsl,imx21-uart", },
        { /* sentinel */ }
    };

    /* other devicetree dependent code */
    [...]
#endif
```

Even if the of_device_id data type is always defined when device tree support is missing, the code wrapped into #ifdef CONFIG_OF ... #endif will be omitted during the build. This is used for conditional compilation. It is not your only choice; there is also the of_match_ptr macro, which simply returns NULL when OF is disabled. Everywhere you'll need to pass your of_match_table as a parameter, it should be wrapped in the of_match_ptr macro, so that it returns NULL when OF is disabled. The macro is defined in include/linux/of.h:

```
#define of_match_ptr(_ptr) (_ptr)  /* When CONFIG_OF is enabled */
#define of_match_ptr(_ptr) NULL    /* When it is not */
```

And we can use it as follows:

```
static int my_probe(struct platform_device *pdev)
{
    const struct of_device_id *match;
    match = of_match_device(of_match_ptr(imx_uart_dt_ids),
                    &pdev->dev);
    [...]
}
static struct platform_driver serial_imx_driver = {
    [...]
    .driver     = {
    .name    = "imx-uart",
    .of_match_table = of_match_ptr(imx_uart_dt_ids),
    },
};
```

This eliminates having a `#ifdef`, returning `NULL` when `OF` is disabled.

Support multiple hardware with per device-specific data

Sometimes, a driver can support different hardware, each with is specific configuration data. That data may be dedicated function tables, specific register values, or anything unique to each hardware. The following example describes a generic approach:

Let us first remember what `struct of_device_id` looks like, in `include/linux/mod_devicetable.h`.

```
/*
 * Struct used for matching a device
 */
struct of_device_id {
        [...]
        char    compatible[128];
const void *data;
};
```

The field we are interested in is `const void *data`, so we can use it to pass any data for each specific device.

Let's say we own three different devices, each with a specific private data. `of_device_id.data` will contain a pointer to specific parameters. This example is inspired by `drivers/tty/serial/imx.c`.

First, we declare private structures:

```
/* i.MX21 type uart runs on all i.mx except i.MX1 and i.MX6q */
enum imx_uart_type {
    IMX1_UART,
    IMX21_UART,
    IMX6Q_UART,
};

/* device type dependent stuff */
struct imx_uart_data {
    unsigned uts_reg;
    enum imx_uart_type devtype;
};
```

The Concept of Device Tree

Then we fill an array with each device-specific data:

```
static struct imx_uart_data imx_uart_devdata[] = {
        [IMX1_UART] = {
                .uts_reg = IMX1_UTS,
                .devtype = IMX1_UART,
        },
        [IMX21_UART] = {
                .uts_reg = IMX21_UTS,
                .devtype = IMX21_UART,
        },
        [IMX6Q_UART] = {
                .uts_reg = IMX21_UTS,
                .devtype = IMX6Q_UART,
        },
};
```

Each compatible entry is tied with a specific array index:

```
static const struct of_device_idimx_uart_dt_ids[] = {
        { .compatible = "fsl,imx6q-uart", .data = &imx_uart_devdata[IMX6Q_UART], },
        { .compatible = "fsl,imx1-uart", .data = &imx_uart_devdata[IMX1_UART], },
        { .compatible = "fsl,imx21-uart", .data = &imx_uart_devdata[IMX21_UART], },
        { /* sentinel */ }
};
MODULE_DEVICE_TABLE(of, imx_uart_dt_ids);

static struct platform_driver serial_imx_driver = {
    [...]
    .driver         = {
        .name   = "imx-uart",
        .of_match_table = of_match_ptr(imx_uart_dt_ids),
    },
};
```

Now in the `probe` function, whatever the match entry is, it will hold a pointer to the device-specific structure:

```
static int imx_probe_dt(struct platform_device *pdev)
{
    struct device_node *np = pdev->dev.of_node;
    const struct of_device_id *of_id =
    of_match_device(of_match_ptr(imx_uart_dt_ids), &pdev->dev);

        if (!of_id)
```

The Concept of Device Tree

```
            /* no device tree device */
            return 1;
    [...]
    sport->devdata = of_id->data; /* Get private data back  */
}
```

In the preceding code, `devdata` is an element of a structure in the original source, and declared like `const struct imx_uart_data *devdata`; we could have stored any specific parameter in the array.

Match style mixing

OF match style can be combined with any other matching mechanism. In the following example, we have a mix of DT and device ID match styles:

We fill an array for the device ID match style, each device having its data:

```
static const struct platform_device_id sdma_devtypes[] = {
    {
        .name = "imx51-sdma",
        .driver_data = (unsigned long)&sdma_imx51,
    }, {
        .name = "imx53-sdma",
        .driver_data = (unsigned long)&sdma_imx53,
    }, {
        .name = "imx6q-sdma",
        .driver_data = (unsigned long)&sdma_imx6q,
    }, {
        .name = "imx7d-sdma",
        .driver_data = (unsigned long)&sdma_imx7d,
    }, {
        /* sentinel */
    }
};
MODULE_DEVICE_TABLE(platform, sdma_devtypes);
```

We do the same for OF match style:

```
static const struct of_device_idsdma_dt_ids[] = {
    { .compatible = "fsl,imx6q-sdma", .data = &sdma_imx6q, },
    { .compatible = "fsl,imx53-sdma", .data = &sdma_imx53, },
        { .compatible = "fsl,imx51-sdma", .data = &sdma_imx51, },
    { .compatible = "fsl,imx7d-sdma", .data = &sdma_imx7d, },
    { /* sentinel */ }
};
MODULE_DEVICE_TABLE(of, sdma_dt_ids);
```

The `probe` function will look as follows:

```
static int sdma_probe(structplatform_device *pdev)
{
conststructof_device_id *of_id =
of_match_device(of_match_ptr(sdma_dt_ids), &pdev->dev);
structdevice_node *np = pdev->dev.of_node;

    /* If devicetree, */
    if (of_id)
drvdata = of_id->data;
    /* else, hard-coded */
    else if (pdev->id_entry)
drvdata = (void *)pdev->id_entry->driver_data;

    if (!drvdata) {
dev_err(&pdev->dev, "unable to find driver data\n");
        return -EINVAL;
    }
    [...]
}
```

Then we declare our platform driver; feed all arrays defined as in the preceding sections:

```
static struct platform_driversdma_driver = {
    .driver = {
    .name    = "imx-sdma",
    .of_match_table = of_match_ptr(sdma_dt_ids),
    },
    .id_table  = sdma_devtypes,
    .remove    = sdma_remove,
    .probe     = sdma_probe,
};
module_platform_driver(sdma_driver);
```

Platform resources and DT

Platform devices can work with the device tree enabled system without any extra modification. It is what we have demonstrated in the section *Handling resources*. By using `platform_xxx` family function, the core also walks through the DT (with `of_xxx` family function) to find the requested resource. The reverse is not true, since `of_xxx` family function is only reserved for the DT. All resource data will be available to the driver in a usual way. The driver now knows whether this device is not initialized with hardcoded parameters in the board file or not. Let us take an example with an uart device node:

```
uart1: serial@02020000 {
    compatible = "fsl,imx6q-uart", "fsl,imx21-uart";
reg = <0x02020000 0x4000>;
    interrupts = <0 26 IRQ_TYPE_LEVEL_HIGH>;
dmas = <&sdma 25 4 0>, <&sdma 26 4 0>;
dma-names = "rx", "tx";
};
```

The following excerpt describes the `probe` function of its driver. In the `probe`, the function `platform_get_resource()` can be used to extract any property which is a resource (memory region, dma, irq), or a specific function, such as `platform_get_irq()`, which extracts the `irq` provided by the `interrupts` property in the DT:

```
static int my_probe(struct platform_device *pdev)
{
struct iio_dev *indio_dev;
struct resource *mem, *dma_res;
struct xadc *xadc;
int irq, ret, dmareq;

    /* irq */
irq = platform_get_irq(pdev, 0);
    if (irq<= 0)
        return -ENXIO;
    [...]

    /* memory region */
mem = platform_get_resource(pdev, IORESOURCE_MEM, 0);
xadc->base = devm_ioremap_resource(&pdev->dev, mem);
    /*
     * We could have used
     *      devm_ioremap(&pdev->dev, mem->start, resource_size(mem));
     * too.
     */
    if (IS_ERR(xadc->base))
        return PTR_ERR(xadc->base);
```

```
    [...]

    /* second dma channel */
dma_res = platform_get_resource(pdev, IORESOURCE_DMA, 1);
dmareq = dma_res->start;

    [...]
}
```

To sum up, for properties such as dma, irq and mem, you have nothing to do in the platform driver to match dtb. If one remembers, this data is of the same type as the data one can pass as a platform resource. To understand why, we just have to look inside these functions; we will see how each of them internally deals with DT functions. The following is an example of the platform_get_irq function:

```
int platform_get_irq(struct platform_device *dev, unsigned int num)
{
    [...]
    struct resource *r;
    if (IS_ENABLED(CONFIG_OF_IRQ) &&dev->dev.of_node) {
        int ret;

        ret = of_irq_get(dev->dev.of_node, num);
        if (ret > 0 || ret == -EPROBE_DEFER)
            return ret;
    }

    r = platform_get_resource(dev, IORESOURCE_IRQ, num);
    if (r && r->flags & IORESOURCE_BITS) {
        struct irq_data *irqd;
        irqd = irq_get_irq_data(r->start);
        if (!irqd)
            return -ENXIO;
        irqd_set_trigger_type(irqd, r->flags & IORESOURCE_BITS);
    }
    return r ? r->start : -ENXIO;
}
```

One may wonder how the `platform_xxx` functions extract resources from the DT. This should have been the `of_xxx` function family. You are right, but during the system boot, the kernel calls `of_platform_device_create_pdata()` on each device node, which will result in creating a platform device with the associated resource, on which you can call the `platform_xxx` family function. Its prototype is as follows:

```
static struct platform_device *of_platform_device_create_pdata(
              struct device_node *np, const char *bus_id,
              void *platform_data, struct device *parent)
```

Platform data versus DT

If your driver expects platform data, you should check the `dev.platform_data` pointer. A non-null value means your driver has been instantiated the old way in the board configuration file, and DT does not enter into it. For drivers instantiated from the DT, `dev.platform_data` will be `NULL`, and your platform device will be given a pointer on the DT entry (node) that corresponds to your device in the `dev.of_node` pointer, from which one can extract the resource and use OF API to parse and extract application data.

There's also a hybrid method that one can use to associate platform data declared in the C files to DT nodes, but that's for special cases only: for DMA, IRQ, and memory. This method is used only when the driver expects only resources, and no application-specific data.

One can transform a legacy declaration of an I2C controller into DT-compatible nodes as follows:

```
#define SIRFSOC_I2C0MOD_PA_BASE 0xcc0e0000
#define SIRFSOC_I2C0MOD_SIZE 0x10000
#define IRQ_I2C0
static struct resource sirfsoc_i2c0_resource[] = {
    {
        .start = SIRFSOC_I2C0MOD_PA_BASE,
        .end = SIRFSOC_I2C0MOD_PA_BASE + SIRFSOC_I2C0MOD_SIZE - 1,
        .flags = IORESOURCE_MEM,
    },{
        .start = IRQ_I2C0,
        .end = IRQ_I2C0,
        .flags = IORESOURCE_IRQ,
    },
};
```

And the DT node:

```
i2c0: i2c@cc0e0000 {
    compatible = "sirf,marco-i2c";
    reg = <0xcc0e0000 0x10000>;
    interrupt-parent = <&phandle_to_interrupt_controller_node>
    interrupts = <0 24 0>;
    #address-cells = <1>;
    #size-cells = <0>;
    status = "disabled";
};
```

Summary

The time to switch from hardcoded device configuration to DT has come. This chapter gave you all you need to handle DTs. Now you have the necessary skills to customize or add whatever node and property you want into the DT, and extract them from within your driver. In the next chapter, we will talk about the I2C driver, and use the DT API to enumerate and configure our I2C devices.

7
I2C Client Drivers

I2C bus, invented by Philips (now NXP) is a two-wire: **Serial Data (SDA)**, **Serial Clock (SCL)** asynchronous serial bus. It is a multi-master bus, though multi-master mode is not widely used. Both SDA and SCL are open drain/open collector, meaning that each of these can drive its output low, but none of these can drive its output high without having pull-up resistors. SCL is generated by the master in order to synchronize data (carried by SDA) transfer over the bus. Both slave and master can send data (not at the same time of course), thus making SDA a bidirectional line. That said the SCL signal is also bidirectional, since slave can stretch the clock by keeping the SCL line low. The bus is controlled by the master, which in our case is a part of the SoC. This bus is frequently used in embedded systems to connect serial EEPROM, RTC chips, GPIO expander, temperature sensors, and so on:

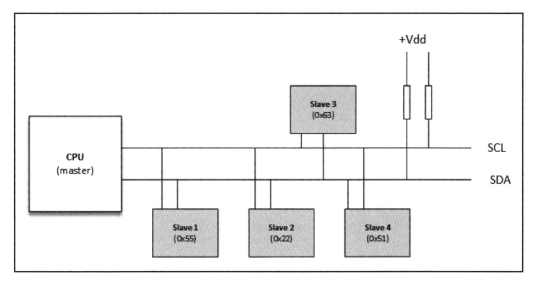

I2C bus and devices

I2C Client Drivers

Clock speed varies from 10 KHz to 100 KHz, and 400 KHz to 2 MHz. We will not cover bus specifications or bus drivers in this book. However, it is up to the bus driver to manage the bus and take care of the specifications. An example of a bus driver for the i.MX6 chip can be found at `drivers/i2C/busses/i2c-imx.c` in kernel source, and I2C specifications can be found at http://www.nxp.com/documents/user_manual/UM10204.pdf.

In this chapter, we are interested in client drivers, in order to handle slave devices seated on the bus. The chapter will cover the following topics:

- I2C client driver architecture
- Accessing the device, thus reading/writing data from/to device
- Declaring clients from DT

The driver architecture

When the device for which you write the driver takes a seat on a physical bus called the *bus controller*, it must rely on the driver of that bus called the *controller driver*, responsible for sharing bus access between devices. The controller driver offers an abstraction layer between your device and the bus. Whenever you perform a transaction (read or write) on an I2C or USB bus for example, the I2C/USB bus controller transparently takes care of that in the background. Every bus controller driver exports a set of functions to ease the development of drivers for devices sitting on that bus. This works for every physical bus (I2C, SPI, USB, PCI, SDIO, and so on).

An I2C driver is represented in the kernel as an instance of `struct i2c_driver`. The I2C client (which represents the device itself) is represented by a `struct i2c_client` structure.

The i2c_driver structure

An I2C driver is declared in kernel as an instance of `struct i2c_driver`, which looks as follows:

```
struct i2c_driver {
    /* Standard driver model interfaces */
    int (*probe)(struct i2c_client *, const struct i2c_device_id *);
    int (*remove)(struct i2c_client *);

    /* driver model interfaces that don't relate to enumeration */
    void (*shutdown)(struct i2c_client *);
```

```
    struct device_driver driver;
    const struct i2c_device_id *id_table;
};
```

The `struct i2c_driver` structure contains and characterizes general access routines, needed to handle the devices claiming the driver, whereas `struct i2c_client` contains device-specific information, like its address. A `struct i2c_client` structure represents and characterizes an I2C device. Later in this chapter, we will see how to populate these structures.

The probe() function

The `probe()` function is a part of the `struct i2c_driver` structure, and is executed any time once an I2C device is instantiated. It is responsible for the following tasks:

- Check whether the device is the one you expected
- Check whether your I2C bus controller of the SoC supports the functionality needed by your device, using the `i2c_check_functionality` function
- Initialize the device
- Set up device specific data
- Register the appropriate kernel framework

The `probe` function's prototype is as follows:

```
static int foo_probe(struct i2c_client *client, const struct
                                             i2c_device_id *id)
```

As you can see, its parameters are:

- `struct i2c_client` pointer: This represents the I2C device itself. This structure inherits from the structure device, and is provided to your `probe` function by the kernel. The client structure is defined in `include/linux/i2c.h`. Its definition is as follows:

```
struct i2c_client {
    unsigned short flags;     /* div., see below         */
    unsigned short addr;      /* chip address - NOTE: 7bit   */
                              /* addresses are stored in the */
                              /* _LOWER_ 7 bits          */
    char name[I2C_NAME_SIZE];
    struct i2c_adapter *adapter; /* the adapter we sit on  */
    struct device dev;        /* the device structure    */
    intirq;                   /* irq issued by device    */
```

I2C Client Drivers

```
        struct list_head detected;
#if IS_ENABLED(CONFIG_I2C_SLAVE)
        i2c_slave_cb_t slave_cb; /* callback for slave mode  */
#endif
};
```

- All fields are filled by the kernel, based on the parameter you provided to register the client. We will see later how to register a device to the kernel.
- `struct i2c_device_id` pointer: This points to the I2C device ID entry that matched the device that is being probed.

Per-device data

The I2C core offers you the possibility to store a pointer to any data structure of your choice, as device-specific data. To store or retrieve the data, use the following function provided by the I2C core:

```
/* set the data */
void i2c_set_clientdata(struct i2c_client *client, void *data);

/* get the data */
void *i2c_get_clientdata(const struct i2c_client *client);
```

These functions internally call `dev_set_drvdata` and `dev_get_drvdata` to update or get the value of the `void *driver_data` field of the `struct device` substructure in the `struct i2c_client` structure.

This is an example of how to use extra client data; an excerpt from `drivers/gpio/gpio-mc9s08dz60.c`:

```
/* This is the device specific data structure */
struct mc9s08dz60 {
    struct i2c_client *client;
    struct gpio_chip chip;
};
static int mc9s08dz60_probe(struct i2c_client *client,
const struct i2c_device_id *id)
{
    struct mc9s08dz60 *mc9s;
    if (!i2c_check_functionality(client->adapter,
            I2C_FUNC_SMBUS_BYTE_DATA))
    return -EIO;
    mc9s = devm_kzalloc(&client->dev, sizeof(*mc9s), GFP_KERNEL);
    if (!mc9s)
        return -ENOMEM;
```

```
    [...]
    mc9s->client = client;
    i2c_set_clientdata(client, mc9s);

    return gpiochip_add(&mc9s->chip);
}
```

Actually, these functions are not really specific to I2C. They do nothing but get/set the void *driver_data pointer that is a member of the struct device, and itself is a member of struct i2c_client. In fact, we could have used dev_get_drvdata and dev_set_drvdata directly. One can see their definitions in linux/include/linux/i2c.h.

The remove() function

The prototype of the remove function looks as follows:

```
static int foo_remove(struct i2c_client *client)
```

The remove() function also provides the same struct i2c_client* as the probe() function, so you can retrieve your private data. For example, you may need to process some cleaning or any other stuff, based on the private data you set up in the probe function:

```
static int mc9s08dz60_remove(struct i2c_client *client)
{
    struct mc9s08dz60 *mc9s;

    /* We retrieve our private data */
    mc9s = i2c_get_clientdata(client);

    /* Wich hold gpiochip we want to work on */
    return gpiochip_remove(&mc9s->chip);
}
```

The remove function has the responsibility to unregister us from the subsystem where we have registered in the probe() function. In the preceding example, we simply remove the gpiochip from the kernel.

I2C Client Drivers

Driver initialization and registration

When one's module gets loaded, one may need to do some initializing. Most of the time, just registering the driver with the I2C core will be sufficient. At the same time, when the module is unloaded, we will usually just need to get ourselves out from the I2C core. In `chapter 5`, *Platform Device Drivers* we saw that it is not worth while to bother ourselves by using init/exit functions, but to use `module_*_driver` functions instead. In this case, the function to use is:

```
module_i2c_driver(foo_driver);
```

Driver and device provisioning

As we have seen in matching mechanisms, we need to provide a `device_id` array in order to expose devices that our driver can manage. Since we are talking about I2C devices, the structure would be `i2c_device_id`. That array will inform the kernel about the devices that we are interested in, in the driver.

Now back to our I2C device driver; having a look in `include/linux/mod_devicetable.h`, you will see how `struct i2c_device_id` is defined:

```
struct i2c_device_id {
    char name[I2C_NAME_SIZE];
    kernel_ulong_tdriver_data;      /* Data private to the driver */
};
```

That said, the `struct i2c_device_id` must be embedded in a `struct i2c_driver`. In order to let the I2C core (for module auto-loading) know about devices we need to handle, we must use the `MODULE_DEVICE_TABLE` macro. The kernel has to be aware of which `probe` or `remove` function to call whenever a match occurs, which is why our `probe` and `remove` functions must also be embedded in the same `i2c_driver` structure:

```
static struct i2c_device_id foo_idtable[] = {
    { "foo", my_id_for_foo },
    { "bar", my_id_for_bar },
    { }
};

MODULE_DEVICE_TABLE(i2c, foo_idtable);

static struct i2c_driver foo_driver = {
    .driver = {
    .name = "foo",
```

```
        },
        .id_table = foo_idtable,
        .probe    = foo_probe,
        .remove   = foo_remove,
}
```

Accessing the client

Serial bus transactions are just a matter of accessing registers to set/get their content. I2C respects that principle. I2C core provides two kind of API, one for plain I2C communications, and another for SMBUS compatible device, which also works with I2C devices, but not the reverse.

Plain I2C communication

The following are essential functions one usually deal with when talking to I2C devices:

```
int i2c_master_send(struct i2c_client *client, const char *buf, int count);
int i2c_master_recv(struct i2c_client *client, char *buf, int count);
```

Almost all I2C communication functions take a `struct i2c_client` as the first parameter. The second parameter contains the bytes to read or write and the third represents the number of bytes to read or write. Like any read/write function, the returned value is the number of bytes being read/written. One can also process message transfers with:

```
int i2c_transfer(struct i2c_adapter *adap, struct i2c_msg *msg,
                 int num);
```

`i2c_transfer` sends a set of messages where each can be either a read or a write operation, and can be mixed in any way. Remember that there is no stop bit between each transaction. Looking at `include/uapi/linux/i2c.h`, a message structure looks as follows:

```
struct i2c_msg {
        __u16 addr;     /* slave address */
        __u16 flags;    /* Message flags */
        __u16 len;      /* msg length */
        __u8 *buf;      /* pointer to msg data */
};
```

I2C Client Drivers

The `i2c_msg` structure describes and characterizes an I2C message. It must contain, for each message, the client address, the number of bytes of the message, and the message payload.

 `msg.len` is a `u16`. It means you must always be less than 2^{16} (64k) with your read/write buffers.

Let us have a look at the `read` function for the microchip I2C 24LC512eeprom character driver; we should understand how things really work. The full code is provided with the source of this book.

```
ssize_t
eep_read(struct file *filp, char __user *buf, size_t count, loff_t *f_pos)
{
    [...]
    int _reg_addr = dev->current_pointer;
    u8 reg_addr[2];
    reg_addr[0] = (u8)(_reg_addr>> 8);
    reg_addr[1] = (u8)(_reg_addr& 0xFF);

    struct i2c_msg msg[2];
    msg[0].addr = dev->client->addr;
    msg[0].flags = 0;                    /* Write */
    msg[0].len = 2;                      /* Address is 2bytes coded */
    msg[0].buf = reg_addr;

    msg[1].addr = dev->client->addr;
    msg[1].flags = I2C_M_RD;             /* We need to read */
    msg[1].len = count;
    msg[1].buf = dev->data;

    if (i2c_transfer(dev->client->adapter, msg, 2) < 0)
        pr_err("ee24lc512: i2c_transfer failed\n");

    if (copy_to_user(buf, dev->data, count) != 0) {
        retval = -EIO;
    goto end_read;
    }
    [...]
}
```

`msg.flags` should be `I2C_M_RD` for a read and 0 for a write transaction. Sometimes, you may not want to create `struct i2c_msg` but just process simple read and write.

System Management Bus (SMBus) compatible functions

SMBus is a two-wire bus developed by Intel, and very similar to I2C. I2C devices are SMBus-compatible, but not the reverse. Therefore, it is better to use SMBus functions if one has a doubt about the chip one is writing the driver for.

The following shows some of the SMBus API:

```
s32 i2c_smbus_read_byte_data(struct i2c_client *client, u8 command);
s32 i2c_smbus_write_byte_data(struct i2c_client *client,
                u8 command, u8 value);
s32 i2c_smbus_read_word_data(struct i2c_client *client, u8 command);
s32 i2c_smbus_write_word_data(struct i2c_client *client,
                u8 command, u16 value);
s32 i2c_smbus_read_block_data(struct i2c_client *client,
                u8 command, u8 *values);
s32 i2c_smbus_write_block_data(struct i2c_client *client,
                u8 command, u8 length, const u8 *values);
```

Have a look in `include/linux/i2c.h` and `drivers/i2c/i2c-core.c` in the kernel sources for more explanation.

The following example shows a simple read/write operation in an I2C gpio expander:

```
struct mcp23016 {
   struct i2c_client   *client;
   structgpio_chip     chip;
   structmutex         lock;
};
[...]
/* This function is called when one needs to change a gpio state */
static int mcp23016_set(struct mcp23016 *mcp,
            unsigned offset, intval)
{
    s32 value;
    unsigned bank = offset / 8 ;
    u8 reg_gpio = (bank == 0) ? GP0 : GP1;
    unsigned bit = offset % 8 ;

    value = i2c_smbus_read_byte_data(mcp->client, reg_gpio);
    if (value >= 0) {
        if (val)
            value |= 1 << bit;
        else
            value &= ~(1 << bit);
```

```
            return i2c_smbus_write_byte_data(mcp->client,
                                      reg_gpio, value);
    } else
        return value;
}
[...]
```

Instantiating I2C devices in the board configuration file (old and depreciated way)

We must inform the kernel about which devices are physically present on the system. There are two ways to achieve that. In the DT, as we will see later in the chapter, or through the board configuration file (which is the old and depreciated way). Let us see how to do that in the board configuration file:

`struct i2c_board_info` is the structure used to represent an I2C device on our board. The structure is defined as follows:

```
struct i2c_board_info {
    char type[I2C_NAME_SIZE];
    unsigned short addr;
    void *platform_data;
    int irq;
};
```

Once again, elements not relevant for us have been removed from the structure.

In the preceding structure, `type` should contain the same value as defined in the device driver in the `i2c_driver.driver.name` field. You will then need to fill an array of `i2c_board_info` and pass it as a parameter to the `i2c_register_board_info` function in the board init routine:

```
int i2c_register_board_info(int busnum, struct i2c_board_info const *info,
  unsigned len)
```

Here, `busnum` is the bus number the devices sit on. This is an old and depreciated method, so I'll not go further into it in this book. Feel free to have a look at *Documentation/i2c/instantiating-devices* in the kernel sources to see how things are done.

I2C and the device tree

As we have seen in the preceding sections, in order to configure I2C devices, there are essentially two steps:

- Define and register the I2C driver
- Define and register the I2C devices

I2C devices belong to nonmemory mapped devices family in the DT, and I2C bus is an addressable bus (by addressable, I mean you can address a specific device on the bus). In this, the `reg` property in the device node represents the device address on the bus.

I2C device nodes are all children of the bus node they seat on. Each device is assigned only an address. There is no length or range involved. Standard properties one needs to declare for I2C devices are `reg`, which represents the address of the device on the bus, and the `compatible` string, which is used to match the device with a driver. For more information on addressing, you can refer to Chapter 6, *The Concept of Device Tree*.

```
&i2c2 { /* Phandle of the bus node */
    pcf8523: rtc@68 {
        compatible = "nxp,pcf8523";
        reg = <0x68>;
    };
    eeprom: ee24lc512@55 { /* eeprom device */
        compatible = "packt,ee24lc512";
        reg = <0x55>;
    };
};
```

The preceding sample declares an HDMI EDID chip at address 0x50, on SoC's I2C bus number 2, and a **real time clock** (**RTC**), at address 0x68 on the same bus.

Defining and registering the I2C driver

What we have seen so far does not change. The extra thing we need is to define a `struct of_device_id`. Struct `of_device_id` defined to match the corresponding node in the `.dts` file:

```
/* no extra data for this device */
static const struct of_device_id foobar_of_match[] = {
        { .compatible = "packtpub,foobar-device" },
        {}
};
MODULE_DEVICE_TABLE(of, foobar_of_match);
```

Now we define the `i2c_driver` as follows:

```
static struct i2c_driver foo_driver = {
    .driver = {
    .name   = "foo",
    .of_match_table = of_match_ptr(foobar_of_match), /* Only this line is added */
    },
    .probe  = foo_probe,
    .id_table = foo_id,
};
```

One can then improve the `probe` function this way:

```
static int my_probe(struct i2c_client *client, const struct i2c_device_id *id)
{
    const struct of_device_id *match;
    match = of_match_device(mcp23s08_i2c_of_match, &client->dev);
    if (match) {
        /* Device tree code goes here */
    } else {
        /*
        * Platform data code comes here.
        * One can use
        *    pdata = dev_get_platdata(&client->dev);
        *
        * or *id*, which is a pointer on the *i2c_device_id* entry that originated
        * the match, in order to use *id->driver_data* to extract the device
        * specific data, as described in platform driver chapter.
        */
    }
```

```
        [...]
}
```

Remark

For kernel versions older than 4.10, if one looks at `drivers/i2c/i2c-core.c`, in the `i2c_device_probe()` function (for information, it is the function the kernel calls every time an I2C device is registered to the I2C core), one will see something like this:

```
if (!driver->probe || !driver->id_table)
        return -ENODEV;
```

This means that even if one does not need to use the `.id_table`, it is mandatory in the driver. In fact, one can use the OF match style only, but cannot get rid of `.id_table`. Kernel developers tried to remove the need for `.id_table` and exclusively use `.of_match_table` for device matching. The patch is available at this URL:
https://git.kernel.org/cgit/linux/kernel/git/torvalds/linux.git/commit/?id=c80f52847c50109ca248c22efbf71ff10553dca4.

Nevertheless, regressions have been found and the commit was reverted. Have a look here for details:
https://git.kernel.org/cgit/linux/kernel/git/torvalds/linux.git/commit/?id=661f6c1cd926c6c973e03c6b5151d161f3a666ed. This has been fixed since kernel version >= 4.10. The fix looks as follows:

```
/*
 * An I2C ID table is not mandatory, if and only if, a suitable Device
 * Tree match table entry is supplied for the probing device.
 */
if (!driver->id_table &&
    !i2c_of_match_device(dev->driver->of_match_table, client))
        return -ENODEV;
```

In other words, one must define both `.id_table` and `.of_match_table` for the I2C driver, otherwise your device will not be probed for kernel version 4.10 or earlier.

Instantiating I2C devices in the device tree - the new way

`struct i2c_client` is the structure used to describe the I2C device. However, with OF style, this structure could not be defined in the board file anymore. The only thing we need to do is provide the device's information in the DT and the kernel will build one from it.

The following code shows how we can declare our I2C `foobar` device node in a `dts` file:

```
&i2c3 {
    status = "okay";
    foo-bar: foo@55 {
    compatible = "packtpub,foobar-device";
reg = &lt;55>;
    };
};
```

Putting it all together

To summarize the steps needed to write I2C client drivers, you need to:

1. Declare device ids supported by the driver. You can do that using `i2c_device_id`. If DT is supported, use `of_device_id` too.
2. Call `MODULE_DEVICE_TABLE(i2c, my_id_table)` to register your device list with the I2C core. If device tree is supported, you must call `MODULE_DEVICE_TABLE(of, your_of_match_table)` to register your device list with the OF core.
3. Write the `probe` and `remove` functions according to their respective prototypes. If needed, write power management functions too. The `probe` function must identify your device, configure it, define per-device (private) data, and register with the appropriate kernel framework. The driver's behavior depends on what you have done in the `probe` function. The `remove` function must undo everything you have done in the `probe` function (free memory and unregister from any framework).

4. Declare and fill a struct i2c_driver structure and set the id_table field with the array of ids you have created. Set .probe and .remove fields with the name of the corresponding function you have written above. In the .driver substructure, set the .owner field to THIS_MODULE, set the driver name, and finally, set the .of_match_table field with the array of of_device_id if DT is supported.
5. Call the module_i2c_driver function with your i2c_driver structure that you just filled above: module_i2c_driver(serial_eeprom_i2c_driver) in order to register your driver with the kernel.

Summary

We just dealt with I2C device drivers. It is time for you to pick any I2C device on the market and write the corresponding driver, with DT support. This chapter talked about the kernel I2C core and associated API, including device tree support, to give you the necessary skills to talk with I2C devices. You should be able to write efficient probe functions and register with the kernel I2C core. In the next chapter, we will use skills we learned here to develop the SPI device driver.

8
SPI Device Drivers

Serial Peripheral Interface (**SPI**) is a (at least) four-wire bus--**Master Input Slave Output** (**MISO**), **Master Output Slave Input** (**MOSI**), **Serial Clock** (**SCK**), and **Chip Select** (**CS**), which is used to connect a serial flash, AD/DA converter. The master always generates the clock. Its speed can reach up to 80 MHz, even if there is no real speed limitation (much faster than I2C as well). The same for the CS line, which is always managed by the master.

Each of these signal names has a synonym:

- Whenever you sees SIMO, SDI, DI, or SDA, they refer to MOSI.
- SOMI, SDO, DO, SDA will refer to MISO.
- SCK, CLK, SCL will refer to SCK.
- S S is the slave select line, also called CS. CSx can be used (where x is an index, CS0, CS1), EN and ENB too, meaning enable. The CS is usually an active low signal:

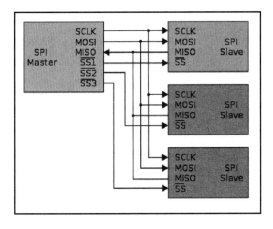

SPI topology (image from wikipedia)

This chapter will walk through SPI driver concepts such as:

- SPI bus description
- Driver architecture and data structure descriptions
- Data sending and receiving in both half and full duplex
- Declaring SPI devices from DT
- Accessing SPI devices from user space, in both half and full duplex

The driver architecture

The required header for SPI stuff in the Linux kernel is `<linux/spi/spi.h>`. Before talking about the driver structure, let us see how SPI devices are defined in the kernel. An SPI device is represented in the kernel as an instance of `spi_device`. The instance of the driver that manages them is `struct spi_driver` structure.

The device structure

`struct spi_device` structure represents an SPI device, and is defined in include/linux/spi/spi.h:

```
struct spi_device {
    struct devicedev;
    struct spi_master*master;
    u32 max_speed_hz;
    u8 chip_select;
    u8 bits_per_word;
    u16 mode;
    int irq;
    [...]
    int cs_gpio;        /* chip select gpio */
};
```

Some fields that are not meaningful for us have been removed. That says, the following is the meaning of elements in the structure:

- `master`: This represents the SPI controller (bus) on which the device is connected.
- `max_speed_hz`: This is the maximum clock rate to be used with this chip (on the current board); this parameter can be changed from within the driver. You can override that parameter using `spi_transfer.speed_hz` for each transfer. We will discuss SPI transfer later.
- `chip_select`: This lets you enable the chip you need to talk to, distinguishing chips handled by the master. The `chip_select` is active low by default. This behavior can be changed in mode, by adding the `SPI_CS_HIGH` flag.
- `mode`: This defines how data should be clocked. The device driver may change this. The data clocking is **Most Significant Bit** (**MSB**) first, by default for each word in a transfer. This behavior can be overridden by specifying `SPI_LSB_FIRST`.
- `irq`: This represents the interrupt number (registered as device resource in your board `init` file or through the DT) you should pass to `request_irq()` to receive interrupts from this device.

A word about SPI modes; they are built using two characteristics:

- CPOL: This is the initial clock polarity:
    - 0: Initial clock state low, and the first edge is rising
    - 1: Initial clock state high, and the first state is falling
- CPHA: This is the clock phase, choosing at which edge the data will be sampled:
    - 0: Data latched at falling edge (high to low transition), whereas output changes at rising edge
    - 1: Data latched at rising edge (low to high transition), and output at falling edge

This allows for four SPI modes, which are defined in the kernel according to the following macro in `include/linux/spi/spi.h`:

```
#define  SPI_CPHA   0x01
#define  SPI_CPOL   0x02
```

You can then produce the following array to summarize things:

Mode	CPOL	CPHA	Kernel macro
0	0	0	#define SPI_MODE_0 (0|0)
1	0	1	#define SPI_MODE_1 (0|SPI_CPHA)
2	1	0	#define SPI_MODE_2 (SPI_CPOL|0)
3	1	1	#define SPI_MODE_3 (SPI_CPOL|SPI_CPHA)

The following is the representation of each SPI mode, as defined in the preceding array. That said, only the MOSI line is represented, but the principle is the same for MISO:

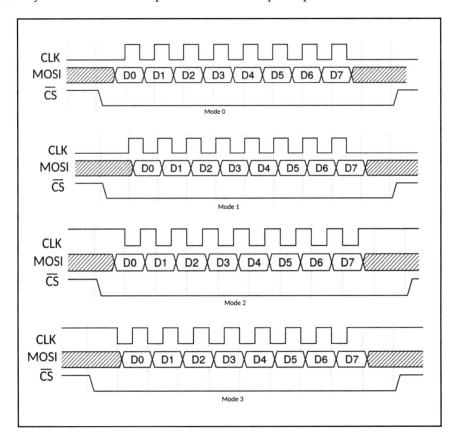

Commonly used modes are SPI_MODE_0 and SPI_MODE_3.

spi_driver structure

struct spi_driver represents the driver you develop to manage your SPI device. Its structure is as follows:

```
struct spi_driver {
   const struct spi_device_id *id_table;
   int (*probe)(struct spi_device *spi);
   int (*remove)(struct spi_device *spi);
   void (*shutdown)(struct spi_device *spi);
   struct device_driver    driver;
};
```

The probe() function

Its prototype is as follows:

```
static int probe(struct spi_device *spi)
```

You may refer to Chapter 7, *I2C Client Drivers* in order to see what is to be done in a probe function. The same steps apply here. Therefore, unlike an I2C driver that has no capability to change the controller bus parameters (CS state, bit per word, clock) at runtime, an SPI driver can. You can set up the bus according to your device properties.

A typical SPI probe function would look like the following:

```
static int my_probe(struct spi_device *spi)
{
    [...] /* declare your variable/structures */

    /* bits_per_word cannot be configured in platform data */
    spi->mode = SPI_MODE_0; /* SPI mode */
    spi->max_speed_hz = 20000000;   /* Max clock for the device */
    spi->bits_per_word = 16;   /* device bit per word */
    ret = spi_setup(spi);
    ret = spi_setup(spi);
    if (ret < 0)
        return ret;

    [...] /* Make some init */
    [...] /* Register with apropriate framework */

    return ret;
}
```

SPI Device Drivers

The `struct spi_device*` is an input parameter, given to the `probe` function by the kernel. It represents the device you are probing. From within your `probe` function, you can get the `spi_device_id` that triggered the match using `spi_get_device_id` (in case of `id_table match`) and extract the driver data:

```
const struct spi_device_id *id = spi_get_device_id(spi);
my_private_data = array_chip_info[id->driver_data];
```

Per-device data

In the `probe` function, it is a common task to track a private (per-device) data to be used during the module lifetime. This has been discussed in Chapter 7, *I2C Client Drivers*.

The following are prototypes of functions one uses for setting/getting per-device data:

```
/* set the data */
void spi_set_drvdata(struct *spi_device, void *data);

/* Get the data back */
 void *spi_get_drvdata(const struct *spi_device);
```

For example:

```
struct mc33880 {
    struct mutex    lock;
    u8       bar;
    struct foo chip;
    struct spi_device *spi;
};

static int mc33880_probe(struct spi_device *spi)
{
    struct mc33880 *mc;
    [...] /* Device set up */

    mc = devm_kzalloc(&spi->dev, sizeof(struct mc33880),
                GFP_KERNEL);
    if (!mc)
        return -ENOMEM;

    mutex_init(&mc->lock);
    spi_set_drvdata(spi, mc);

    mc->spi = spi;
    mc->chip.label = DRIVER_NAME,
    mc->chip.set = mc33880_set;
```

```
        /* Register with appropriate framework */
        [...]
}
```

The remove() function

The `remove` function must release every resource grabbed in the `probe` function. Its structure is as follows:

```
static int  my_remove(struct spi_device *spi);
```

A typical `remove` function may look like the following:

```
static int mc33880_remove(struct spi_device *spi)
{
    struct mc33880 *mc;
    mc = spi_get_drvdata(spi); /* Get our data back */
    if (!mc)
        return -ENODEV;

    /*
     * unregister from frameworks with which we registered in the
     * probe function
     */
    [...]
    mutex_destroy(&mc->lock);
    return 0;
}
```

Driver initialization and registration

For device sitting on a bus, whether it is a physical one or the pseudo platform bus, most of the time, everything is done in the `probe` function. The `init` and `exit` functions are just used to register/unregister the driver with the bus core:

```
static int __init foo_init(void)
{
    [...] /*My init code */
    return spi_register_driver(&foo_driver);
}
module_init(foo_init);

static void __exit foo_cleanup(void)
{
    [...] /* My clean up code */
```

```
        spi_unregister_driver(&foo_driver);
    }
    module_exit(foo_cleanup);
```

That said, if you do not do anything else but register/unregister the driver, the kernel offers a macro:

```
    module_spi_driver(foo_driver);
```

This will internally call `spi_register_driver` and `spi_unregister_driver`. It is exactly the same thing as what we have seen in the previous chapter.

Driver and devices provisioning

As we need `i2c_device_id` for I2C devices, we must use `spi_device_id` for SPI devices, in order to provide a `device_id` array to match our devices. It is defined in include/linux/mod_devicetable.h:

```
    struct spi_device_id {
       char name[SPI_NAME_SIZE];
       kernel_ulong_t driver_data; /* Data private to the driver */
    };
```

We need to embed our array into a `struct spi_device_id` in order to inform the SPI core about the device ID we need to manage in the driver and call MODULE_DEVICE_TABLE macro on the driver structure. Of course, the first parameter of the macro is the name of the bus on which the device sits. In our case, it is SPI:

```
    #define ID_FOR_FOO_DEVICE   0
    #define ID_FOR_BAR_DEVICE   1

    static struct spi_device_id foo_idtable[] = {
        { "foo", ID_FOR_FOO_DEVICE },
        { "bar", ID_FOR_BAR_DEVICE },
        { }
    };
    MODULE_DEVICE_TABLE(spi, foo_idtable);

    static struct spi_driver foo_driver = {
        .driver = {
        .name = "KBUILD_MODULE",
        },

        .id_table   = foo_idtable,
        .probe      = foo_probe,
```

```
        .remove     = foo_remove,
};

module_spi_driver(foo_driver);
```

Instantiate SPI devices in board configuration file – old and depreciated way

Device should be instantiated in board file only if the system does not support device tree. Since device tree has come, this method of instantiating is deprecated. Therefore, let us just remember that the board file resides in `arch/` directory. The structure used to represent an SPI device is `struct spi_board_info`, not the `struct spi_device` we used in the driver. It is only when you have filled and registered the `struct spi_board_info` using the `spi_register_board_info` function that the kernel will build a `struct spi_device` (which will be passed to your driver and register with the SPI core).

Feel free to look at the `struct spi_board_info` field in `include/linux/spi/spi.h`. The definition of `spi_register_board_info` can be found in `drivers/spi/spi.c`. Now let us have a look at some SPI device registration in the board file:

```
/**
 * Our platform data
 */
struct my_platform_data {
   int foo;
   bool bar;
};
static struct my_platform_data mpfd = {
   .foo = 15,
   .bar = true,
};

static struct spi_board_info
   my_board_spi_board_info[] __initdata = {
     {
        /* the modalias must be same as spi device driver name */
         .modalias = "ad7887", /* Name of spi_driver for this device */
         .max_speed_hz = 1000000,  /* max spi clock (SCK) speed in HZ */
         .bus_num = 0, /* Framework bus number */
         .irq = GPIO_IRQ(40),
         .chip_select = 3, /* Framework chip select */
         .platform_data = &mpfd,
         .mode = SPI_MODE_3,
     },{
```

```
            .modalias = "spidev",
            .chip_select = 0,
            .max_speed_hz = 1 * 1000 * 1000,
            .bus_num = 1,
            .mode = SPI_MODE_3,
        },
    };

    static int __init board_init(void)
    {
        [...]
        spi_register_board_info(my_board_spi_board_info,
    ARRAY_SIZE(my_board_spi_board_info));
        [...]

        return 0;
    }
    [...]
```

SPI and device tree

Like I2C devices, SPI devices belong to the non memory mapped devices family in the DT, but are addressable too. Here, the address means the CS index among the list of CS (starting from 0) given to the controller (the master). As an example, we may have three different SPI devices seating on the SPI bus, each with its CS line. The master will be given a set of GPIO, each representing CS to activate a device. If the device X uses the second GPIO line as CS, we must set its address to 1 (as we always start from 0) in reg property.

The following is a real DT listing for SPI devices:

```
ecspi1 {
    fsl,spi-num-chipselects = <3>;
    cs-gpios = <&gpio5 17 0>, <&gpio5 17 0>, <&gpio5 17 0>;
    pinctrl-0 = <&pinctrl_ecspi1 &pinctrl_ecspi1_cs>;
    #address-cells = <1>;
    #size-cells = <0>;
    compatible = "fsl,imx6q-ecspi", "fsl,imx51-ecspi";
    reg = <0x02008000 0x4000>;
    status = "okay";

    ad7606r8_0: ad7606r8@0 {
        compatible = "ad7606-8";
        reg = <0>;
        spi-max-frequency = <1000000>;
        interrupt-parent = <&gpio4>;
        interrupts = <30 0x0>;
```

```
    };
    label: fake_spi_device@1 {
        compatible = "packtpub,foobar-device";
        reg = <1>;
        a-string-param = "stringvalue";
        spi-cs-high;
    };
    mcp2515can: can@2 {
        compatible = "microchip,mcp2515";
        reg = <2>;
        spi-max-frequency = <1000000>;
        clocks = <&clk8m>;
        interrupt-parent = <&gpio4>;
        interrupts = <29 IRQ_TYPE_LEVEL_LOW>;
    };
};
```

There is a new property introduced in SPI device nodes: `spi-max-frequency`. It represents the maximum SPI clocking speed of the device in Hz. Whenever you access the device, the bus controller driver will ensure the clock does not cross this limit. Other properties commonly used are:

- `spi-cpol`: This is a Boolean (empty property) indicating the device requires inverse clock polarity mode. It corresponds to CPOL.
- `spi-cpha`: This is an empty property indicating the device requires shifted clock phase mode. It corresponds to CPHA.
- `spi-cs-high`: By default, SPI devices require CS low to be active. This is a Boolean property indicating the device requires CS active high.

That said, for a complete list of SPI binding elements, you can refer to *Documentation/devicetree/bindings/spi/spi-bus.txt* in the kernel sources.

Instantiate SPI devices in device tree - the new way

By filling our device node in the DT properly, the kernel will build a `struct spi_device` for us, and give it as a parameter to our SPI core functions. The following is just an excerpt from the SPI DT listing defined previously:

```
&ecspi1 {
    status = "okay";
    label: fake_spi_device@1 {
    compatible = "packtpub,foobar-device";
    reg = <1>;
    a-string-param = "stringvalue";
    spi-cs-high;
    };
};
```

Define and register SPI driver

Again the principle is the same as that for I2C drivers. We need to define a `struct of_device_id` to match devices in the DT, and call the `MODULE_DEVICE_TABLE` macro to register with the OF core:

```
static const struct of_device_id foobar_of_match[] = {
        { .compatible = "packtpub,foobar-device" },
        { .compatible = "packtpub,barfoo-device" },
    {}
};
MODULE_DEVICE_TABLE(of, foobar_of_match);
```

Then define our `spi_driver` as the following:

```
static struct spi_driver foo_driver = {
   .driver = {
   .name   = "foo",
       /* The following line adds Device tree */
   .of_match_table = of_match_ptr(foobar_of_match),
   },
   .probe   = my_spi_probe,
   .id_table = foo_id,
};
```

You can then improve the `probe` function this way:

```
static int my_spi_probe(struct spi_device *spi)
{
    const struct of_device_id *match;
    match = of_match_device(of_match_ptr(foobar_of_match), &spi->dev);
    if (match) {
        /* Device tree code goes here */
    } else {
        /*
         * Platform data code comes here.
         * One can use
         *    pdata = dev_get_platdata(&spi->dev);
         *
         * or *id*, which is a pointer on the *spi_device_id* entry that originated
         * the match, in order to use *id->driver_data* to extract the device
         * specific data, as described in Chapter 5, Platform Device Drivers.
         */
    }
    [...]
}
```

Accessing and talking to the client

The SPI I/O model consists of a set of queued messages. We submit one or more `struct spi_message` structures, which are processed and completed synchronously or asynchronously. A single message consists of one or more `struct spi_transfer` objects, each of which represents a full duplex SPI transfer. These are two main structures to exchange data between the driver and the device. They are both defined in `include/linux/spi/spi.h`:

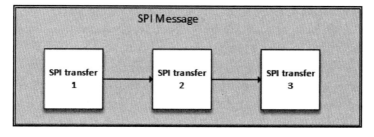

SPI message structure

SPI Device Drivers

`struct spi_transfer` represents a full duplex SPI transfer:

```
struct spi_transfer {
    const void   *tx_buf;
    void *rx_buf;
    unsigned len;

    dma_addr_t tx_dma;
    dma_addr_t rx_dma;

    unsigned cs_change:1;
    unsigned tx_nbits:3;
    unsigned rx_nbits:3;
#define  SPI_NBITS_SINGLE   0x01 /* 1bit transfer */
#define  SPI_NBITS_DUAL     0x02 /* 2bits transfer */
#define  SPI_NBITS_QUAD     0x04 /* 4bits transfer */
    u8 bits_per_word;
    u16 delay_usecs;
    u32 speed_hz;
};
```

The following is the meaning of the structure elements:

- `tx_buf`: This buffer contains the data to be written. It should be NULL or left as it is in case of a read-only transaction. It should be `dma`-safe in the case where you need to perform SPI transactions through **Direct Memory Access** (**DMA**).
- `rx_buf`: This is a buffer for data to be read (with the same properties as `tx_buf`), or NULL in a write-only transaction.
- `tx_dma`: This is the DMA address of `tx_buf`, in case `spi_message.is_dma_mapped` is set to 1. DMA is discussed in Chapter 12, *DMA – Direct Memory Access*.
- `rx_dma`: This is the same as `tx_dma`, but for `rx_buf`.
- `len`: This represents the size of `rx` and `tx` buffers in bytes, meaning they must have the same size if both are used.
- `speed_hz`: This overrides the default speed, specified in `spi_device.max_speed_hz`, but only for the current transfer. If 0, the default value (provided in `struct spi_device` structure) is used.
- `bits_per_word`: Data transfer involves one or more words. A word is a unit of data, whose size in bits may vary according to the need. Here, `bits_per_word` represents the size in bits of a word for this SPI transfer. This override the default value provided in `spi_device.bits_per_word`. If 0, the default (from `spi_device`) is used.

- `cs_change`: This determines the state of the `chip_select` line after this transfer completes.
- `delay_usecs`: This represents the delay (in microseconds) after this transfer before (optionally) changing the `chip_select` status, then starting the next transfer or completing this `spi_message`.

At the other side, the `struct spi_message` is used atomically to wrap one or more SPI transfers. The SPI bus used will be hogged by the driver until every transfer that constitutes the message is completed. SPI message structure is defined in `include/linux/spi/spi.h` too:

```
struct spi_message {
    struct list_head transfers;
    struct spi_device *spi;
    unsigned is_dma_mapped:1;
    /* completion is reported through a callback */
    void (*complete)(void *context);
    void *context;
    unsigned frame_length;
    unsigned actual_length;
    int status;
};
```

- `transfers`: This is the list of transfers that constitutes the message. We will see later how to add a transfer to this list.
- `is_dma_mapped`: This informs the controller whether to use DMA (or not) to perform the transaction. Your code is then responsible in providing DMA and CPU virtual addresses for each transfer buffer.
- `complete`: This is a callback called when the transaction is done, and `context` is the parameter to be given to the callback.
- `frame_length`: This will be set automatically with the total number of bytes in the message.
- `actual_length`: This is the number of bytes transferred in all successful segments.
- `status`: This reports the transfers status. Zero on success, else `-errno`.

`spi_transfer` elements in a message are processed in a FIFO order. Until the message is completed, you have to make sure not to use transfer buffer, in order to avoid data corruption. You make completion call to make sure one can.

Before a message can be submitted to the bus, it has to be initialized with `void spi_message_init(struct spi_message *message)`, which will zero each element in the structure and initialize the `transfers` list. For each transfer to be added to the message, you should call `void spi_message_add_tail(struct spi_transfer *t, struct spi_message *m)` on that transfer, which will result in enqueuing the transfer into `transfers` list. Once done, you have two choices to start the transaction:

- Synchronously, using the `int spi_sync(struct spi_device *spi, struct spi_message *message)` function, which may sleep and which is not to be used in an interrupt context. Completion of the callback is not necessary here. This function is a wrapper around the second function (`spi_async()`).
- Asynchronously, using the `spi_async()` function, which can be used in an atomic context too, and whose prototype is `int spi_async(struct spi_device *spi, struct spi_message *message)`. It is good practice to provide callback here, since it will be executed upon message complete.

The following is what a single transfer SPI message transaction may look like:

```
char tx_buf[] = {
        0xFF, 0xFF, 0xFF, 0xFF, 0xFF,
        0xFF, 0x40, 0x00, 0x00, 0x00,
        0x00, 0x95, 0xEF, 0xBA, 0xAD,
        0xF0, 0x0D,
};

char rx_buf[10] = {0,};
int ret;
struct spi_message single_msg;
struct spi_transfer single_xfer;

single_xfer.tx_buf = tx_buf;
single_xfer.rx_buf = rx_buf;
single_xfer.len    = sizeof(tx_buff);
single_xfer.bits_per_word = 8;

spi_message_init(&msg);
spi_message_add_tail(&xfer, &msg);
ret = spi_sync(spi, &msg);
```

Now let us write a multi-transfer message transaction:

```
struct {
    char buffer[10];
    char cmd[2]
    int foo;
} data;

struct data my_data[3];
initialize_date(my_data, ARRAY_SIZE(my_data));

struct spi_transfer   multi_xfer[3];
struct spi_message    single_msg;
int ret;

multi_xfer[0].rx_buf = data[0].buffer;
multi_xfer[0].len = 5;
multi_xfer[0].cs_change = 1;
/* command A */
multi_xfer[1].tx_buf = data[1].cmd;
multi_xfer[1].len = 2;
multi_xfer[1].cs_change = 1;
/* command B */
multi_xfer[2].rx_buf = data[2].buffer;
multi_xfer[2].len = 10;

spi_message_init(single_msg);
spi_message_add_tail(&multi_xfer[0], &single_msg);
spi_message_add_tail(&multi_xfer[1], &single_msg);
spi_message_add_tail(&multi_xfer[2], &single_msg);
ret = spi_sync(spi, &single_msg);
```

There are other helper functions, all built around `spi_sync()`. Some of them are:

```
int spi_read(struct spi_device *spi, void *buf, size_t len)
int spi_write(struct spi_device *spi, const void *buf, size_t len)
int spi_write_then_read(struct spi_device *spi,
        const void *txbuf, unsigned n_tx,
    void *rxbuf, unsigned n_rx)
```

Please have a look at `include/linux/spi/spi.h` to see the complete list. These wrappers should be used with small amounts of data.

Putting it all together

The steps needed to write SPI client drivers are as follows:

1. Declare device IDs supported by the driver. You can do that using `spi_device_id`. If DT is supported, use `of_device_id` too. You can make an exclusive use of DT.
2. Call `MODULE_DEVICE_TABLE(spi, my_id_table);` to register your device list with the SPI core. If DT is supported, you must call `MODULE_DEVICE_TABLE(of, your_of_match_table);` to register your device list with the `of` core.
3. Write `probe` and `remove` functions according to their respective prototypes. The `probe` function must identify your device, configure it, define per-device (private) data, configure the bus if needed (SPI mode and so on) using `spi_setup` function, and register with the appropriate kernel framework. In the `remove` function, simply undo everything done in the `probe` function.
4. Declare and fill a `struct spi_driver` structure, set the `id_table` field with the array of IDs you have created. Set `.probe` and `.remove` fields with the name of the corresponding functions you have written. In the `.driver` substructure, set the `.owner` field to `THIS_MODULE`, set the driver name, and finally set the `.of_match_table` field with the array of `of_device_id`, if the DT is supported.
5. Call `module_spi_driver` function with your `spi_driver` structure you just filled before `module_spi_driver(serial_eeprom_spi_driver);` in order to register your driver with the kernel.

SPI user mode driver

There are two ways of using the user mode SPI device driver. To be able to do that, you need to enable your device with `spidev` driver. An example would be as follows:

```
spidev@0x00 {
    compatible = "spidev";
    spi-max-frequency = <800000>; /* It depends on your device */
    reg = <0>; /* correspond tochipselect 0 */
};
```

You can call either the read/write functions or an `ioctl()`. With calling read/write you can only read or write at a time. If you need full-duplex read and write, you have to use the **Input Output Control** (**ioctl**) commands . Examples for both are provided. This is the read/write example. You can compile it either with the cross-compiler of the platform or with the native compiler on the board:

```
#include <stdio.h>
#include <fcntl.h>
#include <stdlib.h>

int main(int argc, char **argv)
{
    int i,fd;
    char wr_buf[]={0xff,0x00,0x1f,0x0f};
    char rd_buf[10];

    if (argc<2) {
        printf("Usage:\n%s [device]\n", argv[0]);
        exit(1);
    }
    fd = open(argv[1], O_RDWR);
    if (fd<=0) {
        printf("Failed to open SPI device %s\n",argv[1]);
        exit(1);
    }
    if (write(fd, wr_buf, sizeof(wr_buf)) != sizeof(wr_buf))
        perror("Write Error");
    if (read(fd, rd_buf, sizeof(rd_buf)) != sizeof(rd_buf))
        perror("Read Error");
    else
        for (i = 0; i < sizeof(rd_buf); i++)
            printf("0x%02X ", rd_buf[i]);

    close(fd);
    return 0;
}
```

With IOCTL

The advantage of using IOCTL is that you can work in full duplex. The best example you can find is `documentation/spi/spidev_test.c`, in the kernel source tree, of course.

That said, the preceding example using read/write did not change any SPI configuration. However, the kernel exposes to user space a set of IOCTL commands, which you can use in order to set up the bus according to the need, just like what is done in DT. The following example shows how you can change the bus settings:

```c
#include <stdint.h>
#include <unistd.h>
#include <stdio.h>
#include <stdlib.h>
#include <string.h>
#include <fcntl.h>
#include <sys/ioctl.h>
#include <linux/types.h>
#include <linux/spi/spidev.h>
static int pabort(const char *s)
{
    perror(s);
    return -1;
}

static int spi_device_setup(int fd)
{
    int mode, speed, a, b, i;
    int bits = 8;

    /*
     * spi mode: mode 0
     */
    mode = SPI_MODE_0;
    a = ioctl(fd, SPI_IOC_WR_MODE, &mode); /* write mode */
    b = ioctl(fd, SPI_IOC_RD_MODE, &mode); /* read mode */
    if ((a < 0) || (b < 0)) {
        return pabort("can't set spi mode");
    }

    /*
     * Clock max speed in Hz
     */
    speed = 8000000; /* 8 MHz */
    a = ioctl(fd, SPI_IOC_WR_MAX_SPEED_HZ, &speed); /* Write speed */
    b = ioctl(fd, SPI_IOC_RD_MAX_SPEED_HZ, &speed); /* Read speed */
    if ((a < 0) || (b < 0)) {
```

SPI Device Drivers

```
            return pabort("fail to set max speed hz");
    }
    /*
     * setting SPI to MSB first.
     * Here, 0 means "not to use LSB first".
     * In order to use LSB first, argument should be > 0
     */
    i = 0;
    a = ioctl(dev, SPI_IOC_WR_LSB_FIRST, &i);
    b = ioctl(dev, SPI_IOC_RD_LSB_FIRST, &i);
    if ((a < 0) || (b < 0)) {
        pabort("Fail to set MSB first\n");
    }
    /*
     * setting SPI to 8 bits per word
     */
    bits = 8;
    a = ioctl(dev, SPI_IOC_WR_BITS_PER_WORD, &bits);
    b = ioctl(dev, SPI_IOC_RD_BITS_PER_WORD, &bits);
    if ((a < 0) || (b < 0)) {
        pabort("Fail to set bits per word\n");
    }
    return 0;
}
```

You can have a look at *Documentation/spi/spidev* for more information on spidev ioctl commands. When it comes to sending data over the bus, you can use `SPI_IOC_MESSAGE(N)` request, which offers full-duplex access, and composite operations without chipselect de-activation, thus offering multi-transfer support. It is the equivalent of kernel `spi_sync()`. Here a transfer is represented as an instance of `struct spi_ioc_transfer`, which is the equivalent of kernel `struct spi_transfer`, and whose definition can be found in `include/uapi/linux/spi/spidev.h`. The following is an example of usage:

```
static void do_transfer(int fd)
{
    int ret;
    char txbuf[] = {0x0B, 0x02, 0xB5};
    char rxbuf[3] = {0, };
    char cmd_buff = 0x9f;

    struct spi_ioc_transfer tr[2] = {
        0 = {
            .tx_buf = (unsigned long)&cmd_buff,
            .len = 1,
            .cs_change = 1; /* We need CS to change */
            .delay_usecs = 50, /* wait after this transfer */
```

```c
                .bits_per_word = 8,
        },
        [1] = {
            .tx_buf = (unsigned long)tx,
            .rx_buf = (unsigned long)rx,
            .len = txbuf(tx),
            .bits_per_word = 8,
        },
    };

    ret = ioctl(fd, SPI_IOC_MESSAGE(2), &tr);
    if (ret == 1){
        perror("can't send spi message");
        exit(1);
    }

    for (ret = 0; ret < sizeof(tx); ret++)
        printf("%.2X ", rx[ret]);
    printf("\n");
}


int main(int argc, char **argv)
{
    char *device = "/dev/spidev0.0";
    int fd;
    int error;

    fd = open(device, O_RDWR);
    if (fd < 0)
        return pabort("Can't open device ");

    error = spi_device_setup(fd);
    if (error)
        exit (1);
    do_transfer(fd);
    close(fd);
    return 0;
}
```

Summary

We just dealt with SPI drivers and now can take advantage of this faster serial (and full duplex) bus. We walked through data transfer over SPI, which is the most important section. That said, you may need more abstraction in order not to bother with SPI or I2C APIs. This is where the next chapter comes in, dealing with Regmap API, which offers a higher and unified level of abstraction, so that SPI (or I2C) commands will become transparent to you.

9
Regmap API – A Register Map Abstraction

Before the regmap API was developed, there were redundant codes for the device drivers dealing with SPI core, I2C core, or both. The principle was the same; accessing the register for read/write operations. The following figure shows how either SPI or I2C API were standalone before Regmap was introduced to kernel:

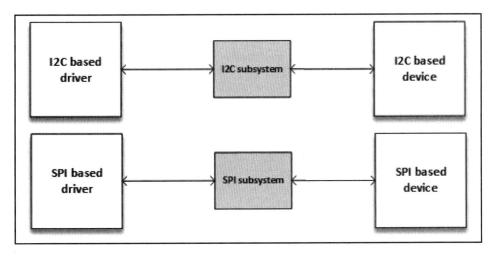

SPI and I2C subsystems before regmap

Regmap API – A Register Map Abstraction

The regmap API was introduced in version 3.1 of the kernel, to factorize and unify the way kernel developers access SPI/I2C devices. It is then just a matter of how to initialize, configure a regmap, and process any read/write/modify operation fluently, whether it is SPI or I2C:

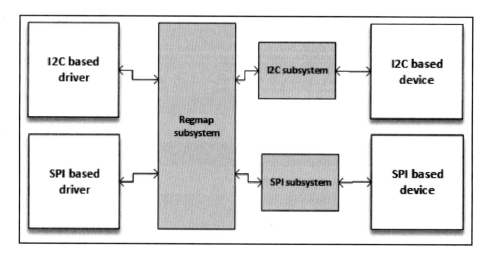

SPI and I2C subsystems after regmap

This chapter will walk through regmap framework by mean of:

- Introducing the main data structures used in by the regmap framework
- Walking through regmap configuration
- Accessing devices using the regmap API
- Introducing the regmap caching system
- Providing a complete driver that summarizes the concepts learned previously

Programming with the regmap API

The regmap API is quite simple. There are only a few structures to know. The two most important structures of this API are `struct regmap_config`, which represents the configuration of the regmap, and `struct regmap`, which is the regmap instance itself. All of the regmap data structures are defined in `include/linux/regmap.h`.

regmap_config structure

`struct regmap_config` stores the configuration of the regmap during the driver's lifetime. What you set here affects read/write operations. It is the most important structure in the regmap API. The source looks like this:

```
struct regmap_config {
    const char *name;

    int reg_bits;
    int reg_stride;
    int pad_bits;
    int val_bits;

    bool (*writeable_reg)(struct device *dev, unsigned int reg);
    bool (*readable_reg)(struct device *dev, unsigned int reg);
    bool (*volatile_reg)(struct device *dev, unsigned int reg);
    bool (*precious_reg)(struct device *dev, unsigned int reg);
    regmap_lock lock;
    regmap_unlock unlock;
    void *lock_arg;

    int (*reg_read)(void *context, unsigned int reg,
                    unsigned int *val);
    int (*reg_write)(void *context, unsigned int reg,
                     unsigned int val);

    bool fast_io;

    unsigned int max_register;
    const struct regmap_access_table *wr_table;
    const struct regmap_access_table *rd_table;
    const struct regmap_access_table *volatile_table;
    const struct regmap_access_table *precious_table;
    const struct reg_default *reg_defaults;
    unsigned int num_reg_defaults;
    enum regcache_type cache_type;
    const void *reg_defaults_raw;
    unsigned int num_reg_defaults_raw;

    u8 read_flag_mask;
    u8 write_flag_mask;

    bool use_single_rw;
    bool can_multi_write;

    enum regmap_endian reg_format_endian;
```

```
            enum regmap_endian val_format_endian;
            const struct regmap_range_cfg *ranges;
            unsigned int num_ranges;
}
```

- `reg_bits`: This mandatory field is the number of bits in a register's address.
- `val_bits`: This represents the number of bits used to store a register's value. It is a mandatory field.
- `writeable_reg`: This is an optional callback function. If provided, it is used by the regmap subsystem when a register needs to be written. Before writing into a register, this function is automatically called to check whether the register can be written to or not:

```
static bool foo_writeable_register(struct device *dev,
                                   unsigned int reg)
{
    switch (reg) {
    case 0x30 ... 0x38:
    case 0x40 ... 0x45:
    case 0x50 ... 0x57:
    case 0x60 ... 0x6e:
    case 0x70 ... 0x75:
    case 0x80 ... 0x85:
    case 0x90 ... 0x95:
    case 0xa0 ... 0xa5:
    case 0xb0 ... 0xb2:
        return true;
    default:
        return false;
    }
}
```

- `readable_reg`: This is the same as `writeable_reg` but for every register read operation.
- `volatile_reg`: This is a callback function called every time a register needs to be read or written through the regmap cache. If the register is volatile, the function should return true. A direct read/write is then performed to the register. If false is returned, it means the register is cacheable. In this case, the cache will be used for a read operation, and the cache will be written in the case of a write operation:

```
static bool foo_volatile_register(struct device *dev,
                                  unsigned int reg)
{
    switch (reg) {
    case 0x24 ... 0x29:
```

```
    case 0xb6 ... 0xb8:
        return true;
    default:
        return false;
    }
}
```

- `wr_table`: Instead of providing a `writeable_reg` callback, one could provide a `regmap_access_table`, which is a structure holding a `yes_range` and a `no_range` field, both pointers to `struct regmap_range`. Any register that belongs to a `yes_range` entry is considered as writeable, and is considered as not writeable if it belongs to a `no_range`.
- `rd_table`: This is same as `wr_table`, but for any read operation.
- `volatile_table`: Instead of `volatile_reg`, one could provide `volatile_table`. The principle is then the same as `wr_table` or `rd_table`, but for the caching mechanism.
- `max_register`: This is optional, it specifies the maximum valid register address, upon which no operation is permitted.
- `reg_read`: Your device may not support simple I2C/SPI read operations. You'll then have no choice but to write your own customized read function. `reg_read` should then point to that function. That said most devices do not need that.
- `reg_write`: This is the same as `reg_read` but for write operations.

I highly recommend you look at `include/linux/regmap.h` for more details on each element.

The following is a kind of initialization of `regmap_config`:

```
static const struct regmap_config regmap_config = {
    .reg_bits     = 8,
    .val_bits     = 8,
    .max_register = LM3533_REG_MAX,
    .readable_reg = lm3533_readable_register,
    .volatile_reg = lm3533_volatile_register,
    .precious_reg = lm3533_precious_register,
};
```

regmap initialization

As we said earlier, the regmap API supports both SPI and I2C protocols. Depending on the protocol you need to support in your driver, you will have to call either `regmap_init_i2c()` or `regmap_init_sp()i` in the `probe` function. To write a generic driver, regmap is the best choice.

The regmap API is generic and homogenous. Only the initialization changes between bus types. Other functions are the same.

 It is a good practice to always initialize the regmap in the `probe` function, and one must always fill the `regmap_config` elements prior to initializing the regmap.

Whether one allocated an I2C or SPI register map, it is freed with `regmap_exit` function:

```
void regmap_exit(struct regmap *map)
```

This function simply release a previously allocated register map.

SPI initialization

Regmap SPI initialization consists of setting the regmap up, so that any device access will internally be translated into SPI commands. The function that does is `regmap_init_spi()`.

```
struct regmap * regmap_init_spi(struct spi_device *spi,
const struct regmap_config);
```

It takes a valid pointer to a `struct spi_device` structure as a parameter, which is the SPI device that will be interacted with, and a `struct regmap_config` that represents the configuration for the regmap. This function returns either a pointer to the allocated struct regmap on success, or a value that will be an `ERR_PTR()` on error.

A full example is as follows:

```
static int foo_spi_probe(struct spi_device *client)
{
    int err;
    struct regmap *my_regmap;
    struct regmap_config bmp085_regmap_config;

        /* fill bmp085_regmap_config somewhere */
        [...]
    client->bits_per_word = 8;
```

```
    my_regmap =
            regmap_init_spi(client,&bmp085_regmap_config);

    if (IS_ERR(my_regmap)) {
        err = PTR_ERR(my_regmap);
        dev_err(&client->dev, "Failed to init regmap: %d\n", err);
        return err;
    }
    [...]
}
```

I2C initialization

On the other hand, I2C regmap initialization consists of calling `regmap_init_i2c()` on the regmap config, which will configure the regmap so that any device access will internally be translated into I2C commands:

```
struct regmap * regmap_init_i2c(struct i2c_client *i2c,
const struct regmap_config);
```

The function takes a `struct i2c_client` structure as parameter, which is the I2C device that will used for interaction, along with a pointer to `struct regmap_config` which represents the configuration for the regmap. This function returns either a pointer to the allocated `struct regmap` on success, or a value that will be an `ERR_PTR()` on error.

A full example is:

```
static int bar_i2c_probe(struct i2c_client *i2c,
const struct i2c_device_id *id)
{
    struct my_struct * bar_struct;
    struct regmap_config regmap_cfg;

        /* fill regmap_cfgsome   where */
        [...]
    bar_struct = kzalloc(&i2c->dev,
sizeof(*my_struct), GFP_KERNEL);
    if (!bar_struct)
        return -ENOMEM;

    i2c_set_clientdata(i2c, bar_struct);

    bar_struct->regmap = regmap_init_i2c(i2c,
&regmap_config);
    if (IS_ERR(bar_struct->regmap))
        return PTR_ERR(bar_struct->regmap);
```

```
        bar_struct->dev = &i2c->dev;
        bar_struct->irq = i2c->irq;
        [...]
}
```

Device access functions

The API handles data parsing, formatting, and transmission. In most cases, device accesses are performed with `regmap_read`, `regmap_write` and `regmap_update_bits`. These are the three most important functions you should always remember when it comes to storing/fetching data into/from the device. Their respective prototypes are:

```
int regmap_read(struct regmap *map, unsigned int reg,
                unsigned int *val);
int regmap_write(struct regmap *map, unsigned int reg,
                unsigned int val);
int regmap_update_bits(struct regmap *map, unsigned int reg,
                unsigned int mask, unsigned int val);
```

- `regmap_write`: This writes data to the device. If set in `regmap_config`, `max_register`, it will be used to check if the register address you need to read from is greater or lower. If the register address passed is lower than or equal to, `max_register`, then the write operation will be performed; otherwise, the regmap core will return invalid I/O error (`-EIO`). Immediately after, the `writeable_reg` callback is called. The callback must return `true` before going on to the next step. If it returns `false`, then `-EIO` is returned and the write operation stopped. If `wr_table` is set instead of `writeable_reg`, then:
  - If the register address lies in `no_range`, `-EIO` is returned.
  - If the register address lies in `yes_range`, the next step is performed.
  - If the register address is present neither in `yes_range` nor `no_range`, then `-EIO` is returned and the operation is terminated.
  - If `cache_type != REGCACHE_NONE`, then cache is enabled. In this case, the cache entry is first updated, and then a write to the hardware is performed; otherwise, a no cache action is performed.
  - If `reg_write` callback is provided, it is used to perform the write operation; otherwise, the generic regmap write function will be executed.

- regmap_read: This reads data from the device. It works exactly like regmap_write with appropriate data structures (readable_reg, and rd_table). Therefore, if provided, reg_read is used to perform the read operation; otherise, the generic remap read function will be performed.

regmap_update_bits function

regmap_update_bits is a three-in-one function. Its prototype is as follows:

```
int regmap_update_bits(struct regmap *map, unsigned int reg,
        unsigned int mask, unsigned int val)
```

It performs a read/modify/write cycle on the register map. It is a wrapper on _regmap_update_bits, which looks as follows:

```
static int _regmap_update_bits(struct regmap *map,
                unsigned int reg, unsigned int mask,
                unsigned int val, bool *change)
{
    int ret;
    unsigned int tmp, orig;

    ret = _regmap_read(map, reg, &orig);
    if (ret != 0)
        return ret;

    tmp = orig& ~mask;
    tmp |= val & mask;

    if (tmp != orig) {
        ret = _regmap_write(map, reg, tmp);
        *change = true;
    } else {
        *change = false;
    }

    return ret;
}
```

This way, bits you need to update must be set to 1 in mask, and the corresponding bits should be set to the value you need to give to them in val.

As an example, to set the first and third bits to 1, the mask should be 0b00000101, and the value should be 0bxxxxx1x1. To clear the seventh bit, mask must be 0b01000000 and the value should be 0bx0xxxxxx, and so on.

Special regmap_multi_reg_write function

The purpose of `remap_multi_reg_write()` function is writing multiple registers to the device. Its prototype looks like as follows:

```
int regmap_multi_reg_write(struct regmap *map,
                const struct reg_sequence *regs, int num_regs)
```

To see how to use that function, you need to know what `struct reg_sequence` is:

```
/**
 * Register/value pairs for sequences of writes with an optional delay in
 * microseconds to be applied after each write.
 *
 * @reg: Register address.
 * @def: Register value.
 * @delay_us: Delay to be applied after the register write in microseconds
 */
struct reg_sequence {
    unsigned int reg;
    unsigned int def;
    unsigned int delay_us;
};
```

And this is how it is used:

```
static const struct reg_sequence foo_default_regs[] = {
    { FOO_REG1,      0xB8 },
    { BAR_REG1,      0x00 },
    { FOO_BAR_REG1,  0x10 },
    { REG_INIT,      0x00 },
    { REG_POWER,     0x00 },
    { REG_BLABLA,    0x00 },
};

staticint probe ( ...)
{
    [...]
    ret = regmap_multi_reg_write(my_regmap, foo_default_regs,
                        ARRAY_SIZE(foo_default_regs));
    [...]
}
```

Other device access functions

`regmap_bulk_read()` and `regmap_bulk_write()` are used to read/write multiple registers from/to the device. Use them with large blocks of data.

```
int regmap_bulk_read(struct regmap *map, unsigned int reg, void
                     *val, size_tval_count);
int regmap_bulk_write(struct regmap *map, unsigned int reg,
                      const void *val, size_t val_count);
```

Feel free to look into the regmap header file in the kernel source to see what choices you have.

regmap and cache

Obviously, regmap supports caching. Whether the cache system is used or not depends on the value of the `cache_type` field in `regmap_config`. Looking at `include/linux/regmap.h`, accepted values are:

```
/* Anenum of all the supported cache types */
enum regcache_type {
    REGCACHE_NONE,
    REGCACHE_RBTREE,
    REGCACHE_COMPRESSED,
    REGCACHE_FLAT,
};
```

It is set to `REGCACHE_NONE` by default, meaning that the cache is disabled. Other values simply define how the cache should be stored.

Your device may have a predefined power-on-reset value in certain registers. Those values can be stored in an array, so that any read operation returns the value contained in the array. However, any write operation affects the real register in the device, and updates the content in the array. It is a kind of cache that we can use to speed up access to the device. That array is `reg_defaults`. Its structure looks like this in the source:

```
/**
 * Default value for a register.  We use an array of structs rather
 * than a simple array as many modern devices have very sparse
 * register maps.
 *
 * @reg: Register address.
 * @def: Register default value.
 */
struct reg_default {
```

```
        unsigned int reg;
        unsigned int def;
};
```

 `reg_defaults` is ignored if `cache_type` is set to none. If no `default_reg` is set but you still enable the cache, the corresponding cache structure will be created for you.

It is quite simple to use. Just declare it and pass it as a parameter to the `regmap_config` structure. Let's have a look at the `LTC3589` regulator driver in `drivers/regulator/ltc3589.c`:

```
static const struct reg_default ltc3589_reg_defaults[] = {
    { LTC3589_SCR1,   0x00 },
    { LTC3589_OVEN,   0x00 },
    { LTC3589_SCR2,   0x00 },
    { LTC3589_VCCR,   0x00 },
    { LTC3589_B1DTV1, 0x19 },
    { LTC3589_B1DTV2, 0x19 },
    { LTC3589_VRRCR,  0xff },
    { LTC3589_B2DTV1, 0x19 },
    { LTC3589_B2DTV2, 0x19 },
    { LTC3589_B3DTV1, 0x19 },
    { LTC3589_B3DTV2, 0x19 },
    { LTC3589_L2DTV1, 0x19 },
    { LTC3589_L2DTV2, 0x19 },
};
static const struct regmap_config ltc3589_regmap_config = {
        .reg_bits = 8,
        .val_bits = 8,
        .writeable_reg = ltc3589_writeable_reg,
        .readable_reg = ltc3589_readable_reg,
        .volatile_reg = ltc3589_volatile_reg,
        .max_register = LTC3589_L2DTV2,
        .reg_defaults = ltc3589_reg_defaults,
        .num_reg_defaults = ARRAY_SIZE(ltc3589_reg_defaults),
        .use_single_rw = true,
        .cache_type = REGCACHE_RBTREE,
};
```

Any read operation on any one of the registers present in the array will immediately return the value in the array. However, a write operation will be performed on the device itself, and updates the affected register in the array. This way, reading the `LTC3589_VRRCR` register will return `0xff`; write any value in that register and it will update its entry in the array so that any new read operation will return the last written value, directly from the cache.

Putting it all together

Perform the following steps to set up regmap subsystem:

1. Set up a struct `regmap_config`, according to your device's characteristic. Set a register range if needed, default values if any, the `cache_type` if needed, and so on. If custom read/write functions are needed, pass them to `reg_read`/`reg_write` fields.
2. In the `probe` function, allocate a regmap using `regmap_init_i2c` or `regmap_init_spi` depending on the bus: I2C or SPI.
3. Whenever you need to read/write from/into registers, call `remap_[read|write]` functions.
4. When you are done with the regmap, call `regmap_exit` to free the register map allocated in `probe`.

A regmap example

To achieve our goal, let's first describe a fake SPI device for which we can write a driver:

- 8-bit register address
- 8-bit register values
- Max register: 0x80
- The write mask is 0x80
- Valid address range:
    - 0x20 to 0x4F
    - 0x60 to 0x7F
- No custom read/write function needed.

Regmap API – A Register Map Abstraction

The following is a fake skeleton:

```
/* mandatory for regmap */
#include <linux/regmap.h>
/* Depending on your need you should include other files */

static struct private_struct
{
    /* Feel free to add whatever you want here */
    struct regmap *map;
    int foo;
};

static const struct regmap_range wr_rd_range[] =
{
    {
            .range_min = 0x20,
            .range_max = 0x4F,
    },{
            .range_min = 0x60,
            .range_max = 0x7F
    },
};

struct regmap_access_table drv_wr_table =
{
        .yes_ranges =   wr_rd_range,
        .n_yes_ranges = ARRAY_SIZE(wr_rd_range),
};

struct regmap_access_table drv_rd_table =
{
        .yes_ranges =   wr_rd_range,
        .n_yes_ranges = ARRAY_SIZE(wr_rd_range),
};

static bool writeable_reg(struct device *dev, unsigned int reg)
{
    if (reg>= 0x20 &&reg<= 0x4F)
        return true;
    if (reg>= 0x60 &&reg<= 0x7F)
        return true;
    return false;
}

static bool readable_reg(struct device *dev, unsigned int reg)
{
    if (reg>= 0x20 &&reg<= 0x4F)
```

```c
        return true;
    if (reg>= 0x60 &&reg<= 0x7F)
        return true;
    return false;
}

static int my_spi_drv_probe(struct spi_device *dev)
{
    struct regmap_config config;
    struct custom_drv_private_struct *priv;
    unsigned char data;

    /* setup the regmap configuration */
    memset(&config, 0, sizeof(config));
    config.reg_bits = 8;
    config.val_bits = 8;
    config.write_flag_mask = 0x80;
    config.max_register = 0x80;
    config.fast_io = true;
    config.writeable_reg = drv_writeable_reg;
    config.readable_reg = drv_readable_reg;

    /*
     * If writeable_reg and readable_reg are set,
     * there is no need to provide wr_table nor rd_table.
     * Uncomment below code only if you do not want to use
     * writeable_reg nor readable_reg.
     */
    //config.wr_table = drv_wr_table;
    //config.rd_table = drv_rd_table;

    /* allocate the private data structures */
    /* priv = kzalloc */

    /* Init the regmap spi configuration */
    priv->map = regmap_init_spi(dev, &config);
    /* Use regmap_init_i2c in case of i2c bus */

    /*
     * Let us write into some register
     * Keep in mind that, below operation will remain same
     * whether you use SPI or I2C. It is and advantage when
     * you use regmap.
     */
    regmap_read(priv->map, 0x30, &data);
    [...] /* Process data */

    data = 0x24;
```

```
        regmap_write(priv->map, 0x23, data); /* write new value */

        /* set bit 2 (starting from 0) and 6 of register 0x44 */
        regmap_update_bits(priv->map, 0x44, 0b00100010, 0xFF);
        [...] /* Lot of stuff */
        return 0;
}
```

Summary

This chapter is all about the regmap API. How easy it is, gives you an idea about how useful and widely used it is. This chapter has told you everything you need to know about the regmap API. Now you should be able to convert any standard SPI/I2C driver into a regmap. The next chapter will cover IIO devices, a framework for an analog to digital converter. Those kinds of device always sit on top of the SPI/I2C buses. It will be a challenge for us, at the end of the next chapter, to write an IIO driver using the regmap API.

10
IIO Framework

Industrial I/O (**IIO**) is a kernel subsystem dedicated to **analogic to digitals converters** (**ADC**) and **digital to analogic converters** (**DAC**). With the growing number of sensors (measurement devices with analogue to digital, or digital to analogue, capabilities) with different code implementations, scattered over the kernel sources, gathering them became necessary. This is what IIO framework does, in a generic and homogeneous way. Jonathan Cameron and the Linux-IIO community have been developing it since 2009.

Accelerometer, Gyroscope, current/voltage measurement chips, light sensors, pressure sensors, and so on all fall into the IIO family of devices.

The IIO model is based on devices and channels architecture:

- Device represents the chip itself. It is the top level of the hierarchy.
- Channel represents a single acquisition line of the device. A device may have one or more channels. For example, an accelerometer is a device with three channels, one for each axis (X, Y, and Z).

The IIO chip is the physical and hardware sensor/converter. It is exposed to the user space as a character device (when triggered buffering is supported), and a **sysfs** directory entry which will contain a set of files, some of which represent the channels. A single channel is represented with a single **sysfs** file entry.

These are the two ways to interact with an IIO driver from the user space:

- /sys/bus/iio/iio:deviceX/: This represents the sensor along with its channels
- /dev/iio:deviceX: This is a character device which exports the device's events and data buffer

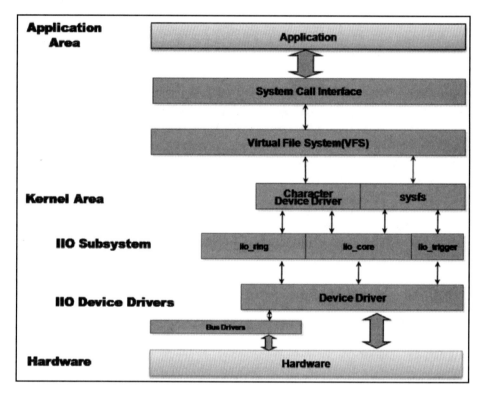

IIO framework architecture and layout

The preceding figure shows how the IIO framework is organized between kernel and user space. The driver manages the hardware and report processing to the IIO core, using a set of facilities and API exposed by the IIO core. The IIO subsystem then abstracts the whole underlying mechanism to the user space by means of the sysfs interface and character device, on top of which users can execute system calls.

IIO APIs are spread over several header files, listed as following:

```
#include <linux/iio/iio.h>    /* mandatory */
#include <linux/iio/sysfs.h>  /* mandatory since sysfs is used */
#include <linux/iio/events.h> /* For advanced users, to manage iio events */
#include <linux/iio/buffer.h> /* mandatory to use triggered buffers */
#include <linux/iio/trigger.h>/* Only if you implement trigger in your driver (rarely used)*/
```

In this chapter, we will describe and handle every concepts of IIO framework, such as

- A walk through its data structure (device, channel, and so on)
- Triggered buffer support and continuous capture, along with its sysfs interface
- Exploring existing IIO triggers
- Capturing data in either one-shot mode or continuous mode
- Listing available tools that can help developers in testing their devices

IIO data structures

An IIO device is represented in the kernel as an instance of the `struct iio_dev`, and described by a `struct iio_info` structure. All of the important IIO structures are defined in `include/linux/iio/iio.h`.

iio_dev structure

This structure represents the IIO device, describing the device, and the driver. It tells us about:

- How many channels are available on the device?
- What modes can the device operate in: one-shot, triggered buffer?

- What hooks are available for this driver?

```
struct iio_dev {
   [...]
   int modes;
   int currentmode;
   struct device dev;

   struct iio_buffer *buffer;
   int scan_bytes;

   const unsigned long *available_scan_masks;
   const unsigned long *active_scan_mask;
   bool scan_timestamp;
   struct iio_trigger *trig;
   struct iio_poll_func *pollfunc;

   struct iio_chan_spec const *channels;
   int num_channels;
   const char *name;
   const struct iio_info *info;
   const struct iio_buffer_setup_ops *setup_ops;
   struct cdev chrdev;
};
```

The complete structure is defined in the IIO header file. Fields that we are not interested in are removed here.

- modes: This represents the different modes supported by the device. Supported modes are:
    - INDIO_DIRECT_MODE which says device provides sysfs type interfaces.
    - INDIO_BUFFER_TRIGGERED says that the device supports hardware triggers. This mode is automatically added to your device when you set up a trigger buffer using the iio_triggered_buffer_setup() function.
    - INDIO_BUFFER_HARDWARE shows the device has a hardware buffer.
    - INDIO_ALL_BUFFER_MODES is the union of the above two.
- currentmode: This represents the mode actually used by the device.
- dev: This represents the struct device (according to Linux device model) the IIO device is tied to.

- `buffer`: This is your data buffer, pushed to the user space when using triggered buffer mode. It is automatically allocated and associated to your device when you enable trigger buffer support using the `iio_triggered_buffer_setup` function.
- `scan_bytes`: This is the number of bytes captured and to be fed to the `buffer`. When using trigger buffer from the user space, the buffer should be at least `indio->scan_bytes` bytes large.
- `available_scan_masks`: This is an optional array of allowed bit masks. When using trigger buffer, one can enable channels to be captured and fed into the IIO buffer. If you do not want to allow some channels to be enabled, you should fill this array with only allowed ones. The following is an example of providing a scan mask for an accelerometer (with X, Y, and Z channels):

    ```
    /*
     * Bitmasks 0x7 (0b111) and 0 (0b000) are allowed.
     * It means one can enable none or all of them.
     * one can't for example enable only channel X and Y
     */
    static const unsigned long my_scan_masks[] = {0x7, 0};
    indio_dev->available_scan_masks = my_scan_masks;
    ```

- `active_scan_mask`: This is a bitmask of enabled channels. Only the data from those channels should be pushed into the `buffer`. For example, for an 8 channels ADC converter, if one only enables the first (0), third (2), and last (7) channels, the bitmask would be 0b10000101 (0x85). `active_scan_mask` will be set to 0x85. The driver can then use the `for_each_set_bit` macro to walk through each set bit, fetch the data according to the channel, and fill the buffer.
- `scan_timestamp`: This tells us whether to push the capture timestamp into the buffer or not. If true, the timestamp will be pushed as the last element of the buffer. The timestamp is 8 bytes (64bits) large.
- `trig`: This is the current device trigger (when buffer mode is supported).
- `pollfunc`: This is the function run on the trigger being received.
- `channels`: This represents the table channel specification structure, to describe every channel the device has.
- `num_channels`: This represents the number of channels specified in `channels`.
- `name`: This represents the device name.
- `info`: Callbacks and constant information from the driver.

- `setup_ops`: Set of callback functions to call before and after the buffer is enabled/disabled. This structure is defined in `include/linux/iio/iio.h` shown as follows:

  ```
  struct iio_buffer_setup_ops {
      int (* preenable) (struct iio_dev *);
      int (* postenable) (struct iio_dev *);
      int (* predisable) (struct iio_dev *);
      int (* postdisable) (struct iio_dev *);
      bool (* validate_scan_mask) (struct iio_dev *indio_dev,
                                   const unsigned long *scan_mask);
  };
  ```

- `setup_ops`: If this is not specified, the IIO core uses the default `iio_triggered_buffer_setup_ops` defined in `drivers/iio/buffer/industrialio-triggered-buffer.c`.
- `chrdev`: This is the associated character device created by the IIO core.

The function used to allocate memory for an IIO device is `iio_device_alloc()`:

```
struct iio_dev *devm_iio_device_alloc(struct device *dev,
                                      int sizeof_priv)
```

`dev` is the device for which `iio_dev` is allocated, and `sizeof_priv` is the memory space used to allocate for any private structure. This way, passing per-device (private) data structure is quite straightforward. The function returns `NULL` if the allocation fails:

```
struct iio_dev *indio_dev;
struct my_private_data *data;
indio_dev = iio_device_alloc(sizeof(*data));
if (!indio_dev)
    return -ENOMEM;
/*data is given the address of reserved momory for private data */
data = iio_priv(indio_dev);
```

After the IIO device memory has been allocated, the next step is to fill different fields. Once done, one has to register the device with the IIO subsystem using `iio_device_register` function:

```
int iio_device_register(struct iio_dev *indio_dev)
```

The device will be ready to accept requests from the user space after this function executes. The reverse operation (usually done in the release function) is `iio_device_unregister()`:

```
void iio_device_unregister(struct iio_dev *indio_dev)
```

Once unregistered, the memory allocated by `iio_device_alloc` can be freed with `iio_device_free`:

```
void iio_device_free(struct iio_dev *iio_dev)
```

Given an IIO device as parameter, one can retrieve the private data in the following manner:

```
struct my_private_data *the_data = iio_priv(indio_dev);
```

iio_info structure

The `struct iio_info` structure is used to declare the hooks used by the IIO core in order to read/write channels/attributes values:

```
struct iio_info {
    struct module *driver_module;
    const struct attribute_group *attrs;

    int (*read_raw)(struct iio_dev *indio_dev,
            struct iio_chan_spec const *chan,
            int *val, int *val2, long mask);

    int (*write_raw)(struct iio_dev *indio_dev,
            struct iio_chan_spec const *chan,
            int val, int val2, long mask);
    [...]
};
```

Fields that we are not interested in have been removed.

- `driver_module`: This is the module structure used to ensure correct ownership of `chrdevs`, usually set to `THIS_MODULE`.
- `attrs`: This represents the devices attributes.
- `read_raw`: This is the callback run when the user reads a device `sysfs` file attribute. The `mask` parameter is a bitmask that allows us to know which type of value is requested. The `channel` parameter lets us know the channel concerned. It can be for the sampling frequency, the scale used to convert the raw value into usable value, or the raw value itself.
- `write_raw`: This is the callback used to write values to the device. One can, for example, use it to set the sampling frequency.

The following code shows how to set up a `struct iio_info` structure:

```
static const struct iio_info iio_dummy_info = {
    .driver_module = THIS_MODULE,
    .read_raw = &iio_dummy_read_raw,
    .write_raw = &iio_dummy_write_raw,
[...]

/*
 * Provide device type specific interface functions and
 * constant data.
 */
indio_dev->info = &iio_dummy_info;
```

IIO channels

A channel represents a single acquisition line. An accelerometer will have, for example, 3 channels (X, Y, Z), since each axis represents a single acquisition line. `struct iio_chan_spec` is the structure that represents and describes a single channel in the kernel:

```
struct iio_chan_spec {
    enum iio_chan_type type;
    int channel;
    int channel2;
    unsigned long address;
    int scan_index;
    struct {
        charsign;
        u8 realbits;
        u8 storagebits;
        u8 shift;
        u8 repeat;
        enum iio_endian endianness;
    } scan_type;
    long info_mask_separate;
    long info_mask_shared_by_type;
    long info_mask_shared_by_dir;
    long info_mask_shared_by_all;
    const struct iio_event_spec *event_spec;
    unsigned int num_event_specs;
    const struct iio_chan_spec_ext_info *ext_info;
    const char *extend_name;
    const char *datasheet_name;
    unsigned modified:1;
    unsigned indexed:1;
    unsigned output:1;
```

```
        unsigned differential:1;
};
```

The following are the meanings of each element in the structure:

- `type`: This specifies which type of measurement the channel makes. In case of voltage measurement, it should be `IIO_VOLTAGE`. For a light sensor, it is `IIO_LIGHT`. For an accelerometer, `IIO_ACCEL` is used. All available types are defined in `include/uapi/linux/iio/types.h`, as `enum iio_chan_type`. To write drivers for a given converter, look into that file to see the type each of your channels falls in.
- `channel`: This specifies the channel index when `.indexed` is set to 1.
- `channel2`: This specifies the channel modifier when `.modified` is set to 1.
- `modified`: This specifies whether a modifier is to be applied to this channel attribute name or not. In that case, the modifier is set in `.channel2`. (For example, `IIO_MOD_X`, `IIO_MOD_Y`, `IIO_MOD_Z` are modifiers for axial-sensors about the xyz-axis). Available modifier list is defined in the kernel IIO header as `enum iio_modifier`. Modifiers only mangle the channel attribute name in `sysfs`, not the value.
- `indexed`: This specifies whether the channel attribute name has an index or not. If yes, the index is specified in the `.channel` field.
- `scan_index` and `scan_type`: These fields are used to identify elements from a buffer, when using buffer triggers. `scan_index` sets the position of the captured channel inside the buffer. Channels with a lower `scan_index` will be placed before channels with a higher index. Setting `.scan_index` to -1 will prevent the channel from buffered capture (no entry in the `scan_elements` directory).

Channel sysfs attributes exposed to user space are specified in the form of bitmasks. Depending on their shared information, attributes can be set into one of the following masks:

- `info_mask_separate` marks the attributes as being specific to this channel.
- `info_mask_shared_by_type` marks the attribute as being shared by all channels of the same type. The information exported is shared by all channels of the same type.
- `info_mask_shared_by_dir` marks the attribute as being shared by all channels of the same direction. The information exported is shared by all channels of the same direction.

- info_mask_shared_by_all marks the attribute as being shared by all channels, whatever their type or direction may be. The information exported is shared by all channels. Bitmasks for enumeration of those attributes are all defined in include/linux/iio/iio.h:

```
enum iio_chan_info_enum {
    IIO_CHAN_INFO_RAW = 0,
    IIO_CHAN_INFO_PROCESSED,
    IIO_CHAN_INFO_SCALE,
    IIO_CHAN_INFO_OFFSET,
    IIO_CHAN_INFO_CALIBSCALE,
    [...]
    IIO_CHAN_INFO_SAMP_FREQ,
    IIO_CHAN_INFO_FREQUENCY,
    IIO_CHAN_INFO_PHASE,
    IIO_CHAN_INFO_HARDWAREGAIN,
    IIO_CHAN_INFO_HYSTERESIS,
    [...]
};
```

The endianness field should be one of:

```
enum iio_endian {
    IIO_CPU,
    IIO_BE,
    IIO_LE,
};
```

Channel attribute naming conventions

The attribute's name is automatically generated by the IIO core with the following pattern: {direction}_{type}_{index}_{modifier}_{info_mask}:

- direction corresponds to the attribute direction, according to the struct iio_direction structure in drivers/iio/industrialio-core.c:

```
static const char * const iio_direction[] = {
    [0] = "in",
    [1] = "out",
};
```

- `type` corresponds to the channel type, according to the char array `const iio_chan_type_name_spec`:

  ```
  static const char * const iio_chan_type_name_spec[] = {
      [IIO_VOLTAGE] = "voltage",
      [IIO_CURRENT] = "current",
      [IIO_POWER] = "power",
      [IIO_ACCEL] = "accel",
      [...]
      [IIO_UVINDEX] = "uvindex",
      [IIO_ELECTRICALCONDUCTIVITY] = "electricalconductivity",
      [IIO_COUNT] = "count",
      [IIO_INDEX] = "index",
      [IIO_GRAVITY] = "gravity",
  };
  ```

- `index` pattern depends on the channel `.indexed` field being set or not. If set, the index will be taken from the `.channel` field in order to replace the `{index}` pattern.
- `modifier` pattern depends on the channel `.modified` field being set or not. If set, the modifier will be taken from the `.channel2` field, and the `{modifier}` pattern will be replaced according to the char array `struct iio_modifier_names` structure:

  ```
  static const char * const iio_modifier_names[] = {
      [IIO_MOD_X] = "x",
      [IIO_MOD_Y] = "y",
      [IIO_MOD_Z] = "z",
      [IIO_MOD_X_AND_Y] = "x&y",
      [IIO_MOD_X_AND_Z] = "x&z",
      [IIO_MOD_Y_AND_Z] = "y&z",
      [...]
      [IIO_MOD_CO2] = "co2",
      [IIO_MOD_VOC] = "voc",
  };
  ```

- `info_mask` depends on the channel info mask, private or shared, indexing value in the char array `iio_chan_info_postfix`:

  ```
  /* relies on pairs of these shared then separate */
  static const char * const iio_chan_info_postfix[] = {
      [IIO_CHAN_INFO_RAW] = "raw",
      [IIO_CHAN_INFO_PROCESSED] = "input",
      [IIO_CHAN_INFO_SCALE] = "scale",
      [IIO_CHAN_INFO_CALIBBIAS] = "calibbias",
      [...]
  ```

```
    [IIO_CHAN_INFO_SAMP_FREQ] = "sampling_frequency",
    [IIO_CHAN_INFO_FREQUENCY] = "frequency",
    [...]
};
```

Distinguishing channels

You may find yourself in trouble when there are multiple data channels per channel type. The dilemma would be: how to identify them. There are two solutions for that: indexes and modifiers.

Using indexes: Given an ADC device with one channel line, indexation is not needed. Its channel definition would be:

```
static const struct iio_chan_spec adc_channels[] = {
    {
        .type = IIO_VOLTAGE,
        .info_mask_separate = BIT(IIO_CHAN_INFO_RAW),
    },
}
```

The attribute name resulting from the preceding channel described will be in_voltage_raw.

/sys/bus/iio/iio:deviceX/in_voltage_raw

Now let us say the converter has 4 or even 8 channels. How do we identify them? The solution is to use indexes. Setting the .indexed field to 1 will mangle the channel attribute name with the .channel value replacing the {index} pattern:

```
static const struct iio_chan_spec adc_channels[] = {
    {
        .type = IIO_VOLTAGE,
        .indexed = 1,
        .channel = 0,
        .info_mask_separate = BIT(IIO_CHAN_INFO_RAW),
    },
    {
        .type = IIO_VOLTAGE,
        .indexed = 1,
        .channel = 1,
        .info_mask_separate = BIT(IIO_CHAN_INFO_RAW),
    },
    {
        .type = IIO_VOLTAGE,
        .indexed = 1,
        .channel = 2,
```

```
                    .info_mask_separate = BIT(IIO_CHAN_INFO_RAW),
        },
        {
                    .type = IIO_VOLTAGE,
                    .indexed = 1,
                    .channel = 3,
                    .info_mask_separate = BIT(IIO_CHAN_INFO_RAW),
        },
}
```

The resulting channel attributes are:

/sys/bus/iio/iio:deviceX/in_voltage0_raw
/sys/bus/iio/iio:deviceX/in_voltage1_raw
/sys/bus/iio/iio:deviceX/in_voltage2_raw
/sys/bus/iio/iio:deviceX/in_voltage3_raw

Using modifiers: Given a light sensor with two channels—one for infrared light and one for both infrared and visible light, without index or modifier, an attribute name would be in_intensity_raw. Using indexes here can be error-prone, because it makes no sense to have in_intensity0_ir_raw, and in_intensity1_ir_raw. Using modifiers will help to provide meaningful attribute names. The channel's definition could look like as follows:

```
static const struct iio_chan_spec mylight_channels[] = {
        {
                    .type = IIO_INTENSITY,
                    .modified = 1,
                    .channel2 = IIO_MOD_LIGHT_IR,
                    .info_mask_separate = BIT(IIO_CHAN_INFO_RAW),
                    .info_mask_shared = BIT(IIO_CHAN_INFO_SAMP_FREQ),
        },
        {
                    .type = IIO_INTENSITY,
                    .modified = 1,
                    .channel2 = IIO_MOD_LIGHT_BOTH,
                    .info_mask_separate = BIT(IIO_CHAN_INFO_RAW),
                    .info_mask_shared = BIT(IIO_CHAN_INFO_SAMP_FREQ),
        },
        {
                    .type = IIO_LIGHT,
                    .info_mask_separate = BIT(IIO_CHAN_INFO_PROCESSED),
                    .info_mask_shared = BIT(IIO_CHAN_INFO_SAMP_FREQ),
        },
}
```

IIO Framework

Resulting attributes will be:

- `/sys/bus/iio/iio:deviceX/in_intensity_ir_raw` for the channel measuring IR intensity
- `/sys/bus/iio/iio:deviceX/in_intensity_both_raw` for the channel measuring both infrared and visible light
- `/sys/bus/iio/iio:deviceX/in_illuminance_input` for the processed data
- `/sys/bus/iio/iio:deviceX/sampling_frequency` for the sampling frequency, shared by all

This is valid with accelerometer too, as we will see further on in the case study. For now, let's summarize what we have discussed so far in a dummy IIO driver.

Putting it all together

Let us summarize what we have seen so far in a simple dummy driver, which will expose four voltage channels. We will ignore `read()` or `write()` functions:

```
#include <linux/init.h>
#include <linux/module.h>
#include <linux/kernel.h>
#include <linux/platform_device.h>
#include <linux/interrupt.h>
#include <linux/of.h>
#include <linux/iio/iio.h>
#include <linux/iio/sysfs.h>
#include <linux/iio/events.h>
#include <linux/iio/buffer.h>

#define FAKE_VOLTAGE_CHANNEL(num)                              \
    {                                                          \
        .type = IIO_VOLTAGE,                                   \
        .indexed = 1,                                          \
        .channel = (num),                                      \
        .address = (num),                                      \
        .info_mask_separate = BIT(IIO_CHAN_INFO_RAW),          \
        .info_mask_shared_by_type = BIT(IIO_CHAN_INFO_SCALE)   \
    }

struct my_private_data {
    int foo;
    int bar;
    struct mutex lock;
};
```

```c
static int fake_read_raw(struct iio_dev *indio_dev,
                struct iio_chan_spec const *channel, int *val,
                int *val2, long mask)
{
    return 0;
}

static int fake_write_raw(struct iio_dev *indio_dev,
                struct iio_chan_spec const *chan,
                int val, int val2, long mask)
{
    return 0;
}

static const struct iio_chan_spec fake_channels[] = {
    FAKE_VOLTAGE_CHANNEL(0),
    FAKE_VOLTAGE_CHANNEL(1),
    FAKE_VOLTAGE_CHANNEL(2),
    FAKE_VOLTAGE_CHANNEL(3),
};

static const struct of_device_id iio_dummy_ids[] = {
    { .compatible = "packt,iio-dummy-random", },
    { /* sentinel */ }
};

static const struct iio_info fake_iio_info = {
    .read_raw      = fake_read_raw,
    .write_raw     = fake_write_raw,
    .driver_module = THIS_MODULE,
};

static int my_pdrv_probe (struct platform_device *pdev)
{
    struct iio_dev *indio_dev;
    struct my_private_data *data;

    indio_dev = devm_iio_device_alloc(&pdev->dev, sizeof(*data));
    if (!indio_dev) {
        dev_err(&pdev->dev, "iio allocation failed!\n");
        return -ENOMEM;
    }

    data = iio_priv(indio_dev);
    mutex_init(&data->lock);
    indio_dev->dev.parent = &pdev->dev;
    indio_dev->info = &fake_iio_info;
    indio_dev->name = KBUILD_MODNAME;
```

```c
        indio_dev->modes = INDIO_DIRECT_MODE;
        indio_dev->channels = fake_channels;
        indio_dev->num_channels = ARRAY_SIZE(fake_channels);
        indio_dev->available_scan_masks = 0xF;

        iio_device_register(indio_dev);
        platform_set_drvdata(pdev, indio_dev);
        return 0;
    }

    static void my_pdrv_remove(struct platform_device *pdev)
    {
        struct iio_dev *indio_dev = platform_get_drvdata(pdev);
        iio_device_unregister(indio_dev);
    }

    static struct platform_driver mypdrv = {
        .probe      = my_pdrv_probe,
        .remove     = my_pdrv_remove,
        .driver     = {
            .name         = "iio-dummy-random",
            .of_match_table = of_match_ptr(iio_dummy_ids),
            .owner        = THIS_MODULE,
        },
    };
    module_platform_driver(mypdrv);
    MODULE_AUTHOR("John Madieu <john.madieu@gmail.com>");
    MODULE_LICENSE("GPL");
```

After loading the module above, we will have the following output, showing that our device really corresponds to the platform device we have registered:

```
~# ls -l /sys/bus/iio/devices/
lrwxrwxrwx 1 root root 0 Jul 31 20:26 iio:device0 ->
../../../devices/platform/iio-dummy-random.0/iio:device0
lrwxrwxrwx 1 root root 0 Jul 31 20:23 iio_sysfs_trigger ->
../../../devices/iio_sysfs_trigger
```

The following listing shows the channels that this device has, along with its name, which correspond exactly to what we have described in the driver:

```
~# ls /sys/bus/iio/devices/iio\:device0/
dev in_voltage2_raw name uevent
in_voltage0_raw in_voltage3_raw power
in_voltage1_raw in_voltage_scale subsystem
~# cat /sys/bus/iio/devices/iio:device0/name
iio_dummy_random
```

Triggered buffer support

In many data analysis applications, it is useful to be able to capture data based on some external signal (trigger). These triggers might be:

- A data ready signal
- An IRQ line connected to some external system (GPIO or something else)
- On-processor periodic interrupt
- User space reading/writing a specific file in sysfs

IIO device drivers are completely unrelated to triggers. A trigger may initialize data capture on one or many devices. These triggers are used to fill buffers, exposed to user space as character devices.

One can develop one's own trigger driver, but that is beyond the scope of this book. We will try to focus on existing ones only. These are:

- `iio-trig-interrupt`: This provides support for using any IRQ as IIO triggers. In old kernel versions, it used to be `iio-trig-gpio`. The kernel option to enable this trigger mode is `CONFIG_IIO_INTERRUPT_TRIGGER`. If built as a module, the module would be called `iio-trig-interrupt`.
- `iio-trig-hrtimer`: This provides a frequency-based IIO trigger using HRT as the interrupt source (since kernel v4.5). In older kernel versions, it used to be `iio-trig-rtc`. The kernel option responsible for this trigger mode is `IIO_HRTIMER_TRIGGER`. If built as a module, the module would be called `iio-trig-hrtimer`.
- `iio-trig-sysfs`: This allow us to use sysfs entry to trigger data capture. `CONFIG_IIO_SYSFS_TRIGGER` is the kernel option to add the support of this trigger mode.
- `iio-trig-bfin-timer`: This allows us to use a blackfin timer as IIO triggers (still in staging).

IIO Framework

IIO exposes API so that we can:

- Declare any given number of triggers
- Choose which channels will have their data pushed into buffer

When your IIO device provides the support of the trigger buffer, you must set `iio_dev.pollfunc`, which is executed when the trigger fires. This handler has the responsibility to find enabled channels through `indio_dev->active_scan_mask`, retrieve their data, and feed them into `indio_dev->buffer` using the `iio_push_to_buffers_with_timestamp` function. As such, buffers and triggers are very connected in the IIO subsystem.

The IIO core provides a set of helper functions to set up triggered buffers that one can find in `drivers/iio/industrialio-triggered-buffer.c`.

The following are the steps to support triggered buffers from within your driver:

1. Fill an `iio_buffer_setup_ops` structure if needed:

   ```
   const struct iio_buffer_setup_ops sensor_buffer_setup_ops = {
       .preenable    = my_sensor_buffer_preenable,
       .postenable   = my_sensor_buffer_postenable,
       .postdisable  = my_sensor_buffer_postdisable,
       .predisable   = my_sensor_buffer_predisable,
   };
   ```

2. Write the top half associated to the trigger. In 99% of cases, one has to just feed the timestamp associated with the capture:

   ```
   irqreturn_t sensor_iio_pollfunc(int irq, void *p)
   {
       pf->timestamp = iio_get_time_ns((struct indio_dev *)p);
       return IRQ_WAKE_THREAD;
   }
   ```

3. Write the trigger bottom half, which will fetch data from each enabled channel, and feed them into the buffer:

   ```
   irqreturn_t sensor_trigger_handler(int irq, void *p)
   {
       u16 buf[8];
       int bit, i = 0;
       struct iio_poll_func *pf = p;
       struct iio_dev *indio_dev = pf->indio_dev;
   ```

```
    /* one can use lock here to protect the buffer */
    /* mutex_lock(&my_mutex); */

    /* read data for each active channel */
    for_each_set_bit(bit, indio_dev->active_scan_mask,
                     indio_dev->masklength)
        buf[i++] = sensor_get_data(bit)

    /*
     * If iio_dev.scan_timestamp = true, the capture timestamp
     * will be pushed and stored too, as the last element in the
     * sample data buffer before pushing it to the device buffers.
     */
    iio_push_to_buffers_with_timestamp(indio_dev, buf, timestamp);

    /* Please unlock any lock */
    /* mutex_unlock(&my_mutex); */

    /* Notify trigger */
    iio_trigger_notify_done(indio_dev->trig);
    return IRQ_HANDLED;
}
```

4. Finally, in the `probe` function, one has to set up the buffer itself, prior to registering the device with `iio_device_register()`:

```
iio_triggered_buffer_setup(indio_dev, sensor_iio_polfunc,
                           sensor_trigger_handler,
                           sensor_buffer_setup_ops);
```

The magic function here is `iio_triggered_buffer_setup`. This will also give the `INDIO_DIRECT_MODE` capability to your device. When a trigger is given (from user space) to your device, you have no way of knowing when capture will be fired.

While continuous buffered capture is active, one should prevent (by returning an error) the driver from performing sysfs per-channel data capture (performed by the `read_raw()` hook) in order to avoid undetermined behavior, since both the trigger handler and `read_raw()` hook will try to access the device at the same time. The function used to check whether buffered mode is actually used is `iio_buffer_enabled()`. The hook will look like this:

```
static int my_read_raw(struct iio_dev *indio_dev,
                       const struct iio_chan_spec *chan,
                       int *val, int *val2, long mask)
{
    [...]
```

```
        switch (mask) {
        case IIO_CHAN_INFO_RAW:
            if (iio_buffer_enabled(indio_dev))
                return -EBUSY;
        [...]
}
```

The `iio_buffer_enabled()` function simply tests if the buffer is enabled for a given IIO device.

Let us describe some important things used in the preceding section:

- `iio_buffer_setup_ops` provides buffer setup functions to be called at fixed step of the buffer configuration sequence (before/after enable/disable). If not specified, the default `iio_triggered_buffer_setup_ops` will be given to your device by the IIO core.
- `sensor_iio_pollfunc` is the trigger's top half. As with every top half, it runs in interrupt context and must do as little processing as possible. In 99% of cases, you just have to feed the timestamp associated with the capture. Once again, one can use the default IIO `iio_pollfunc_store_time` function.
- `sensor_trigger_handler` is the bottom half, which runs in a kernel thread, allowing us to do any processing including even acquiring mutex or sleep. The heavy processing should take place here. It usually reads data from the device and stores it in the internal buffer together with the timestamp recorded in the top half, and pushes it to your IIO device buffer.

A trigger is mandatory for triggered buffering. It tells the driver when to read the sample from the device and put it into the buffer. Triggered buffering is not mandatory to write IIO device drivers. One can use single shot capture through sysfs too, by reading raw attributesof the channel, which will only perform a single conversion (for the channel attribute being read). Buffer mode allows continuous conversions, thus capturing more than one channel in a single shot.

IIO trigger and sysfs (user space)

There are two locations in sysfs related to triggers:

- `/sys/bus/iio/devices/triggerY/` which is created once an IIO trigger is registered with the IIO core and corresponds to triggers with index `Y`. There is at least one attribute in the directory:
  - `name` which is the trigger name that can be later used for association with a device
- `/sys/bus/iio/devices/iio:deviceX/trigger/*` directory will be automatically created if your device supports a triggered buffer. One can associate a trigger with our device by writing the trigger's name in the `current_trigger` file.

Sysfs trigger interface

The sysfs trigger is enabled in the kernel by the `CONFIG_IIO_SYSFS_TRIGGER=y` config option, which will result in the `/sys/bus/iio/devices/iio_sysfs_trigger/` folder being automatically created, and can be used for sysfs trigger management. There will be two files in the directory, `add_trigger` and `remove_trigger`. Its driver is in `drivers/iio/trigger/iio-trig-sysfs.c`.

add_trigger file

This is used to create a new sysfs trigger. You can create a new trigger by writing a positive value (which will be used as a trigger ID) into that file. It will create the new sysfs trigger, accessible at `/sys/bus/iio/devices/triggerX`, where `X` is the trigger number.

For example:

```
# echo 2 > add_trigger
```

This will create a new sysfs trigger, accessible at `/sys/bus/iio/devices/trigger2`. If the trigger with the specified ID is already present in the system, an invalid argument message will be returned. The sysfs trigger name pattern is `sysfstrig{ID}`. The command `echo 2 > add_trigger` will create the trigger `/sys/bus/iio/devices/trigger2` whose name is `sysfstrig2`:

```
$ cat /sys/bus/iio/devices/trigger2/name
sysfstrig2
```

Each sysfs trigger contains at least one file: `trigger_now`. Writing 1 into that file will instruct all devices having the corresponding trigger name in their `current_trigger` to start capture, and push data into their respective buffer. Each device buffer must have its size set, and must be enabled (echo 1 > /sys/bus/iio/devices/iio:deviceX/buffer/enable).

remove_trigger file

To remove a trigger, the following command is used:

```
# echo 2 > remove_trigger
```

Tying a device with a trigger

Associating a device with a given trigger consists of writing the name of the trigger to the `current_trigger` file available under the device's trigger directory. For example, let us say we need to tie a device with the trigger that has index 2:

```
# set trigger2 as current trigger for device0
# echo sysfstrig2 >
/sys/bus/iio/devices/iio:device0/trigger/current_trigger
```

To detach the trigger from the device, one should write an empty string to the `current_trigger` file of the device trigger directory, shown as follows:

```
# echo "" > iio:device0/trigger/current_trigger
```

We will see further on in the chapter a practical example dealing with the sysfs trigger for data capture.

The interrupt trigger interface

Consider the following sample:

```
static struct resource iio_irq_trigger_resources[] = {
    [0] = {
        .start = IRQ_NR_FOR_YOUR_IRQ,
        .flags = IORESOURCE_IRQ | IORESOURCE_IRQ_LOWEDGE,
    },
};

static struct platform_device iio_irq_trigger = {
    .name = "iio_interrupt_trigger",
    .num_resources = ARRAY_SIZE(iio_irq_trigger_resources),
```

```
    .resource = iio_irq_trigger_resources,
};
platform_device_register(&iio_irq_trigger);
```

Declare our IRQ trigger and it will result in the IRQ trigger standalone module being loaded. If its `probe` function succeeds, there will be a directory corresponding to the trigger. IRQ trigger names have the form `irqtrigX`, where X corresponds to the virtual IRQ you just passed, the one you will see in `/proc/interrupt`:

```
$ cd /sys/bus/iio/devices/trigger0/
$ cat name
```

`irqtrig85`: As we have done with other triggers, you just have to assign that trigger to your device, by writing its name into your device `current_trigger` file.

```
# echo "irqtrig85" >
/sys/bus/iio/devices/iio:device0/trigger/current_trigger
```

Now, every time the interrupt will be fired, device data will be captured.

> The IRQ trigger driver does not support DT yet, which is the reason why we used our board `init` file. But it does not matter; since the driver requires a resource, we can use DT without any code change.

The following is an example of device tree node declaring the IRQ trigger interface:

```
mylabel: my_trigger@0{
    compatible = "iio_interrupt_trigger";
    interrupt-parent = <&gpio4>;
    interrupts = <30 0x0>;
};
```

The example supposes the IRQ line is the GPIO#30 that belongs to the GPIO controller node `gpio4`. This consists of using a GPIO as an interrupt source, so that whenever the GPIO changes to a given state, the interrupt is raised, thus triggering the capture.

The hrtimer trigger interface

The `hrtimer` trigger relies on the configfs file system (see *Documentation/iio/iio_configfs.txt* in kernel sources), which can be enabled through the `CONFIG_IIO_CONFIGFS` config option, and mounted on our system (usually under the `/config` directory):

```
# mkdir /config
# mount -t configfs none /config
```

Now, loading the module `iio-trig-hrtimer` will create IIO groups accessible under `/config/iio`, allowing users to create hrtimer triggers under `/config/iio/triggers/hrtimer`.

For example:

```
# create a hrtimer trigger
$ mkdir /config/iio/triggers/hrtimer/my_trigger_name
# remove the trigger
$ rmdir /config/iio/triggers/hrtimer/my_trigger_name
```

Each hrtimer trigger contains a single `sampling_frequency` attribute in the trigger directory. A full and working example is provided further in the chapter in the section *Data capture using hrtimer trigger*.

IIO buffers

The IIO buffer offers continuous data capture, where more than one data channel can be read at once. The buffer is accessible from the user space through the `/dev/iio:device` character device node. From within the trigger handler, the function used to fill the buffer is `iio_push_to_buffers_with_timestamp`. The function responsible to allocate the trigger buffer for your device is `iio_triggered_buffer_setup()`.

IIO buffer sysfs interface

An IIO buffer has an associated attributes directory under `/sys/bus/iio/iio:deviceX/buffer/*`. Here are some of the existing attributes:

- `length`: The total number of data samples (capacity) that can be stored by the buffer. This is the number of scans contained by the buffer.
- `enable`: This activates buffer capture, start the buffer capture.
- `watermark`: This attribute has been available since kernel version v4.2. It is a positive number which specifies how many scan elements a blocking read should wait for. If using `poll` for example, it will block until the watermark is reached. It makes sense only if the watermark is greater than the requested amount of reads. It does not affect non-blocking reads. One can block on poll with a timeout and read the available samples after the timeout expires, and thus have a maximum delay guarantee.

IIO buffer setup

A channel whose data is to be read and pushed into the buffer is called a scan element. Their configurations are accessible from the user space through the /sys/bus/iio/iio:deviceX/scan_elements/* directory, containing the following attributes:

- en (actually a suffix for attribute name), is used to enable the channel. If and only if its attribute is non-zero, then a triggered capture will contain data samples for this channel. For example, in_voltage0_en, in_voltage1_en and so on.
- type describes the scan element data storage within the buffer, and hence the form in which it is read from user space. For example, in_voltage0_type. The format is [be|le]:[s|u]bits/storagebitsXrepeat[>>shift].
    - be or le specifies the endianness (big or little).
    - s or u specifies the sign, either signed (2's complement) or unsigned.
    - bits is the number of valid data bits.
    - storagebits is the number of bits this channel occupies in the buffer. That said, a value may be really coded in 12 bits (**bits**), but occupies 16 bits (**storagebits**) in the buffer. One must therefore shift the data four times to the right to obtain the actual value. This parameter depends on the device, and one should refer to its data sheet.
    - shift represents the number of times one should shift the data value prior to masking out unused bits. This parameter is not always needed. If the number of valid bit (**bits**) is equal to the number of storage bits, the shift will be 0. One can also find this parameter in the device data sheet.
    - repeat specifies the number of bit/storagebit repetitions. When the repeat element is 0 or 1, then the repeat value is omitted.

The best way to explain this section is by an excerpt of kernel doc, which can find here: https://www.kernel.org/doc/html/latest/driver-api/iio/buffers.html. For example, a driver for a 3-axis accelerometer, with 12-bit resolution where data is stored in two 8-bit registers, is as follows:

```
  7   6   5   4   3   2   1   0
+---+---+---+---+---+---+---+---+
|D3 |D2 |D1 |D0 | X | X | X | X |  (LOW byte, address 0x06)
+---+---+---+---+---+---+---+---+

  7   6   5   4   3   2   1   0
+---+---+---+---+---+---+---+---+
|D11|D10|D9 |D8 |D7 |D6 |D5 |D4 |  (HIGH byte, address 0x07)
+---+---+---+---+---+---+---+---+
```

will have the following scan element type for each axis:

```
$ cat /sys/bus/iio/devices/iio:device0/scan_elements/in_accel_y_type
le:s12/16>>4
```

One should interpret this as being little endian-signed data, 16 bits-sized, which needs to be shifted right by 4 bits before masking out the 12 valid bits of data.

The element in `struct iio_chan_spec` that is responsible for determining how a channel's value should be stored into the buffer is `scant_type`.

```
struct iio_chan_spec {
        [...]
        struct {
            char sign; /* Should be 'u' or 's' as explained above */
            u8 realbits;
            u8 storagebits;
            u8 shift;
            u8 repeat;
            enum iio_endian endianness;
        } scan_type;
        [...]
};
```

This structure absolutely matches
`[be|le]:[s|u]bits/storagebitsXrepeat[>>shift]`, which is the pattern described in the previous section. Let us have a look at each member of the structure:

- `sign` represents the sign of the data and matches `[s|u]` in the pattern
- `realbits` corresponds to `bits` in the pattern
- `storagebits` matches the same name in the pattern
- `shift` corresponds to shift in the pattern, same for `repeat`
- `iio_indian` represents the endianness and matches `[be|le]` in the pattern

At this point, one is able to write the IIO channel structure that corresponds to the type previously explained:

```
struct struct iio_chan_spec accel_channels[] = {
        {
                .type = IIO_ACCEL,
                .modified = 1,
                .channel2 = IIO_MOD_X,
                /* other stuff here */
                .scan_index = 0,
                .scan_type = {
                        .sign = 's',
                        .realbits = 12,
                        .storagebits = 16,
                        .shift = 4,
                        .endianness = IIO_LE,
                },
        }
        /* similar for Y (with channel2 = IIO_MOD_Y, scan_index = 1)
         * and Z (with channel2 = IIO_MOD_Z, scan_index = 2) axis
         */
}
```

Putting it all together

Let us have a closer look at the digital triaxial acceleration sensor BMA220 from BOSH. This is an SPI/I2C-compatible device, with 8 bit-sized registers, along with an on-chip motion-triggered interrupt controller, which actually senses tilt, motion, and shock vibration. Its data sheet is available at: `http://www.mouser.fr/pdfdocs/BSTBMA220DS00308.PDF`, and its driver has been introduced since kernel v4.8 (`CONFIG_BMA200`). Let us walk through it:

Firstly, we declare our IIO channels using `struct iio_chan_spec`. Once the triggered buffer is used, then we need to fill the `.scan_index` and `.scan_type` fields:

```c
#define BMA220_DATA_SHIFT 2
#define BMA220_DEVICE_NAME "bma220"
#define BMA220_SCALE_AVAILABLE "0.623 1.248 2.491 4.983"

#define BMA220_ACCEL_CHANNEL(index, reg, axis) {                \
    .type = IIO_ACCEL,                                          \
    .address = reg,                                             \
    .modified = 1,                                              \
    .channel2 = IIO_MOD_##axis,                                 \
    .info_mask_separate = BIT(IIO_CHAN_INFO_RAW),               \
    .info_mask_shared_by_type = BIT(IIO_CHAN_INFO_SCALE),       \
    .scan_index = index,                                        \
    .scan_type = {                                              \
        .sign = 's',                                            \
        .realbits = 6,                                          \
        .storagebits = 8,                                       \
        .shift = BMA220_DATA_SHIFT,                             \
        .endianness = IIO_CPU,                                  \
    },                                                          \
}

static const struct iio_chan_spec bma220_channels[] = {
    BMA220_ACCEL_CHANNEL(0, BMA220_REG_ACCEL_X, X),
    BMA220_ACCEL_CHANNEL(1, BMA220_REG_ACCEL_Y, Y),
    BMA220_ACCEL_CHANNEL(2, BMA220_REG_ACCEL_Z, Z),
};
```

`.info_mask_separate = BIT(IIO_CHAN_INFO_RAW)` says there will be a `*_raw` sysfs entry (attribute) for each channel, and `.info_mask_shared_by_type = BIT(IIO_CHAN_INFO_SCALE)` says that there is only a `*_scale` sysfs entry for all channels of the same type:

```
jma@jma:~$ ls -l /sys/bus/iio/devices/iio:device0/
(...)
# without modifier, a channel name would have in_accel_raw (bad)
-rw-r--r-- 1 root root 4096 jul 20 14:13 in_accel_scale
-rw-r--r-- 1 root root 4096 jul 20 14:13 in_accel_x_raw
-rw-r--r-- 1 root root 4096 jul 20 14:13 in_accel_y_raw
-rw-r--r-- 1 root root 4096 jul 20 14:13 in_accel_z_raw
(...)
```

Reading `in_accel_scale` calls the `read_raw()` hook with the mask set to `IIO_CHAN_INFO_SCALE`. Reading `in_accel_x_raw` calls the `read_raw()` hook with the mask set to `IIO_CHAN_INFO_RAW`. The real value is therefore `raw_value * scale`.

What `.scan_type` says is that the value returned by each channel is, 8 bit-sized (will occupy 8 bits in the buffer), but the useful payload only occupies 6 bits, and the data must be right-shifted 2 times prior to masking out unused bits. Any scan element type will look like this:

```
$ cat /sys/bus/iio/devices/iio:device0/scan_elements/in_accel_x_type
le:s6/8>>2
```

The following is our `pollfunc` (actually it is the bottom half), which reads samples from the device and pushes read values into the buffer (`iio_push_to_buffers_with_timestamp()`). Once done, we inform the core (`iio_trigger_notify_done()`):

```
static irqreturn_t bma220_trigger_handler(int irq, void *p)
{
    int ret;
    struct iio_poll_func *pf = p;
    struct iio_dev *indio_dev = pf->indio_dev;
    struct bma220_data *data = iio_priv(indio_dev);
    struct spi_device *spi = data->spi_device;

    mutex_lock(&data->lock);
    data->tx_buf[0] = BMA220_REG_ACCEL_X | BMA220_READ_MASK;
    ret = spi_write_then_read(spi, data->tx_buf, 1, data->buffer,
                    ARRAY_SIZE(bma220_channels) - 1);
    if (ret < 0)
        goto err;

    iio_push_to_buffers_with_timestamp(indio_dev, data->buffer,
                    pf->timestamp);
err:
    mutex_unlock(&data->lock);
    iio_trigger_notify_done(indio_dev->trig);

    return IRQ_HANDLED;
}
```

The following is the `read` function. It is a hook, called every time one reads a sysfs entry of the device:

```
static int bma220_read_raw(struct iio_dev *indio_dev,
                struct iio_chan_spec const *chan,
                int *val, int *val2, long mask)
{
   int ret;
   u8 range_idx;
   struct bma220_data *data = iio_priv(indio_dev);

   switch (mask) {
   case IIO_CHAN_INFO_RAW:
           /* If buffer mode enabled, do not process single-channel read */
           if (iio_buffer_enabled(indio_dev))
                   return -EBUSY;
           /* Else we read the channel */
           ret = bma220_read_reg(data->spi_device, chan->address);
           if (ret < 0)
                   return -EINVAL;
           *val = sign_extend32(ret >> BMA220_DATA_SHIFT, 5);
           return IIO_VAL_INT;
   case IIO_CHAN_INFO_SCALE:
           ret = bma220_read_reg(data->spi_device, BMA220_REG_RANGE);
           if (ret < 0)
                   return ret;
           range_idx = ret & BMA220_RANGE_MASK;
           *val = bma220_scale_table[range_idx][0];
           *val2 = bma220_scale_table[range_idx][1];
           return IIO_VAL_INT_PLUS_MICRO;
   }

   return -EINVAL;
}
```

When one reads a *raw sysfs file, the hook is called, given IIO_CHAN_INFO_RAW in the mask parameter, and the corresponding channel in the *chan parameter. *val and val2 are actually output parameters. They must be set with the raw value (read from the device). Any read performed on the *scale sysfs file will call the hook with IIO_CHAN_INFO_SCALE in mask parameter, and so on for each attribute mask.

This is also the case with the `write` function, used to write value into the device. There is an 80% chance your driver does not require a `write` function. This `write` hook lets the user change the device's scale:

```
static int bma220_write_raw(struct iio_dev *indio_dev,
                struct iio_chan_spec const *chan,
                int val, int val2, long mask)
{
    int i;
    int ret;
    int index = -1;
    struct bma220_data *data = iio_priv(indio_dev);

    switch (mask) {
    case IIO_CHAN_INFO_SCALE:
        for (i = 0; i < ARRAY_SIZE(bma220_scale_table); i++)
            if (val == bma220_scale_table[i][0] &&
                val2 == bma220_scale_table[i][1]) {
                index = i;
                break;
            }
        if (index < 0)
            return -EINVAL;

        mutex_lock(&data->lock);
        data->tx_buf[0] = BMA220_REG_RANGE;
        data->tx_buf[1] = index;
        ret = spi_write(data->spi_device, data->tx_buf,
                sizeof(data->tx_buf));
        if (ret < 0)
            dev_err(&data->spi_device->dev,
                "failed to set measurement range\n");
        mutex_unlock(&data->lock);

        return 0;
    }

    return -EINVAL;
}
```

IIO Framework

This function is called whenever one writes a value to the device. Frequently changed parameters are the scale. An example could be: echo <desired-scale> > /sys/bus/iio/devices/iio;devices0/in_accel_scale.

Now, it comes to fill a `struct iio_info` structure, to be given to our `iio_device`:

```
static const struct iio_info bma220_info = {
    .driver_module   = THIS_MODULE,
    .read_raw        = bma220_read_raw,
    .write_raw       = bma220_write_raw, /* Only if your driver need it */
};
```

In the `probe` function, we allocate and set up a `struct iio_dev` IIO device. Memory for private data is reserved too:

```
/*
 * We provide only two mask possibility, allowing to select none or every
 * channels.
 */
static const unsigned long bma220_accel_scan_masks[] = {
   BIT(AXIS_X) | BIT(AXIS_Y) | BIT(AXIS_Z),
   0
};

static int bma220_probe(struct spi_device *spi)
{
   int ret;
   struct iio_dev *indio_dev;
   struct bma220_data *data;

   indio_dev = devm_iio_device_alloc(&spi->dev, sizeof(*data));
   if (!indio_dev) {
         dev_err(&spi->dev, "iio allocation failed!\n");
         return -ENOMEM;
   }

   data = iio_priv(indio_dev);
   data->spi_device = spi;
   spi_set_drvdata(spi, indio_dev);
   mutex_init(&data->lock);

   indio_dev->dev.parent = &spi->dev;
   indio_dev->info = &bma220_info;
   indio_dev->name = BMA220_DEVICE_NAME;
   indio_dev->modes = INDIO_DIRECT_MODE;
   indio_dev->channels = bma220_channels;
   indio_dev->num_channels = ARRAY_SIZE(bma220_channels);
```

```
        indio_dev->available_scan_masks = bma220_accel_scan_masks;

        ret = bma220_init(data->spi_device);
        if (ret < 0)
               return ret;

        /* this call will enable trigger buffer support for the device */
        ret = iio_triggered_buffer_setup(indio_dev, iio_pollfunc_store_time,
                              bma220_trigger_handler, NULL);
        if (ret < 0) {
               dev_err(&spi->dev, "iio triggered buffer setup failed\n");
               goto err_suspend;
        }

        ret = iio_device_register(indio_dev);
        if (ret < 0) {
               dev_err(&spi->dev, "iio_device_register failed\n");
               iio_triggered_buffer_cleanup(indio_dev);
               goto err_suspend;
        }

        return 0;

err_suspend:
     return bma220_deinit(spi);
}
```

One can enable this driver by means of the CONFIG_BMA220 kernel option. That said, this is available only from v4.8 onwards in kernel. The closest device one can use for this on older kernel versions is BMA180, which one can enable using the CONFIG_BMA180 option.

IIO data access

You may have guessed that there are only two ways to access data with the IIO framework; one-shot capture through sysfs channels, or continuous mode (triggered buffer) through an IIO character device.

One-shot capture

One-shot data capture is done through sysfs interface. By reading the sysfs entry that corresponds to a channel, you'll capture only the data specific to that channel. Given a temp sensor with two channels: one for the ambient temp, and the other for the thermocouple temp:

```
# cd /sys/bus/iio/devices/iio:device0
# cat in_voltage3_raw
6646

# cat in_voltage_scale
0.305175781
```

Processed value is obtained by multiplying the scale by the raw value.

`Voltage value:` 6646 * 0.305175781 = 2028.19824053

The device datasheet says the process value is given in MV. In our case, it corresponds to 2.02819V.

Buffer data access

To get a triggered acquisition working, the trigger support must have been implemented in your driver. Then, to acquire data from within user space, one must: create a trigger, assign it, enable the ADC channels, set the dimension of the buffer, and enable it). Here is the code for this:

Capturing using the sysfs trigger

Capturing data using the sysfs trigger consists of sending a set of command few sysfs files. Let us enumerate what we should do to achieve that:

1. **Create the trigger**: Before the trigger can be assigned to any device, it should be created:

    ```
    # echo 0 > /sys/devices/iio_sysfs_trigger/add_trigger
    ```

Here, 0 corresponds to the index we need to assign to the trigger. After this command, the trigger directory will be available under /sys/bus/iio/devices/, as trigger0.

2. **Assign the trigger to the device**: A trigger is uniquely identified by its name, which we can use in order to tie device to the trigger. Since we used 0 as index, the trigger will be named sysfstrig0:

   ```
   # echo sysfstrig0 >
   /sys/bus/iio/devices/iio:device0/trigger/current_trigger
   ```

We could have used this command too: cat /sys/bus/iio/devices/trigger0/name > /sys/bus/iio/devices/iio:device0/trigger/current_trigger. That said, if the value we wrote does not correspond to an existing trigger name, nothing will happen. To make sure we really defined a trigger, we can use cat /sys/bus/iio/devices/iio:device0/trigger/current_trigger.

3. **Enable some scan elements**: This step consists of choosing which channels should have their data value pushed into the buffer. One should pay attention to available_scan_masks in the driver:

   ```
   # echo 1 >
   /sys/bus/iio/devices/iio:device0/scan_elements/in_voltage4_en
   # echo 1 >
   /sys/bus/iio/devices/iio:device0/scan_elements/in_voltage5_en
   # echo 1 >
   /sys/bus/iio/devices/iio:device0/scan_elements/in_voltage6_en
   # echo 1 >
   /sys/bus/iio/devices/iio:device0/scan_elements/in_voltage7_en
   ```

4. **Setup the buffer size**: Here one should set the number of sample sets that may be held by the buffer:

   ```
   # echo 100 > /sys/bus/iio/devices/iio:device0/buffer/length
   ```

5. **Enable the buffer**: This step consists of marking the buffer as being ready to receive pushed data:

   ```
   # echo 1 > /sys/bus/iio/devices/iio:device0/buffer/enable
   ```

IIO Framework

To stop the capture, we'll have to write 0 in the same file.

6. **Fire the trigger**: Launch acquisition:

   ```
   # echo 1 > /sys/bus/iio/devices/trigger0/trigger_now
   ```

Now acquisition is done, we can:

7. Disable the buffer:

   ```
   # echo 0 > /sys/bus/iio/devices/iio:device0/buffer/enable
   ```

8. Detach the trigger:

   ```
   # echo "" > /sys/bus/iio/devices/iio:device0/trigger/current_trigger
   ```

9. Dump the content of our IIO character device:

   ```
   # cat /dev/iio\:device0 | xxd -
   ```

Capturing using the hrtimer trigger

The following is the set of commands that allow to capture data using hrtimer trigger:

```
# echo /sys/kernel/config/iio/triggers/hrtimer/trigger0
# echo 50 > /sys/bus/iio/devices/trigger0/sampling_frequency
# echo 1 > /sys/bus/iio/devices/iio:device0/scan_elements/in_voltage4_en
# echo 1 > /sys/bus/iio/devices/iio:device0/scan_elements/in_voltage5_en
# echo 1 > /sys/bus/iio/devices/iio:device0/scan_elements/in_voltage6_en
# echo 1 > /sys/bus/iio/devices/iio:device0/scan_elements/in_voltage7_en
# echo 1 > /sys/bus/iio/devices/iio:device0/buffer/enable
# cat /dev/iio:device0 | xxd -
0000000: 0188 1a30 0000 0000 8312 68a8 c24f 5a14 ...0......h..OZ.
0000010: 0188 1a30 0000 0000 192d 98a9 c24f 5a14 ...0.....-...OZ.
[...]
```

And, we look at the type to figure out how to process data:

```
$ cat /sys/bus/iio/devices/iio:device0/scan_elements/in_voltage_type
be:s14/16>>2
```

Voltage processing: $0x188 >> 2 = 98 * 250 = 24500 = 24.5$ v

IIO tools

There are some useful tools you can use in order to ease and speed up your app's development with IIO devices. They are available in `tools/iio` in the kernel tree:

- `lsiio.c`: To enumerate IIO triggers, devices, and channels
- `iio_event_monitor.c`: Monitor an IIO device's ioctl interface for IIO events
- `generic_buffer.c`: To retrieve, process, and print data received from an IIO device's buffer
- `libiio`: A powerful library developed by analog device to interface IIO devices, and available at https://github.com/analogdevicesinc/libiio.

Summary

By the end of this chapter, you should now be familiar with IIO framework and vocabulary. You know what channels, device, and triggers are. You can even play with your IIO device from the user space, through sysfs or character device. The time to write your own IIO driver has come. There are a lot of available existing drivers not supporting trigger buffers. You can try to add such features in one of them. In the next chapter, we will play with the most useful/used resource on a system: the memory. Be strong, the game has just started.

11
Kernel Memory Management

On Linux systems, every memory address is virtual. They do not point to any address in the RAM directly. Whenever one accesses a memory location, a translation mechanism is performed in order to match the corresponding physical memory.

Let us start with a short story to introduce the virtual memory concept. Given a hotel, there can be a phone in each room, having a private number. Any installed phone, of course belongs to the hotel. None of them can be joined directly from outside the hotel.

If you need to contact an occupant of a room, let us say your friend, he must have given you the hotel's switchboard number and the room number in which he stays. Once you call the switchboard and give the room number of the occupant you need to talk to, just at this moment, the receptionist redirects your call to the real private phone of the room. Only the receptionist and the room occupant know the private number mapping:

```
(switchboard number + room number) <=> private (real) phone number
```

Kernel Memory Management

Every time someone in the city (or over the world) wants to contact a room occupant, he has to pass by the hotline. He needs to know the right hotline number of the hotel, and the room number. This way, `switchboard number + room number` = virtual address, whereas `private phone number` corresponds to the physical address. There are some rules related to hotels that apply on Linux as well:

Hotel	Linux
You cannot contact an occupant who has no private phone in the room. There is not even a way to attempt to do this. Your call will be ended suddenly.	You cannot access a non-existing memory in your address space. This will cause a segmentation fault.
You cannot contact an occupant who does not exist, or whose check-in the hotel is not aware of, or whose information is not found by the switchboard.	If you access unmapped memory, the CPU raises a page fault and the OS handles it.
You can't contact an occupant whose stay is over.	You cannot access freed memory. Maybe it has been allocated to another process
Many hotels may have the same brand, but located at different places, each of them having a different hotline number. If you make a mistake with the hotline number.	Different processes may have the same virtual addresses mapped in their address space, but pointing to another different physical addresses.
There is a book (or software with a database) holding the mapping between the room number and the private phone number, and consulted by the receptionist on demand.	Virtual addresses are mapped to the physical memory by page tables, which are maintained by the operating system kernel and consulted by the processor.

That is how one can imagine the virtual addresses work on a Linux system.

In this chapter, we will deal with the whole Linux memory management system covering following topics:

- Memory layout along with address translation and MMU
- Memory allocation mechanisms (page allocator, slab allocator, kmalloc allocator, and so on)
- I/O memory access
- Mapping kernel memory to user space and implementing `mmap()` callback function

- Introducing Linux caching system
- Introducing the device managed resource framework (devres)

System memory layout - kernel space and user space

Throughout this chapter, terms such as kernel space and user space will refer to their virtual address space. On Linux systems, each process owns a virtual address space. It is a kind of memory sandbox during the process life. That address space is 4 GB in size on 32-bits systems (even on a system with physical memory less than 4 GB). For each process, that 4 GB address space is split in two parts:

- User space virtual addresses
- Kernel space virtual addresses

The way the split is done depends on a special kernel configuration option, `CONFIG_PAGE_OFFSET`, which defines where the kernel addresses section starts in a process address space. The common value is `0xC0000000` by default on 32-bit systems, but this may be changed, as it is the case for i.MX6 family processors from NXP, which uses `0x80000000`. In the whole chapter, we will consider `0xC0000000` by default. This is called 3G/1G split, where the user space is given the lower 3 GB of virtual address space, and the kernel uses the upper remaining 1 GB. A typical process's virtual address space layout looks like:

Both addresses used in the kernel and the user space are virtual addresses. The difference is that accessing a kernel address needs a privileged mode. Privileged mode has extended privileges. When the CPU runs the user space side code, the active process is said to be running in the user mode; when the CPU runs the kernel space side code, the active process is said to be running in the kernel mode.

Given an address (virtual of course), one can distinguish whether it is a kernel space or a user space address by using process layout shown above. Every address falling into 0-3 GB, comes from the user space; otherwise, it is from the kernel.

There is a reason why the kernel shares its address space with every process: because every single process at a given moment uses system calls, which will involve the kernel. Mapping the kernel's virtual memory address into each process's virtual address space allow us to avoid the cost of switching out the memory address space on each entry to (and exit from) the kernel. It is the reason why the kernel address space is permanently mapped on top of each process in order to speed up kernel access through system calls.

The memory management unit organizes memory into units of fixed size called pages. A page consists of 4,096 bytes (4 KB). Even if this size may differ on other systems, it is fixed on ARM and x86, which are architectures we are interested in:

- A memory page, virtual page, or simply page are terms one uses to refer to a fixed-length contiguous block of virtual memory. The same name `page` is used as a kernel data structure to represent a memory page.
- On the other hand, a frame (or page frame) refers to a fixed-length contiguous block of physical memory on top of which the operating system maps a memory page. Each page frame is given a number, called **page frame number** (**PFN**). Given a page, one can easily get its PFN and vice versa, using the `page_to_pfn` and `pfn_to_page` macros, which will be discussed in detail in the next sections.
- A page table is the kernel and architecture data structure used to store the mapping between virtual addresses and physical addresses. The key pair page/frame describes a single entry in the page table. This represents a mapping.

Since a memory page is mapped to a page frame, it goes without saying that pages and page frames have the same sizes, 4 K in our case. The size of a page is defined in the kernel through the `PAGE_SIZE` macro.

 There are situations where one needs memory to be page-aligned. One says a memory is page-aligned if its address starts exactly at the beginning of a page. For example, on a 4 K page size system, 4,096, 20,480, and 409,600 are instances of page-aligned memory addresses. In other words, any memory whose address is a multiple of the system page size is said to be page-aligned.

Kernel addresses – concept of low and high memory

The Linux kernel has its own virtual address space as every user mode process does. The virtual address space of the kernel (1 GB sized in 3G/1G split) is divided into two parts:

- Low memory or LOWMEM, which is the first 896 MB
- High Memory or HIGHMEM, represented by the top 128 MB

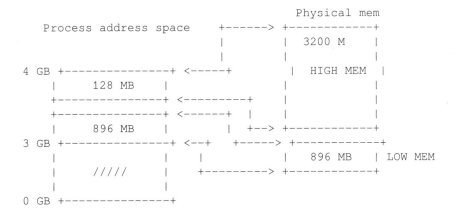

Low memory

The first 896 MB of kernel address space constitutes the low memory region. Early at boot, the kernel permanently maps those 896 MB. Addresses that result from that mapping are called **logical addresses**. These are virtual addresses, but can be translated into physical addresses by subtracting a fixed offset, since the mapping is permanent and known in advance. Low memory match with lower bound of physical addresses. One could define low memory as being the memory for which logical addresses exist in the kernel space. Most of the kernel memory function returns low memory. In fact, to serve different purposes, kernel memory is divided into a zone. Actually, the first 16 MB of LOWMEM is reserved for DMA usage. Because of hardware limitations, the kernel cannot treat all pages as identical. We can then identify three different memory zones in the kernel space:

- `ZONE_DMA`: This contains page frames of memory below 16 MB, reserved for **Direct Memory Access (DMA)**
- `ZONE_NORMAL`: This contains page frames of memory above 16 MB and below 896 MB, for normal use
- `ZONE_HIGHMEM`: This contains page frames of memory at and above 896 MB

That says on a 512 MB system, there will be no `ZONE_HIGHMEM`, 16 MB for `ZONE_DMA`, and 496 MB for `ZONE_NORMAL`.

Another definition of logical addresses: addresses in kernel space, mapped linearly on physical addresses, which can be converted into physical addresses just with an offset, or applying a bitmask. One can convert a physical address into a logical address using the `__pa(address)` macro, and then revert with the `__va(address)` macro.

High memory

The top 128 MB of the kernel address space is called the high memory region. It is used by the kernel to temporarily map physical memory above 1 G. When physical memory above 1 GB (or more precisely, 896 MB), needs to be accessed, the kernel uses those 128 MB to create temporary mapping to its virtual address space, thus achieving the goal of being able to access all physical pages. One could define high memory as being memory for which logical addresses do not exist, and which is not mapped permanently into kernel address space. The physical memory above 896 MB is mapped on demand to the 128 MB of the HIGHMEM region.

Mapping to access high memory is created on the fly by the kernel, and destroyed when done. This makes high memory access slower. That said, the concept of high memory does not exist on the 64-bits systems, due to the huge address range (2^{64}), where the 3G/1G split does not make sense anymore.

User space addresses

In this section, we will deal with the user space by means of processes. Each process is represented in the kernel as an instance of struct task_struct (see include/linux/sched.h), which characterizes and describes a process. Each process is given a table of memory mapping, stored in a variable of type struct mm_struct (see include/linux/mm_types.h). You can then guess that there is at least one mm_struct field embedded in each task_struct. The following line is the part of struct task_struct definition that we are interested in:

```
struct task_struct{
    [...]
    struct mm_struct *mm, *active_mm;
    [...]
}
```

The kernel global variable current, points to the current process. The field *mm, points to its memory mapping table. By definition, current->mm points to the current process memory mappings table.

Now let us see what a struct mm_struct looks like:

```
struct mm_struct {
        struct vm_area_struct *mmap;
        struct rb_root mm_rb;
        unsigned long mmap_base;
        unsigned long task_size;
        unsigned long highest_vm_end;
        pgd_t * pgd;
        atomic_t mm_users;
        atomic_t mm_count;
        atomic_long_t nr_ptes;
#if CONFIG_PGTABLE_LEVELS > 2
        atomic_long_t nr_pmds;
#endif
        int map_count;
        spinlock_t page_table_lock;
        struct rw_semaphore mmap_sem;
        unsigned long hiwater_rss;
```

Kernel Memory Management

```
            unsigned long hiwater_vm;
            unsigned long total_vm;
            unsigned long locked_vm;
            unsigned long pinned_vm;
            unsigned long data_vm;
            unsigned long exec_vm;
            unsigned long stack_vm;
            unsigned long def_flags;
            unsigned long start_code, end_code, start_data, end_data;
            unsigned long start_brk, brk, start_stack;
            unsigned long arg_start, arg_end, env_start, env_end;

            /* Architecture-specific MM context */
            mm_context_t context;

            unsigned long flags;
            struct core_state *core_state;
#ifdef CONFIG_MEMCG
            /*
             * "owner" points to a task that is regarded as the canonical
             * user/owner of this mm. All of the following must be true in
             * order for it to be changed:
             *
             * current == mm->owner
             * current->mm != mm
             * new_owner->mm == mm
             * new_owner->alloc_lock is held
             */
            struct task_struct __rcu *owner;
#endif
            struct user_namespace *user_ns;
            /* store ref to file /proc/<pid>/exe symlink points to */
            struct file __rcu *exe_file;
};
```

I intentionally removed some fields we are not interested in. There are some fields we will talk about later: pgd for example, which is a pointer to the process's base (first entry) level 1 table (PGD), written in the translation table base address of the CPU at context switching. Anyway, before going further, let us see the representation of a process address space:

Kernel Memory Management

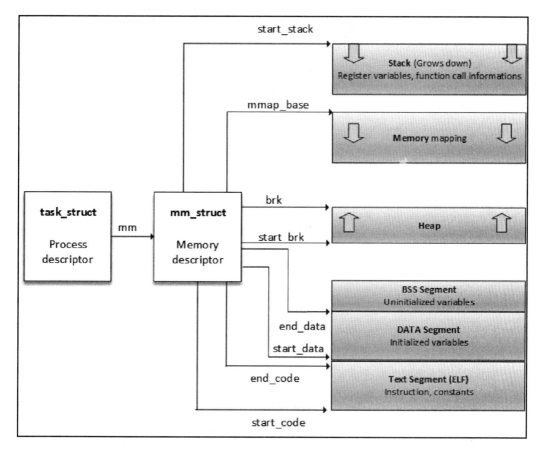

Process memory layout

From the process point of view, a memory mapping can be seen as nothing but a set of page table entries dedicated to a consecutive virtual address range. That *consecutive virtual address range* is called memory area, or **virtual memory area** (**VMA**). Each memory mapping is described by a start address and length, permissions (such as whether the program can read, write, or execute from that memory), and associated resources (such as physical pages, swap pages, file contents, and so on).

A mm_struct has two ways to store process regions (VMA):

1. In a red-black tree, whose root element is pointed by the field mm_struct->mm_rb.
2. In a linked list, where the first element is pointed by the field mm_struct->mmap.

Virtual Memory Area (VMA)

The kernel uses virtual memory areas to keep track of the processes memory mappings, for example, a process having one VMA for its code, one VMA for each type of data, one VMA for each distinct memory mapping (if any), and so on. VMAs are processor-independent structures, with permissions and access control flags. Each VMA has a start address, a length, and their sizes are always a multiple of page size (PAGE_SIZE). A VMA consists of a number of pages, each of which has an entry in the page table.

 Memory regions described by VMA are always virtually contiguous, not physically. One can check all VMAs associated with a process through the /proc/<pid>/maps file, or using the pmap command on a process ID.

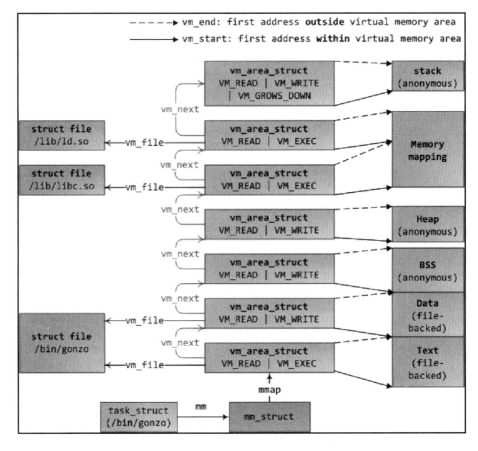

Image source: http://duartes.org/gustavo/blog/post/how-the-kernel-manages-your-memory/

```
# cat /proc/1073/maps
00400000-00403000 r-xp 00000000 b3:04 6438 /usr/sbin/net-listener
00602000-00603000 rw-p 00002000 b3:04 6438 /usr/sbin/net-listener
00603000-00624000 rw-p 00000000 00:00 0 [heap]
7f0eebe4d000-7f0eebe54000 r-xp 00000000 b3:04 11717
/usr/lib/libffi.so.6.0.4
7f0eebe54000-7f0eec054000 ---p 00007000 b3:04 11717
/usr/lib/libffi.so.6.0.4
7f0eec054000-7f0eec055000 rw-p 00007000 b3:04 11717
/usr/lib/libffi.so.6.0.4
7f0eec055000-7f0eec069000 r-xp 00000000 b3:04 21629 /lib/libresolv-2.22.so
7f0eec069000-7f0eec268000 ---p 00014000 b3:04 21629 /lib/libresolv-2.22.so
[...]
7f0eee1e7000-7f0eee1e8000 rw-s 00000000 00:12 12532 /dev/shm/sem.thk-
mcp-231016-sema
[...]
```

Each line in the preceding excerpt represents a VMA, and fields map the following pattern: `{address (start-end)} {permissions} {offset} {device (major:minor)} {inode} {pathname (image)}`:

- `address`: This represents the starting and ending address of the VMA.
- `permissions`: This describes access right of the region: `r` (read), `w` (write), and `x` (execute), including `p` (if the mapping is private) and `s` (for shared mapping).
- `Offset`: In the case of file mapping (`mmap` system call), it is the offset in the file where the mapping takes place. It is 0 otherwise.
- `major:minor`: In case of file mapping, these represent the major and minor number of the devices in which the file is stored (device holding the file).
- `inode`: In the case of mapping from a file, the inode number of the mapped file.
- `pathname`: This is the name of the mapped file, or left blank otherwise. There are other region name such as `[heap]`, `[stack]`, or `[vdso]`, which stands for virtual dynamic shared object, which is a shared library mapped by the kernel into every process address space, in other to reduce performance penalties when system calls switch to kernel mode.

Each page allocated to a process belongs to an area; thus, any page that does not live in the VMA does not exist and cannot be referenced by the process.

 High memory is perfect for user space because user space's virtual address must be explicitly mapped. Thus, most high memory is consumed by user applications. `__GFP_HIGHMEM` and `GFP_HIGHUSER` are the flags for requesting the allocation of (potentially) high memory. Without these flags, all kernel allocations return only low memory. There is no way to allocate contiguous physical memory from user space in Linux.

One can use the `find_vma` function to find the VMA that corresponds to a given virtual address. `find_vma` is declared in `linux/mm.h`:

```
* Look up the first VMA which satisfies  addr < vm_end,  NULL if none. */
extern struct vm_area_struct * find_vma(struct mm_struct * mm, unsigned
long addr);
```

This is an example:

```
struct vm_area_struct *vma = find_vma(task->mm, 0x13000);
if (vma == NULL) /* Not found ? */
    return -EFAULT;
if (0x13000 >= vma->vm_end) /* Beyond the end of returned VMA ? */
    return -EFAULT;
```

The whole process of memory mapping can be obtained by reading files: `/proc/<PID>/map`, `/proc/<PID>/smap`, and `/proc/<PID>/pagemap`.

Address translation and MMU

Virtual memory is a concept, an illusion given to a process so it thinks it has large and almost infinite memory, and sometimes more than the system really has. It is up to the CPU to make the conversion from virtual to physical address every time one accesses a memory location. That mechanism is called address translation, and is performed by the **Memory Management Unit (MMU)**, which is a part of the CPU.

MMU protects memory from unauthorized access. Given a process, any page that needs to be accessed must exist in one of the process VMAs, and thus, must live in the process page table (every process has its own).

Kernel Memory Management

Memory is organized by chunks of fixed size named **pages** for virtual memory, and **frames** for physical memory, sized 4 KB in our case. Anyway, you do not need to guess the page size of the system you write the driver for. It is defined and accessible with the `PAGE_SIZE` macro in the kernel. Remember therefore, page size is imposed by the hardware (CPU). Considering a 4 KB page sized system, bytes 0 to 4095 fall in page 0, bytes 4096-8191 fall in page 1, and so on.

The concept of page table is introduced to manage mapping between pages and frames. Pages are spread over tables, so that each PTE corresponds to a mapping between a page and a frame. Each process is then given a set of page tables to describe its whole memory space.

In order to walk through pages, each page is assigned an index (like an array), called the page number. When it comes to frame, it is PFN. This way, virtual memory addresses are composed of two parts: a page number and an offset. The offset represents the 12 less significant bits of the address, whereas 13 less significant bits represent it on 8 KB page size systems:

0xBAADF	0x0DA
Virtual Page Number	Offset

Virtual address representation

How do the OS or CPU know which physical address corresponds to a given virtual address? They use the page table as the translation table, and know that each entry's index is a virtual page number, and the value is the PFN. To access physical memory given a virtual memory, the OS first extracts the offset, the virtual page number, and then walks through the process's page tables in order to match virtual page number to physical page. Once a match occurs, it is then possible to access data into that page frame:

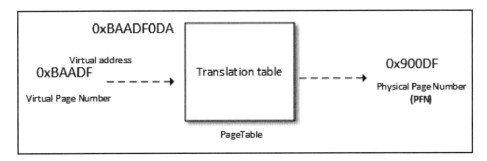

Address translation

Kernel Memory Management

The offset is used to point to the right location into the frame. Page table does not only hold mapping between physical and virtual page number, but also access control information (read/write access, privileges, and so on).

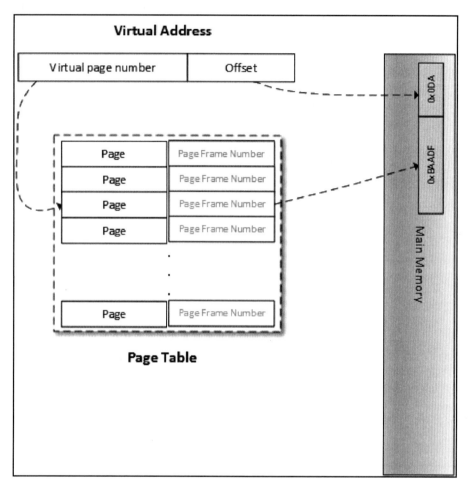

Virtual to physical address translation

The number of bits used to represent the offset is defined by the kernel macro `PAGE_SHIFT`. `PAGE_SHIFT` is the number of bits to shift one bit left to obtain the `PAGE_SIZE` value. It is also the number of bits to right-shift to convert the virtual address to the page number and the physical address to the page frame number. The following are the definitions of these macros from `/include/asm-generic/page.h` in the kernel sources:

```
#define PAGE_SHIFT      12
#ifdef __ASSEMBLY__
#define PAGE_SIZE       (1 << PAGE_SHIFT)
#else
#define PAGE_SIZE       (1UL << PAGE_SHIFT)
#endif
```

Page table is a partial solution. Let us see why. Most architecture requires 32 bits (4 bytes) to represent a PTE. Each process having its private 3 GB user space address, we need 786,432 entries to characterize and cover a process address space. It represents too much physical memory spent per process, just to characterize the memory mappings. In fact, a process generally uses a small but scattered portion of its virtual address space. To resolve that issue, the concept of *level* is introduced. Page tables are hierarchized by level (page level). The space necessary to store a multi-level page table only depends on the virtual address space actually in use, instead of being proportional to the maximum size of the virtual address space. This way, unused memory is no longer represented, and the page table walk through time is reduced. This way, each table entry in level N will point to an entry in table of level N+1. Level 1 is the higher level.

Linux uses a four-level paging model:

- **Page Global Directory** (**PGD**): It is the first level (level 1) page table. Each entry's type is `pgd_t` in kernel (generally an `unsigned long`), and point on an entry in table at the second level. In kernel, the structure `tastk_struct` represents a process's description, which in turn has a member (mm) whose type is `mm_struct`, and that characterizes and represents the process's memory space. In the `mm_struct`, there is a processor-specific field `pgd`, which is a pointer on the first entry (entry 0) of the process's level-1 (PGD) page table. Each process has one and only one PGD, which may contain up to 1024 entries.
- **Page Upper Directory** (**PUD**): This exist only on architectures using four-level tables. It represent the socong level of indirection.

Kernel Memory Management

- **Page Middle Directory (PMD)**: This is the third indirection level, and exists only on architectures using four-level tables.
- **Page Table (PTE)**: Leaves of the tree. It is an array of pte_t, where each entry points to the physical page.

All levels are not always used. The i.MX6's MMU only supports a 2 level page table (PGD and PTE), it is the case for almost all 32-bit CPUs) In this case, PUD and PMD are simply ignored.

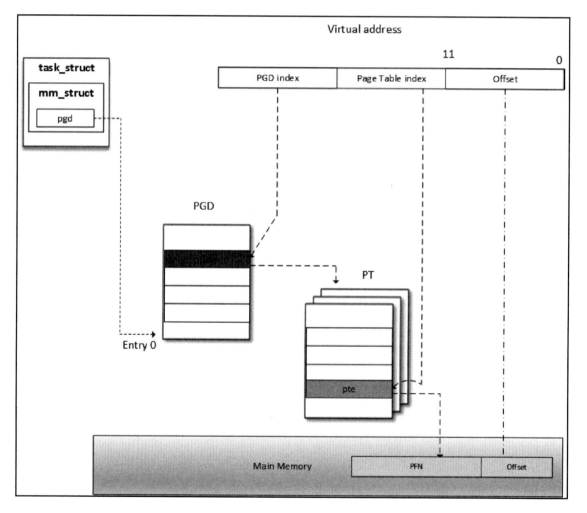

Two-level tables overview

Kernel Memory Management

You might ask how MMU is aware of the process page table. It is simple, MMU does not store any address. Instead, there is a special register in the CPU, called **page table base register** (**PTBR**) or **Translation Table Base Register 0** (**TTBR0**), which points to the base (entry 0) of the level-1 (top level) page table (PGD) of the process. It is exactly where the field `pdg` of `struct mm_struct` points: `current->mm.pgd == TTBR0`.

At context switch (when a new process is scheduled and given the CPU), the kernel immediately configures the MMU, and updates the PTBR with the new process's `pgd`. Now when a virtual address is given to MMU, it uses the PTBR's content to locate the process's level-1 page table (PGD), and then it uses the level-1 index, extracted from the **most significant bits** (**MSBs**) of the virtual address, to find the appropriate table entry, which contains a pointer to the base address of the appropriate level-2 page table. Then, from that base address, it uses the level-2 index to find the appropriate entry and so on until it reaches the PTE. ARM architecture (i.MX6 in our case) has a 2-level page table. In this case, the level-2 entry is a PTE, and points to the physical page (PFN). Only the physical page is found at this step. To access the exact memory location in the page, the MMU extracts the memory offset, also part of the virtual address, and points on the same offset in the physical page.

When a process needs to read from or write into a memory location (of course we're talking about virtual memory), the MMU performs a translation into that process's page table, to find the right entry (`PTE`). The virtual page number is extracted (from the virtual address) and used by the processor as an index into the processes page table to retrieve its page table entry. If there is a valid page table entry at that offset, the processor takes the page frame number from this entry. If not, it means the process accessed an unmapped area of its virtual memory. A page fault is then raised and the OS should handle it.

In the real world, address translation requires a page table walk, and it is not always a one-shot operation. There are at least as many memory accesses as there are table levels. A four-level page table would require four memory accesses. In other words, every virtual access would result in five physical memory accesses. The virtual memory concept would be useless if its access were four times slower than a physical access. Fortunately, SoC manufacturers worked hard to find a clever trick to address this performance issue: modern CPUs use a small associative and very fast memory called **translation lookaside buffer** (**TLB**), in order to cache the PTEs of recently accessed virtual pages.

Page look up and TLB

Before the MMU proceeds to address translation, there is another step involved. As there is a cache for recently accessed data, there is also a cache for recently translated addresses. As a data cache speeds up the data accessing process, TLB speeds up virtual address translation (yes, address translation is a tricky task. It is content-addressable memory, abbreviated (**CAM**), where the key is the virtual address and the value is the physical address. In other words, the TLB is a cache for the MMU. At each memory access, the MMU first checks for recently used pages in the TLB, which contains a few of the virtual address ranges to which physical pages are currently assigned.

How does TLB work

On a virtual memory access, the CPU walks through the TLB trying to find the virtual page number of the page that is being accessed. This step is called TLB lookup. When a TLB entry is found (a match occurred), one says there is a **TLB hit** and the CPU just keeps running and uses the PFN found in the TLB entry to calculate the target physical address. There is no page fault when a TLB hit occurs. As one can see, as long as a translation can be found in the TLB, virtual memory access will be as fast as a physical access. If no TLB entry is found (no match occured), one says there is a **TLB miss**.

On a TLB miss event, there are two possibilities, depending on the processor type, TLB miss events can be handled by the software, or by the hardware, through the MMU:

- **Software handling**: The CPU raises a TLB miss interruption, caught by the OS. The OS then walks through the process's page table to find the right PTE. If there is a matching and valid entry, then the CPU installs the new translation in the TLB. Otherwise, the page fault handler is executed.
- **Hardware handling**: It is up to the CPU (the MMU in fact) to walk through the process's page table in hardware. If there is a matching and valid entry, the CPU adds the new translation in the TLB. Otherwise, the CPU raises a page fault interruption, handled by the OS.

In both cases, the page fault handler is the same: the `do_page_fault()` function is executed, which is architecture-dependent. For ARM, the `do_page_fault` is defined in `arch/arm/mm/fault.c`:

Kernel Memory Management

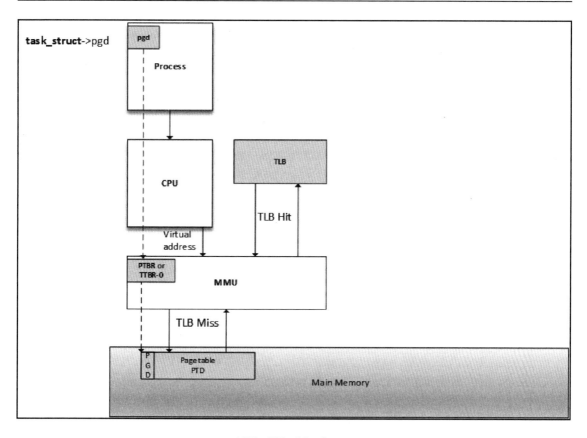

MMU and TLB walkthrough process

 Page table and Page directory entries are architecture-dependent. It is up to the Operating system to ensure that the structure of the table corresponds to a structure recognized by the MMU. On the ARM processor, you must write the location of the translation table in CP15 (coprocessor 15) register c2, and then enable the caches and the MMU by writing to the CP15 register c1. Have a look at both http://infocenter.arm.com/help/index.jsp?topic=/com.arm.doc.dui0056d/BABHJIBH.htm and http://infocenter.arm.com/help/index.jsp?topic=/com.arm.doc.ddi0433c/CIHFDBEJ.html for detailed information.

Memory allocation mechanism

Let us look at the following figure, showing us different memory allocators existing on a Linux-based system, and discuss it later:

Inspired from: `http://free-electrons.com/doc/training/linux-kernel/linux-kernel-slides.pdf`.

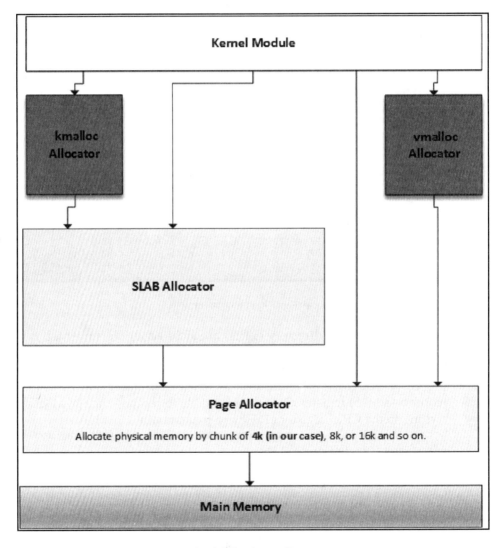

Overview of kernel memory allocator

There is an allocation mechanism to satisfy any kind of memory request. Depending on what you need memory for, you can choose the one closer to your goal. The main allocator is the **Page Allocator**, which only works with pages (a page being the smallest memory unit it can deliver). Then comes the **SLAB Allocator** that is built on top of the page allocator, getting pages from it and returning smaller memory entities (by mean of slabs and caches). This is the allocator on which the **kmalloc Allocator** relies.

Page allocator

Page allocator is the low-level allocator on the Linux system, the one on which other allocators rely on. System's physical memory is made up of fixed-size blocks (called page frames). A page frame is represented in the kernel as an instance of the struct page structure. A page is the smallest unit of memory that the OS will give to any memory request at low level.

Page allocation API

You will have understood that the kernel page allocator allocates and deallocates blocks of pages using the buddy algorithm. Pages are allocated in blocks that are powers of 2 in size (in order to get the best from the buddy algorithm). That means that it can allocate a block 1 page, 2 pages, 4 pages, 8, 16, and so on:

1. alloc_pages(mask, order) allocates 2^{order} pages and returns an instance of struct page which represents the first page of the reserved block. To allocate only one page, order should be 0. It is what alloc_page(mask) does:

   ```
   struct page *alloc_pages(gfp_t mask, unsigned int order)
   #define alloc_page(gfp_mask) alloc_pages(gfp_mask, 0)
   ```

   __free_pages() is used to free memory allocated with alloc_pages() function. It takes a pointer to the allocated page(s) as a parameter, with the same order as was used for allocation.

   ```
   void __free_pages(struct page *page, unsigned int order);
   ```

Kernel Memory Management

2. There are other functions working in the same way, but instead of an instance of struct page, they return the address (virtual of course) of the reserved block. These are __get_free_pages(mask, order) and __get_free_page(mask):

   ```
   unsigned long __get_free_pages(gfp_t mask, unsigned int order);
   unsigned long get_zeroed_page(gfp_t mask);
   ```

   free_pages() is used to free page allocated with __get_free_pages(). It takes the kernel address representing the start region of allocated page(s), along with the order, which should be the same as that used for allocation:

   ```
   free_pages(unsigned long addr, unsigned int order);
   ```

In either case, mask specifies details about the request, which are the memory zones and the behavior of allocators. Choices available are:

- GFP_USER, for user memory allocation.
- GFP_KERNEL, the commonly used flag for kernel allocation.
- GFP_HIGHMEM, which requests memory from the HIGH_MEM zone.
- GFP_ATOMIC, which allocates memory in an atomic manner that cannot sleep. Used when one needs to allocate memory from an interrupt context.

There is a warning on using GFP_HIGHMEM, which should not be used with __get_free_pages() (or __get_free_page()). Since HIGHMEM memory is not guaranteed to be contiguous, you can't return an address of a memory allocated from that zone. Globally only a subset of GFP_* is allowed in memory-related functions:

```
unsigned long __get_free_pages(gfp_t gfp_mask, unsigned int order)
{
    struct page *page;

    /*
     * __get_free_pages() returns a 32-bit address, which cannot represent
     * a highmem page
     */
    VM_BUG_ON((gfp_mask & __GFP_HIGHMEM) != 0);

    page = alloc_pages(gfp_mask, order);
    if (!page)
        return 0;
    return (unsigned long) page_address(page);
}
```

Kernel Memory Management

 The maximum number of pages one can allocate is 1024. It means that on a 4 Kb sized system, you can allocate up to 1024*4 Kb = 4 MB at most. It is the same for `kmalloc`.

Conversion functions

The `page_to_virt()` function is used to convert the struct page (as returned by `alloc_pages()` for example) into the kernel address. `virt_to_page()` takes a kernel virtual address and returns its associated struct page instance (as if it was allocated using the `alloc_pages()` function). Both `virt_to_page()` and `page_to_virt()` are defined in <asm/page.h>:

```
struct page *virt_to_page(void *kaddr);
void *page_to_virt(struct page *pg)
```

The macro `page_address()` can be used to return the virtual address that corresponds to the beginning address (the logical address of course) of a struct page instance:

```
void *page_address(const struct page *page)
```

We can see how it is used in the `get_zeroed_page()` function:

```
unsigned long get_zeroed_page(unsigned int gfp_mask)
{
    struct page * page;

    page = alloc_pages(gfp_mask, 0);
    if (page) {
        void *address = page_address(page);
        clear_page(address);
        return (unsigned long) address;
    }
    return 0;
}
```

`__free_pages()` and `free_pages()` can be mixed. The main difference between them is that `free_page()` takes a virtual address as a parameter, whereas `__free_page()` takes a struct page structure.

Slab allocator

Slab allocator is the one on which `kmalloc()` relies. Its main purpose is to eliminate the fragmentation caused by memory (de)allocation that would be caused by the buddy system in the case of small size memory allocation, and speed up memory allocation for commonly used objects.

The buddy algorithm

To allocate memory, the requested size is round up to a power of two, and the buddy allocator searches the appropriate list. If no entries exist on the requested list, an entry from the next upper list (which has blocks of twice the size of the previous list) is split into two halves (called **buddies**). The allocator uses the first half, while the other is added to the next list down. This is a recursive approach, which stops when either the buddy allocator successfully finds a block which we can be split, or reaches the largest size of block and there are no free blocks available.

The following case study is heavily inspired from http://dysphoria.net/OperatingSystems1/4_allocation_buddy_system.html. For example, if the minimum allocation size is 1 KB, and the memory size is 1 MB, the buddy allocator will create an empty list for 1 KB holes, empty list for 2 KB holes, one for 4 KB holes, 8 KB, 16 KB, 32 KB, 64 KB, 128 KB, 256 KB, 512 KB, and one list for 1 MB holes. All of them are initially empty, except for the 1 MB list which has only one hole.

Now let us imagine a scenario where we want to allocate a **70K** block. The buddy allocator will round it up to **128K**, and end up splitting the 1 MB into two **512K** blocks, then **256K**, and finally **128K**, then it will allocate one of the **128K** blocks to the user. The following are schemes that summarize this scenario:

Kernel Memory Management

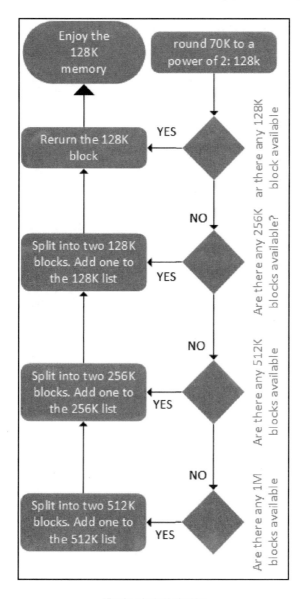

Allocation using buddy algorithm

Kernel Memory Management

The deallocation is as fast as allocation. The following figure summarize the deallocation algorithm:

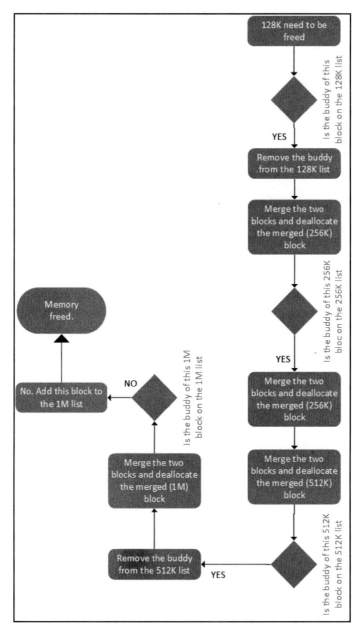

Deallocation using buddy algorithm

A journey into the slab allocator

Before we introduce the slab allocator, let us define some terms it uses:

- **Slab**: This is a contiguous piece of physical memory made of several page frames. Each slab is divided into equal chunks of the same size, used to store specific types of kernel object, such as inodes, mutexes, and so on. Each slab is then an array of objects.
- **Cache**: It is made of one or more slabs in a linked list, and they are represented in the kernel as instances the of `struct kmem_cache_t` structure. The cache only stores objects of the same type (for example, inodes only, or only address space structures)

Slabs may be in one of the following states:

- **Empty**: This is where all objects (chunks) on the slab are marked as free
- **Partial**: Both used and free objects exist in the slab
- **Full**: All objects on the slab are marked as used

It is up to the memory allocator to build caches. Initially, each slab is marked as empty. When one (code) allocates memory for a kernel object, the system looks for a free location for that object on a partial/free slab in a cache for that type of object. If not found, the system allocates a new slab and adds it into the cache. The new object gets allocated from this slab, and the slab is marked as **partial**. When the code is done with the memory (memory freed), the object is simply returned to the slab cache in its initialized state.

Kernel Memory Management

It is the reason why the kernel also provides helper functions to obtain zeroed initialized memory, in order to get rid of the previous content. The slab keeps a reference count of how many of its objects are being used, so that when all slabs in a cache are full and another object is requested, the slab allocator is responsible for adding new slabs:

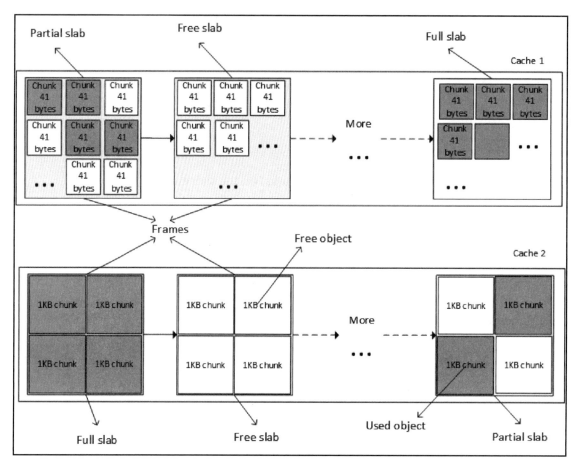

Slab cache overview

It is a bit like creating a per-object allocator. The system allocate one cache per type of object, and only objects of the same type can be stored in a cache (For example, only `task_struct` structure).

There are different kinds of slab allocator in the kernel, depending on whether or not one needs compactness, cache-friendliness, or raw speed:

- The **SLOB**, which is as compact as possible
- The **SLAB**, which is as cache-friendly as possible
- The **SLUB**, which is quite simple and requires fewer instruction cost counts

kmalloc family allocation

kmalloc is a kernel memory allocation function, such as malloc() in user space. Memory returned by kmalloc is contiguous in physical memory and in virtual memory:

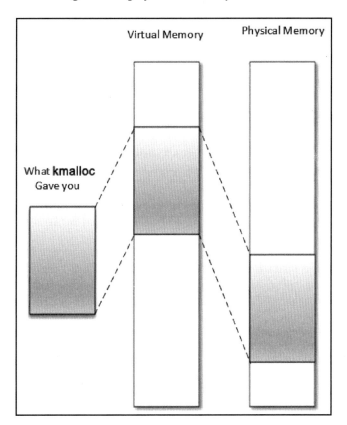

Kernel Memory Management

The kmalloc allocator is the general and higher-level memory allocator in the kernel, which relies on the SLAB allocator. Memory returned from kmalloc has a kernel logical address because it is allocated from the LOW_MEM region, unless HIGH_MEM is specified. It is declared in <linux/slab.h>, which is the header to include when using kmalloc in your driver. The following is the prototype:

```
void *kmalloc(size_t size, int flags);
```

size specifies the size of the memory to be allocated (in bytes). flag determines how and where memory should be allocated. Available flags are the same as the page allocator (GFP_KERNEL, GFP_ATOMIC, GFP_DMA, and so on).

- GFP_KERNEL: This is the standard flag. We cannot use this flag in the interrupt handler because its code may sleep. It always returns memory from LOM_MEM zone (hence a logical address).
- GFP_ATOMIC: This guarantees the atomicity of the allocation. The only flag to use when we are in the interrupt context. Please do not abuse this, since it uses an emergence pool of memory.
- GFP_USER: This allocates memory to a user space process. Memory is then distinct and separated from that allocated to the kernel.
- GFP_HIGHUSER: This allocates memory from HIGH_MEMORY zone
- GFP_DMA: This allocates memory from DMA_ZONE.

On successful allocation of memory, kmalloc returns the virtual address of the chunk allocated, guaranteed to be physically contiguous. On error, it returns NULL.

Kmalloc relies on SLAB caches when allocating small size memories. In this case, the kernel rounds the allocated area size up to the size of the smallest SLAB cache in which it can fit. Always use it as your default memory allocator. In architectures used in this book (ARM and x86), the maximum size per allocation is 4 MB, and 128 MB for total allocations. Have a look at https://kaiwantech.wordpress.com/2011/08/17/kmalloc-and-vmalloc-linux-kernel-memory-allocation-api-limits/.

The kfree function is used to free the memory allocated by kmalloc. The following is the prototype of kfree();

```
void kfree(const void *ptr)
```

Kernel Memory Management

Let us see an example:

```
#include <linux/init.h>
#include <linux/module.h>
#include <linux/slab.h>
#include <linux/mm.h>

MODULE_LICENSE("GPL");
MODULE_AUTHOR("John Madieu");

void *ptr;

static int
alloc_init(void)
{
    size_t size = 1024; /* allocate 1024 bytes */
    ptr = kmalloc(size,GFP_KERNEL);
    if(!ptr) {
        /* handle error */
        pr_err("memory allocation failed\n");
        return -ENOMEM;
    }
    else
        pr_info("Memory allocated successfully\n");
    return 0;
}

static void alloc_exit(void)
{
    kfree(ptr);
    pr_info("Memory freed\n");
}

module_init(alloc_init);
module_exit(alloc_exit);
```

Other family-like functions are:

```
void kzalloc(size_t size, gfp_t flags);
void kzfree(const void *p);
void *kcalloc(size_t n, size_t size, gfp_t flags);
void *krealloc(const void *p, size_t new_size, gfp_t flags);
```

Kernel Memory Management

krealloc() is the kernel equivalent of the user space realloc() function. Because memory returned by kmalloc() retains the contents from its previous incarnation, there could be a security risk if it's exposed to user space. To get zeroed kmalloc'ed memory, one should use kzalloc. kzfree() is the freeing function for kzalloc(), whereas kcalloc() allocates memory for an array, and its parameters n and size represent respectively the number of elements in the array and the size of an element.

Since kmalloc() returns a memory area in the kernel permanent mapping (which mean physically contiguous), the memory address can be translated to a physical address using virt_to_phys(), or to a IO bus address using virt_to_bus(). These macros internally call either __pa() or __va() if necessary. The physical address (virt_to_phys(kmalloc'ed address)), downshifted by PAGE_SHIFT, will produce a PFN of the first page from which the chunk is allocated.

vmalloc allocator

vmalloc() is the last kernel allocator we will discuss in the book. It returns memory only contiguous on the virtual space (not physically contiguous):

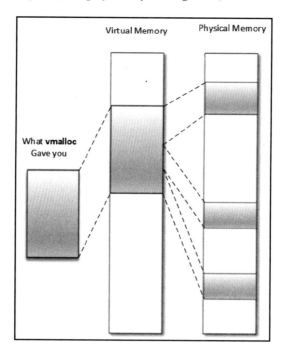

The returned memory always comes from HIGH_MEM zone. Addresses returned cannot be translated into a physical one or into bus address, because one cannot assert that the memory is physically contiguous. It means memory returned by vmalloc() can't be used outside the microprocessor (you cannot easily use it for DMA purposes). It is correct to use vmalloc() to allocate memory for a large (it does not make sense to use it to allocate one page for example) sequential that exists only in software, for example, a network buffer. It is important to note that vmalloc() is slower than kmalloc() or page allocator functions, because it must retrieve the memory, build the page tables, or even remap into a virtually contiguous range, whereas kmalloc() never does that.

Before using this vmalloc API, you should include this header in the code:

```
#include <linux/vmalloc.h>
```

The following are the vmalloc family prototype:

```
void *vmalloc(unsigned long size);
void *vzalloc(unsigned long size);
void vfree( void *addr);
```

size is the size of memory you need to allocate. Upon successful allocation of memory, it returns the address of the first byte of the allocated memory block. On failure, it returns a NULL. vfree function, which is used to free the memory allocated by vmalloc().

The following is an example of using vmalloc:

```
#include<linux/init.h>
#include<linux/module.h>
#include <linux/vmalloc.h>

void *ptr;
static int alloc_init(void)
{
    unsigned long size = 8192;
    ptr = vmalloc(size);
    if(!ptr)
    {
        /* handle error */
        printk("memory allocation failed\n");
        return -ENOMEM;
    }
    else
        pr_info("Memory allocated successfully\n");
    return 0;
}
static void my_vmalloc_exit(void) /* function called at the time of rmmod
```

```
*/
{
    vfree(ptr); //free the allocated memory
    printk("Memory freed\n");
}
module_init(my_vmalloc_init);
module_exit(my_vmalloc_exit);

MODULE_LICENSE("GPL");
MODULE_AUTHOR("john Madieu, john.madieu@gmail.com");
```

One can use `/proc/vmallocinfo` to display all vmalloc'ed memory on the system. `VMALLOC_START` and `VMALLOC_END` are two symbols that delimit the vmalloc address range. They are architecture-dependent and defined in `<asm/pgtable.h>`.

Process memory allocation under the hood

Let us focus on the lower level allocator, which allocates pages of memory. The kernel will report allocation of frame pages (physical pages) until really necessary (when those are actually accessed, by reading or writing). This on-demand allocation is called **lazy-allocation**, eliminating the risk of allocating pages that will never be used.

Whenever a page is requested, only the page table is updated, in most of the cases, a new entry is created, which means only virtual memory is allocated. Only when you access the page, an interrupt called **page fault** is raised. This interrupt has a dedicated handler, called the page fault handler, and is called by the MMU in response to an attempt to access virtual memory, which did not immediately succeed.

Actually, a page fault interrupt is raised whatever the access type is (read, write, execute), to a page whose entry in the page table has not got the appropriate permission bits set to allow that type of access. The response to that interrupt falls in one of the following three ways:

- **The hard fault**: The page does not reside anywhere (neither in the physical memory nor a memory-mapped file), which means the handler cannot immediately resolve the fault. The handler will perform I/O operations in order to prepare the physical page needed to resolve the fault, and may suspend the interrupted process and switch to another while the system works to resolve the issue.
- **The soft fault**: The page resides elsewhere in memory (in the working set of another process). It means the fault handler may resolve the fault by immediately attaching a page of physical memory to the appropriate page table entry, adjusting the entry, and resuming the interrupted instruction.

- **The fault cannot be resolved**: This will result in a bus error or segv. SIGSEGV is sent to the faulty process, killing it (the default behavior) unless a signal handler has been installed for SIGSEV to change the default behavior.

Memory mappings generally start out with no physical pages attached, by defining the virtual address ranges without any associated physical memory. The actual physical memory is allocated later in response to a page fault exception, when the memory is accessed, since the kernel provides some flags to determine whether the attempted access was legal, and specify the behavior of the page fault handler. Thus, the user space brk(), mmap() and similar allocate (virtual) space, but physical memory is attached later.

A page fault occurring in the interrupt context causes a **double fault** interrupt, which usually panics the kernel (calling the panic() function). It is the reason why memory allocated in the interrupt context is taken from a memory pool, which does not raise page fault interrupts. If an interrupt occurs when a double fault is being handled, a triple fault exception is generated, causing the CPU to shut down and the OS immediately reboots. This behavior is actually arc-dependent.

The copy-on-write (CoW) case

The CoW (heavily used with fork()) is a kernel feature that does not allocate several time the memory for a data shared by two or more processes, until a process touches it (write into it); in this case memory is allocated for its private copy. The following shows how a page fault handler manages CoW (one-page case study):

- A PTE is added to the process page table, and marked as un-writable.
- The mapping will result in a VMA creation in the process VMA list. The page is added to that VMA and that VMA is marked as writable.
- On page access (at the first write), the fault handler notices the difference, which means: **this is a Copy on write**. It will then allocate a physical page, which is assigned to the PTE added above, update the PTE flags, flush the TLB entry, and execute the do_wp_page() function, which can copy the content from the shared address to the new location.

Work with I/O memory to talk with hardware

Apart from performing data RAM-oriented operations, one can perform I/O memory transactions, to talk with the hardware. When it comes to the access device's register, the kernel offers two possibilities depending on the system architecture:

- **Through the I/O ports**: This is also called **Port Input Output** (**PIO**). Registers are accessible through a dedicated bus, and specific instructions (`in` and `out`, in assembler generally) are needed to access those registers. It is the case on x86 architectures.
- **Memory Mapped Input Output** (**MMIO**): This is the most common and most used method. The device's registers are mapped to memory. Simply read and write to a particular address to write to the registers of the device. It is the case on ARM architectures.

PIO devices access

On a system on which PIO is used, there are two different address spaces, one for memory, which we have already discussed, and the other one for I/O ports, called the port address space, limited to 65,536 ports only. This is a old way, and very uncommon nowadays.

The kernel exports a few functions (symbols) to handle I/O port. Prior to accessing any port regions, we must first inform the kernel that we are using a range of ports using the `request_region()` function, which will return `NULL` on error. Once done with the region, one must call `release_region()`. These are both declared in `linux/ioport.h`. Their prototypes are:

```
struct resource *request_region(unsigned long start,
                                unsigned long len, char *name);
void release_region(unsigned long start, unsigned long len);
```

Those functions inform the kernel about your intention to use/release of a region `len` ports, starting from `start`. The `name` parameter should be set with the name of your device. Their use is not mandatory. This is a kind of politeness, which prevents two or more drivers from referencing the same range of ports. One can display information about the ports actually in use on the system by reading the content of `/proc/ioports` files.

Once one is done with region reservation, one can access the port using the following functions:

```
u8 inb(unsigned long addr)
u16 inw(unsigned long addr)
u32 inl(unsigned long addr)
```

which respectively access (read) 8-, 16-, or 32-bits sized (wide) ports, and the following functions:

```
void outb(u8 b, unsigned long addr)
void outw(u16 b, unsigned long addr)
void outl(u32 b, unsigned long addr)
```

which write b data, 8-, 16-, or 32-bits sized, into addr port.

The fact that PIO uses a different set of instruction to access I/O ports or MMIO is a disadvantage because PIO requires more instructions than normal memory to accomplish the same task. For instance, 1-bit testing has only one instruction in MMIO, whereas PIO requires reading the data into a register before testing the bit, which is more than one instruction.

MMIO devices access

Memory-mapped I/O reside same address space than memory. The kernel uses part of the address space normally used by RAM (HIGH_MEM actually) to map the devices registers, so that instead of having real memory (that is, RAM) at that address, I/O device take place. Thus, communicating to an I/O device becomes like reading and writing to memory addresses devoted to that I/O device.

Like PIO, there are MMIO functions, to inform the kernel about our intention to use a memory region. Remember it is a pure reservation only. These are request_mem_region() and release_mem_region():

```
struct resource* request_mem_region(unsigned long start,
                                    unsigned long len, char *name)
void release_mem_region(unsigned long start, unsigned long len)
```

It is also a politeness.

One can display memory regions actually in use on the system by reading the content of the /proc/iomem file.

Kernel Memory Management

Prior to accessing a memory region (and after you successfully request it), the region must be mapped into kernel address space by calling special architecture-dependent functions (which make use of MMU to build the page table, and thus cannot be called from the interrupt handler). These are ioremap() and iounmap(), which handle cache coherency too:

```
void __iomem *ioremap(unsigned long phys_add, unsigned long size)
void iounmap(void __iomem *addr)
```

ioremap() returns a __iomem void pointer to the start of the mapped region. Do not be tempted to deference (get/set the value by reading/writing to the pointer) such pointers. The kernel provides functions to access ioremap'ed memories. These are:

```
unsigned int ioread8(void __iomem *addr);
unsigned int ioread16(void __iomem *addr);
unsigned int ioread32(void __iomem *addr);
void iowrite8(u8 value, void __iomem *addr);
void iowrite16(u16 value, void __iomem *addr);
void iowrite32(u32 value, void __iomem *addr);
```

ioremap builds new page tables, just as vmalloc does. However, it does not actually allocate any memory but instead, returns a special virtual address that one can use to access the specified physical address range.

On 32-bit systems, the fact that MMIO steals physical memory address space to create mapping for memory-mapped I/O devices is a disadvantage, since it prevents the system from using the stolen memory for general RAM purpose.

__iomem cookie

__iomem is a kernel cookie used by Sparse, a semantic checker used by the kernel to find possible coding faults. To take advantage of the features offered by Sparse, it should be enabled at kernel compile time; if not, __iomem cookie will be ignored anyway.

The C=1 in the command line will enable Sparse for you, but parse should be installed first on your system:

```
sudo apt-get install sparse
```

For example, when building a module, use:

```
make -C $KPATH M=$PWD C=1 modules
```

Alternatively, if the makefile is well written, just type:

```
make C=1
```

The following shows how __iomem is defined in the kernel:

```
#define __iomem    __attribute__((noderef, address_space(2)))
```

It prevents us from faulty drivers performing I/O memory access. Adding the __iomem for all I/O accesses is a way to be stricter too. Since even I/O access is done through virtual memory (on systems with MMU), this cookie prevents us from using absolute physical addresses, and requires us to use `ioremap()`, which will return a virtual address tagged with __iomem cookie:

```
void __iomem *ioremap(phys_addr_t offset, unsigned long size);
```

So we can use dedicated functions, such as `ioread23()` and `iowrite32()`. You may wonder why one does not use the `readl()`/`writel()` function. Those are deprecated, since these do not make sanity checks and are less secure (no __iomem required), than `ioreadX()`/`iowriteX()` family functions, which accept only __iomem addresses.

In addition, `noderef` is an attribute used by Sparse to make sure programmers do not dereference a __iomem pointer. Even though it could work on some architecture, you are not encouraged to do that. Use the special `ioreadX()`/`iowriteX()` function instead. It is portable and works on every architecture. Now let us see how Sparse will warn us when dereferencing a __iomem pointer:

```
#define BASE_ADDR 0x20E01F8
void * _addrTX = ioremap(BASE_ADDR, 8);
```

First, Sparse is not happy because of the wrong type initializer:

```
warning: incorrect type in initializer (different address spaces)
expected void *_addrTX
got void [noderef] <asn:2>*
```

Or:

```
u32 __iomem* _addrTX = ioremap(BASE_ADDR, 8);
*_addrTX = 0xAABBCCDD; /* bad. No dereference */
pr_info("%x\n", *_addrTX); /* bad. No dereference */
```

Sparse is still not happy:

```
Warning: dereference of noderef expression
```

This last example makes Sparse happy:

```
void __iomem* _addrTX = ioremap(BASE_ADDR, 8);
iowrite32(0xAABBCCDD, _addrTX);
pr_info("%x\n", ioread32(_addrTX));
```

The two rules that you must remember are:

- Always use `__iomem` where it is required whether it is as a return type or as a parameter type, and use Sparse to make sure you did so
- Do not dereference a `__iomem` pointer; use a dedicated function instead

Memory (re)mapping

Kernel memory sometimes needs to be remapped, either from kernel to user space, or from kernel to kernel space. The common use case is remapping the kernel memory to user space, but there are other cases, when one need to access high memory for example.

kmap

Linux kernel permanently maps 896 MB of its address space to the lower 896 MB of the physical memory (low memory). On a 4 GB system, there is only 128 MB left to the kernel to map the remaining 3.2 GB of physical memory (high memory). Low memory is directly addressable by the kernel because of the permanent and one-to-one mapping. When it comes to high memory (memory above 896 MB), the kernel has to map the requested region of high memory into its address space, and the 128 MB mentioned before are especially reserved for this. The function used to perform this trick, `kmap()`. `kmap()`, is used to map a given page into the kernel address space.

```
void *kmap(struct page *page);
```

`page` is a pointer to the `struct page` structure to map. When a high memory page is allocated, it is not directly addressable. `kmap()` is the function one must call to temporarily map high memory into the kernel address space. The mapping will last until `kunmap()` is called:

```
void kunmap(struct page *page);
```

Kernel Memory Management

By temporarily, I mean the mapping should be undone as soon as it is not needed anymore. Remember, 128 MB is not enough to map 3.2 GB. The best programming practice is to unmap high memory mappings when no longer required. It is why the `kmap()` - `kunmap()` sequence has to be entered around every access to the high memory page. .

This function works on both high memory and low memory. That says, if the page structure resides in low memory, then just the virtual address of the page is returned (because low memory pages already have permanent mappings). If the page belongs to high memory, a permanent mapping is created in the kernel's page tables and the address is returned:

```
void *kmap(struct page *page)
{
   BUG_ON(in_interrupt());
   if (!PageHighMem(page))
        return page_address(page);

   return kmap_high(page);
}
```

Mapping kernel memory to user space

Mapping physical addresses is one of the most useful functionalities, especially in embedded systems. Sometime you may want to share part of kernel memory with user space. As said earlier, CPU runs in unprivileged mode when running in user space. To let a process access a kernel memory region, we need to remap that region into the process address space.

Using remap_pfn_range

`remap_pfn_range()` maps physical memory (by means of kernel logical address) to a user space process. It is particularly useful for implementing the `mmap()` system call.

After calling the `mmap()` system call on a file (whether it is a device file or not), the CPU will switch to privileged mode, and run the corresponding `file_operations.mmap()` kernel function, which in turn will call `remap_pfn_range()`. The kernel PTE of the mapped region will be derived, and given to the process, of course, with different protection flags. The process's VMA list is updated with a new VMA entry (with appropriate attributes), which will use PTE to access the same memory.

Kernel Memory Management

Thus, instead of wasting memory by copying, the kernel just duplicates the PTEs. However, kernel and user space PTE have different attributes. `remap_pfn_range()` has the following prototype:

```
int remap_pfn_range(struct vm_area_struct *vma, unsigned long addr,
            unsigned long pfn, unsigned long size, pgprot_t flags);
```

A successful call will return 0, and a negative error code on failure. Most of the arguments for `remap_pfn_range()` are provided when the `mmap()` method is called.

- `vma`: This is the virtual memory area provided by the kernel in the case of a `file_operations.mmap()` call. It corresponds to the user process `vma` into which the mapping should be done.
- `addr`: This is the user virtual address where VMA should start (`vma->vm_start`), which will result in a mapping from a virtual address range between `addr` and `addr + size`.
- `pfn`: This represents the page frame number of the kernel memory region to map. It corresponds to the physical address right-shifted by `PAGE_SHIFT` bits. The `vma` offset (offset into the object where the mapping must start) should be taken into account to produce the PFN. Since the `vm_pgoff` field of the VMA structure contains the offset value in the form of the number of pages, it is precisely what you need (with a `PAGE_SHIFT` left-shifting) to extract the offset in the form of bytes: `offset = vma->vm_pgoff << PAGE_SHIFT`). Finally, `pfn = virt_to_phys(buffer + offset) >> PAGE_SHIFT`.
- `size`: This is the dimension, in bytes, of the area being remapped.
- `prot`: This represents the protection requested for the new VMA. The driver can mangle the default value, but should use the value found in `vma->vm_page_prot` as the skeleton using the OR operator, since some of its bits are already set by user space. Some of these flags are:
    - `VM_IO`, which specifies a device's memory mapped I/O
    - `VM_DONTCOPY`, which tells the kernel not to copy this `vma` on fork
    - `VM_DONTEXPAND`, which prevents `vma` from expanding with `mremap(2)`
    - `VM_DONTDUMP`, prevents the `vma` from being included in the core dump

One may need to modify this value in order to disable caching if using this with I/O memory (`vma->vm_page_prot = pgprot_noncached(vma->vm_page_prot);`).

Using io_remap_pfn_range

The `remap_pfn_range()` function discussed does not apply anymore when it comes to mapping I/O memory to user space. In this case, the appropriate function is `io_remap_pfn_range()`, whose parameters are the same. The only thing that changes is where the PFN comes from. Its prototype looks like:

```
int io_remap_page_range(struct vm_area_struct *vma,
                        unsigned long virt_addr,
                        unsigned long phys_addr,
                        unsigned long size, pgprot_t prot);
```

There is no need to use `ioremap()` when at tempting to map I/O memory to user space. - `ioremap()` is intended for kernel purposes (mapping I/O memory into kernel address space), where as `io_remap_pfn_range` is for user space purposes.

Just pass your real physical I/O address (downshifted by `PAGE_SHIFT` to produce a PFN) directly to `io_remap_pfn_range()`. Even if there are some architectures where `io_remap_pfn_range()` is defined as being `remap_pfn_range()`, there are other architectures where it is not the case. For portability reasons, you should only use `remap_pfn_range()` in situations where the PFN parameter points to RAM, and `io_remap_pfn_range()` in situations where `phys_addr` refers to I/O memory.

The mmap file operation

Kernel `mmap` function is part of `struct file_operations` structure, which is executed when the user executes the system call `mmap(2)`, used to maps physical memory into a user virtual address. The kernel translates any access to that mapped region of memory through the usual pointer dereferences into a file operation. It is even possible to map device physical memory directly to user space (see `/dev/mem`). Essentially writing to memory becomes like writing into a file. It is just a more convenient way of calling `write()`.

Normally, user space processes cannot access device memory directly for security purposes. Therefore, user space processes use the `mmap()` system call to ask kernel to map the device into the virtual address space of the calling process. After the mapping, the user space process can write directly into the device memory through the returned address.

The mmap system call is declared as follows:

```
mmap (void *addr, size_t len, int prot,
      int flags, int fd, ff_t offset);
```

Kernel Memory Management

The driver should have defined the mmap file operation (`file_operations.mmap`) in order to support `mmap(2)`. From the kernel side, the mmap field in the driver's file operation structure (`struct file_operations` structure) has the following prototype:

```
int (*mmap) (struct file *filp, struct vm_area_struct *vma);
```

where:

- `filp` is a pointer to the open device file for the driver that results from the translation of the fd parameter.
- `vma` is allocated and given as a parameter by the kernel. It is a pointer to the user process's vma where the mapping should go. To understand how the kernel creates the new vma, let's recall the `mmap(2)` system call's prototype:

```
void *mmap(void *addr, size_t length, int prot, int flags, int fd, off_t offset);
```

The parameters of this function somehow affect some fields of the vma:

- `addr`: is the user space's virtual address where the mapping should start. It has an impact on `vma>vm_start`. If NULL (the most portable way) was specified, automatically determinate the correct address.
- `length`: This specifies the length of the mapping, and indirectly has an impact on `vma->vm_end`. Remember, the size of a `vma` is always a multiple of `PAGE_SIZE`. In other words, `PAGE_SIZE` is always the smallest size a `vma` can have. The kernel will always alter the size of the `vma` so that is is a multiple of `PAGE_SIZE`.

```
If length <= PAGE_SIZE
    vma->vm_end - vma->vm_start == PAGE_SIZE.
If PAGE_SIZE < length <= (N * PAGE_SIZE)
        vma->vm_end - vma->vm_start == (N * PAGE_SIZE)
```

- `prot`: This affects the permissions of the VMA, which the driver can find in `vma->vm_pro`. As discussed earlier, the driver can update these values, but not alter them.
- `flags`: This determine the type of mapping that the driver can find in `vma->vm_flags`. The mapping can be private or shared.
- `offset`: This specifies the offset within the mapped region, thus mangling the value of `vma->vm_pgoff`.

Implementing mmap in the kernel

Since user space code cannot access kernel memory, the purpose of the mmap() function is to derive one or more protected kernel page table entries (which correspond to the memory to be mapped) and duplicate the user space page tables, remove the kernel flag protection, and set permission flags that will allow the user to access the same memory as the kernel without needing special privileges.

The steps to write a mmap file operation are as follows:

1. Get the mapping offset and check whether it is beyond our buffer size or not:

    ```
    unsigned long offset = vma->vm_pgoff << PAGE_SHIFT;
    if (offset >= buffer_size)
            return -EINVAL;
    ```

2. Check if the mapping size is bigger than our buffer size:

    ```
    unsigned long size = vma->vm_end - vma->vm_start;
    if (size > (buffer_size - offset))
        return -EINVAL;
    ```

3. Get the PFN which corresponds to the PFN of the page where the offset position of our buffer falls:

    ```
    unsigned long pfn;
    /* we can use page_to_pfn on the struct page structure
     * returned by virt_to_page
     */
    /* pfn = page_to_pfn (virt_to_page (buffer + offset)); */

    /* Or make PAGE_SHIFT bits right-shift on the physical
     * address returned by virt_to_phys
     */
    pfn = virt_to_phys(buffer + offset) >> PAGE_SHIFT;
    ```

4. Set the appropriate flag, whether I/O memory is present or not:

    - Disable caching using vma->vm_page_prot = pgprot_noncached(vma->vm_page_prot).
    - Set the VM_IO flag: vma->vm_flags |= VM_IO.
    - Prevent the VMA from swapping out: vma->vm_flags |= VM_DONTEXPAND | VM_DONTDUMP. In kernel versions older than 3.7, you should use only the VM_RESERVED flag instead.

5. Call `remap_pfn_range` with the PFN calculated, the size, and the protection flags:

```
if (remap_pfn_range(vma, vma->vm_start, pfn, size,
vma->vm_page_prot)) {
    return -EAGAIN;
}
return 0;
```

6. Pass your mmap function to the `struct file_operations` structure:

```
static const struct file_operations my_fops = {
    .owner = THIS_MODULE,
    [...]
    .mmap = my_mmap,
    [...]
};
```

Linux caching system

Caching is the process by which frequently accessed or newly written data is fetched from, or written to a small and faster memory, called a **cache**.

Dirty memory is data-backed (for example, file-backed) memory whose content has been modified (typically in a cache) but not written back to the disk yet. The cached version of the data is newer than the on-disk version, meaning that both versions are out of sync. The mechanism by which cached data is written back on the disk (back store) is called **writeback**. We will eventually update the on-disk version, bringing the two in sync. *Clean memory* is file-backed memory in which the contents are in sync with the disk.

Linux delays write operations in order to speed up the read process, and reduces disk wear leveling by writing data only when necessary. A typical example is the `dd` command. Its complete execution does not mean that the data is written to the target device; this is the reason why `dd` in most cases is chained to a `sync` command.

What is a cache?

A cache is temporary, small, and fast memory used to keep copies of data from larger and often very slow memory, typically placed in systems where there is a working set of data accessed far more often than the rest (for example, hard drive, memory).

When the first read occurs, let us say a process requests some data from the large and slower disk, the requested data is returned to the process, and a copy of accessed data is tracked and cached as well. Any consequent read will fetch data from the cache. Any data modification will be applied in the cache, not on the main disk. Then, the cache region whose content has been modified and differs (is newer than) from the on-disk version will be tagged as **dirty**. When the cache runs full, and since cached data is tacked, new data begins to evict the data that has not been accessed and has been sitting idle for the longest, so that if it is needed again, it will have to be fetched from the large/slow storage again.

CPU cache – memory caching

There are three cache memories on the modern CPU, ordered by size and access speed:

- The **L1** cache that has the smallest amount of memory (often between 1k and 64k) is directly accessible by the CPU in a single clock cycle, which makes it the fastest as well. Frequently used things are in L1 and remain in L1 until some other thing's usage becomes more frequent than the existing one and there is less space in L1. If so, it is moved to a bigger space L2.
- The **L2** cache is the middle level, with a larger amount of memory (up to several megabytes) adjacent to the processor, which can be accessed in a small number of clock cycles. This applies when moving things from L2 to L3.
- The **L3** cache, even slower than L1 and L2, may be two times faster than the main memory (RAM). Each core may have its own L1 and L2 cache; therefore, they all share the L3 cache. Size and speed are the main criteria that change between each cache level: L1 < L2 < L3. Whereas original memory access may be 100 ns for example, the L1 cache access can be 0.5 ns.

A real-life example is how a library may put several copies of the most popular titles on display for easy and fast access, but have a large-scale archive with a far greater collection available, at the inconvenience of having to wait for a librarian to go get it for you. The display cases would be analogous to a cache, and the archive would be the large, slow memory.

The main issue that a CPU cache addresses is latency, which indirectly increases the throughput, because access to uncached memory may take a while.

The Linux page cache – disk caching

The page cache, as its name suggests, is a cache of pages in RAM, containing chunks of recently accessed files. The RAM acts as a cache for pages that resides on the disk. In other words, it is the kernel cache of file contents. Cached data may be regular filesystem files, block device files, or memory-mapped files. Whenever a `read()` operation is invoked, the kernel first checks whether the data resides in the page cache, and immediately returns it if found. Otherwise, the data will be read from the disk.

If a process needs to write data without any caching involved, it has to use the `O_SYNC` flag, which guarantees the `write()` command will not return before all data has been transferred to the disk, or the `O_DIRECT`, flag, which only guarantees that no caching will be used for data transfer. That says, `O_DIRECT` actually depends on filesystem used and is not recommended.

Specialized caches (user space caching)

- **Web browser cache**: This stores frequently accessed web pages and images onto the disk, instead of fetching them from the web. Whereas the first access to online data may last for more than hundreds of milliseconds, the second access will fetch data from the cache (which is a disk in this case) in only 10 ms.
- **libc or user-app cache**: Memory and disk cache implementations will try to guess what you need to use next, while browser caches keep a local copy in case you need to use it again.

Why delay writing data to disk?

There are essentially two reasons to that:

- Better usage of the disk characteristics; this is efficiency
- Allows the application to continue immediately after a write; this is performance

For example, delaying disk access and processing data only when it reaches a certain size may improve disk performance, and reduce wear leveling of eMMC (on embedded systems). Every chunk write is merged into a single and contiguous write operation. Additionally, written data is cached, allowing the process to return immediately so that any subsequent read will fetch the data from the cache, resulting in a more responsive program. Storage devices prefer a small number of large operations instead of several small operations.

By reporting write operation on the permanent storage later, we can get rid of latency issues introduced by these disks, which are relatively slow.

Write caching strategies

Depending on the cache strategy, several benefits may be enumerated:

- Reduced latency on data accessing, thus increasing application performance
- Improved storage lifetime
- Reduced system work load
- Reduced risk of data loss

Caching algorithms usually fall into one of the following three different strategies:

1. The **write-throughcache**, where any write operation will automatically update both the memory cache and the permanent storage. This strategy is preferred for applications where data loss cannot be tolerated, and applications that write and then frequently re-read data (since data is stored in the cache and results in low read latency).
2. The **write-aroundcache**, which is similar to write-through, with the difference that it immediately invalidates the cache (which is also costly for the system since any write results in automatic cache invalidation). The main consequence is that any subsequent read will fetch data from the disk, which is slow, thus increasing latency. It prevents the cache from being flooded with data that will not be subsequently read.

3. Linux employs the third and last strategy, called **write back cache**, which can write data to the cache every time a change occurs without updating the corresponding location in the main memory. Instead, the corresponding pages in the page cache are marked as **dirty** (this task is done by MMU using TLB) and added to a so-called list, maintained by the kernel. The data is written into the corresponding location in the permanent storage only at specified intervals or under certain conditions. When the data in the pages is up to date with the data in the page cache, the kernel removes the pages from the list, and they are not marked dirty.
4. On Linux systems, you can find this from `/proc/meminfo` under `Dirty`:

```
cat /proc/meminfo | grep Dirty
```

The flusher threads

The write back cache defers I/O data operations in the page cache. A set or kernel threads, called flusher threads, are responsible for that. Dirty page write back occurs when any one of the following situations is satisfied:

1. When free memory falls below a specified threshold to regain memory consumed by dirty pages.
2. When dirty data lasts until a specific period. The oldest data is written back to the disk to ensure that dirty data does not remain dirty indefinitely.
3. When a user process invokes the `sync()` and `fsync()` system calls. This is an on demand write back.

Device-managed resources – Devres

Devres is a kernel facility helping the developer by automatically freeing the allocated resource in a driver. It simplifies errors handling in `init/probe/open` functions. With devres, each resource allocator has its managed version that will take care of resource release and freeing for you.

> This section heavily relies on the *Documentation/driver-model/devres.txt* file in the kernel source tree, which deals with devres API and lists supported functions along with their descriptions.

The memory allocated with resource-managed functions is associated with the device. devres consists of a linked list of arbitrarily sized memory areas associated with a `struct device`. Each devers resource allocator inserts the allocated resource in the list. The resource remains available until it is manually freed by the code, when the device is detached from the system, or when the driver is unloaded. Each devres entry is associated with a `release` function. There are different ways to release a devres. No matter what, all devres entries are released on driver detach. On release, the associated release function is invoked and then the devres entry is freed.

The following is the list of resources available for a driver:

- Memory for private data structures
- Interrutps (IRQs)
- Memory region allocation (`request_mem_region()`)
- I/O mapping of memory regions (`ioremap()`)
- Buffer memory (possibly with DMA mapping)
- Different framework data structures: Clocks, GPIOs, PWMs, USB phy, regulators, DMA, and so on

Almost every function discussed in this chapter has its managed version. In the majority of cases, the name given to the managed version of a function is obtained by prefixing the original function name with `devm`. For example, `devm_kzalloc()` is the managed version of `kzalloc()`. Additionally, parameters remain unchanged, but are shifted to the right, since the first parameter is the struct device for which the resource is allocated. There is an exception for functions for which the non-managed version is already given a struct device in its parameters:

```
void *kmalloc(size_t size, gfp_t flags)
void * devm_kmalloc(struct device *dev, size_t size, gfp_t gfp)
```

When the device is detached from the system or the driver for the device is unloaded, that memory is freed automatically. It is possible to free the memory with `devm_kfree()` if it's no longer needed.

The old way:

```
ret = request_irq(irq, my_isr, 0, my_name, my_data);
if(ret) {
    dev_err(dev, "Failed to register IRQ.\n");
    ret = -ENODEV;
    goto failed_register_irq; /* Unroll */
}
```

The right way:

```
ret = devm_request_irq(dev, irq, my_isr, 0, my_name, my_data);
if(ret) {
    dev_err(dev, "Failed to register IRQ.\n");
    return -ENODEV; /* Automatic unroll */
}
```

Summary

This chapter is one of the most important chapters. It demystifies memory management and allocation (how and where) in the kernel. Every memory aspect is discussed and detailed, as well as dvres is also explained. The caching mechanism is briefly discussed in order to give an overview of what goes on under the hood during I/O operations. It is a strong base from which introduce and understand the next chapter, which deals with DMA.

12
DMA – Direct Memory Access

DMA is a feature of computer systems that allows devices to access the main system memory RAM without CPU intervention, which then allows them to devote themselves to other tasks. One typically uses it for accelerating network traffic, but it supports any kind of copy.

The DMA controller is the peripheral responsible for DMA management. One mostly finds it in modern processors and microcontrollers. DMA is a feature used to perform memory read and write operations without stealing CPU cycles. When one needs to transfer a block of data, the processor feeds the DMA controller with the source and destination addresses and the total number of bytes. The DMA controller then transfers the data from the source to the destination automatically, without stealing CPU cycles. When the number of bytes remaining reaches zero, the block transfer ends.

In this chapter, we will cover the following topics:

- Coherent and non-coherent DMA mappings, as well as coherency issues
- DMA engine API
- DMA and DT binding

Setting up DMA mappings

For any type of DMA transfer, one needs to provide source and destination addresses, as well as the number of words to transfer. In the case of a peripheral DMA, the peripheral's FIFO serves as either the source or the destination. When the peripheral serves as the source, a memory location (internal or external) serves as the destination address. When the peripheral serves as the destination, a memory location (internal or external) serves as the source address.

With a peripheral DMA, we specify either the source or the destination, depending on the direction of the transfer. In others words, a DMA transfer requires suitable memory mappings. This is what we will discuss in the following sections.

Cache coherency and DMA

As discussed in `Chapter 11`, *Kernel Memory Management*, copies of recently accessed memory areas are stored in the cache. This applies to DMA memory too. The reality is that memory shared between two independent devices is generally the source of cache coherency problems. Cache incoherence is an issue coming from the fact that other devices may not be aware of an update from a writing device. On the other hand, cache coherency ensures that every write operation appears to occur instantaneously, so that all devices sharing the same memory region see exactly the same sequence of changes.

A well-explained situation of the coherency issue is illustrated in the following excerpt from LDD3:

> *Let us imagine a CPU equipped with a cache and an external memory that can be accessed directly by devices using DMA. When the CPU accesses location X in the memory, the current value will be stored in the cache. Subsequent operations on X will update the cached copy of X, but not the external memory version of X, assuming a write-back cache. If the cache is not flushed to the memory before the next time a device tries to access X, the device will receive a stale value of X. Similarly, if the cached copy of X is not invalidated when a device writes a new value to the memory, then the CPU will operate on a stale value of X.*

There are actually two ways to address this issue:

- A hardware-based solution. Such systems are **coherent systems**.
- A software-based solution, where the OS is responsible for ensuring cache coherency. One calls such systems **non-coherent systems**.

DMA mappings

Any suitable DMA transfer requires suitable memory mapping. A DMA mapping consists of allocating a DMA buffer and generating a bus address for it. Devices actually use bus addresses. Bus addresses are each instance of the `dma_addr_t` type.

One distinguishes two types of mapping: **coherent DMA mappings** and **streaming DMA mappings**. One can use the former over several transfers, which automatically addresses cache coherency issues. Therefore, it is too expensive. The streaming mapping has a lot of constraints and does not automatically address coherency issues, although, there is a solution for that, which consists of several function calls between each transfer. Coherent mapping usually exists for the life of the driver, whereas one streaming mapping is usually unmapped once the DMA transfer completes.

One should use streaming mapping when one can and coherent mapping when one must.

Back to the code; the main header should include the following to handle DMA mapping:

```
#include <linux/dma-mapping.h>
```

Coherent mapping

The following function sets up a coherent mapping:

```
void *dma_alloc_coherent(struct device *dev, size_t size,
                dma_addr_t *dma_handle, gfp_t flag)
```

This function handles both the allocation and the mapping of the buffer, and returns a kernel virtual address for that buffer, which is `size` bytes wide and accessible by the CPU. `dev` is your device structure. The third argument is an output parameter that points to the associated bus address. Memory allocated for the mapping is guaranteed to be physically contiguous, and `flag` determines how memory should be allocated, which is usually `GFP_KERNEL`, or `GFP_ATOMIC` (if we are in an atomic context).

Do note that this mapping is said to be:

- **Consistent (coherent)**, since it allocates uncached unbuffered memory for a device for performing DMA
- **Synchronous**, because a write by either the device or the CPU can be immediately read by either without worrying about cache coherency

DMA – Direct Memory Access

In order to free a mapping, one can use the following function:

```
void dma_free_coherent(struct device *dev, size_t size,
                void *cpu_addr, dma_addr_t dma_handle);
```

Here `cpu_addr` corresponds to the kernel virtual address returned by `dma_alloc_coherent()`. This mapping is expensive, and the minimum it can allocate is a page. In fact, it only allocates the number of pages that is the power of 2. The order of pages is obtained with `int order = get_order(size)`. One should use this mapping for buffers that last the life of the device.

Streaming DMA mapping

Streaming mapping has more constraints, and is different from coherent mapping for the following reasons:

- Mappings need to work with a buffer that has already been allocated.
- Mappings may accept several non-contiguous and scattered buffers.
- A mapped buffer belongs to the device and not to the CPU anymore. Before the CPU can use the buffer, it should be unmapped first (after `dma_unmap_single()` or `dma_unmap_sg()`). This is for caching purposes.
- For write transactions (CPU to device), the driver should place data in the buffer before the mapping.
- The direction the data should move into has to be specified, and the data should only be used based on this direction.

One may wonder why one should not access the buffer until it is unmapped. The reason is simple: CPU mapping is cacheable. The `dma_map_*()` family functions, which are used for streaming mapping, will first clean/invalidate the caches related to the buffer and rely on the CPU not to access it until the corresponding `dma_unmap_*()`. That will then invalidate (if necessary) the caches again, in case of any speculative fetches in the meantime, before the CPU may read any data written to memory by the device. Now the CPU can access the buffer.

There are actually two forms of streaming mapping:

- Single buffer mapping, which allow only one-page mapping
- Scatter/gather mapping, which allows passing several buffers (scattered over memory)

For either mapping, direction should be specified, by a symbol of type `enum dma_data_direction`, defined in `include/linux/dma-direction.h`:

```
enum dma_data_direction {
    DMA_BIDIRECTIONAL = 0,
    DMA_TO_DEVICE = 1,
    DMA_FROM_DEVICE = 2,
    DMA_NONE = 3,
};
```

Single buffer mapping

This is for occasional mapping. One can set up a single buffer with this:

```
dma_addr_t dma_map_single(struct device *dev, void *ptr,
        size_t size, enum dma_data_direction direction);
```

The direction should be `DMA_TO_DEVICE`, `DMA_FROM_DEVICE`, or `DMA_BIDIRECTIONAL`, as described in the preceding code. `ptr` is the kernel virtual address of the buffer, and `dma_addr_t` is the returned bus address for the device. Make sure to use the direction that really fits your need, not just always `DMA_BIDIRECTIONAL`.

One should free the mapping with this:

```
void dma_unmap_single(struct device *dev, dma_addr_t dma_addr,
        size_t size, enum dma_data_direction direction);
```

Scatter/gather mapping

Scatter/gather mappings are a special type of streaming DMA mapping where one can transfer several buffer regions in a single shot, instead of mapping each buffer individually and transferring them one by one. Suppose you have several buffers that might not be physically contiguous, all of which need to be transferred at the same time to or from the device. This situation may occur due to:

- A readv or writev system call
- A disk I/O request
- Or simply a list of pages in a mapped kernel I/O buffer

The kernel represents the scatterlist as a coherent structure, `struct scatterlist`:

```
struct scatterlist {
    unsigned long page_link;
    unsigned int offset;
    unsigned int length;
    dma_addr_t dma_address;
    unsigned int dma_length;
};
```

In order to set up a scatterlist mapping, one should:

- Allocate your scattered buffers.
- Create an array of the scatter list and fill it with allocated memory using `sg_set_buf()`. Note that scatterlist entries must be of page size (except ends).
- Call `dma_map_sg()` on the scatterlist.
- Once done with DMA, call `dma_unmap_sg()` to unmap the scatterlist.

While one can send contents of several buffers over DMA one at a time by individually mapping each of them, scatter/gather can send them all at once by sending the pointer to the scatterlist to the device, along with a length, which is the number of entries in the list:

```
u32 *wbuf, *wbuf2, *wbuf3;
wbuf = kzalloc(SDMA_BUF_SIZE, GFP_DMA);
wbuf2 = kzalloc(SDMA_BUF_SIZE, GFP_DMA);
wbuf3 = kzalloc(SDMA_BUF_SIZE/2, GFP_DMA);

struct scatterlist sg[3];
sg_init_table(sg, 3);
sg_set_buf(&sg[0], wbuf, SDMA_BUF_SIZE);
sg_set_buf(&sg[1], wbuf2, SDMA_BUF_SIZE);
sg_set_buf(&sg[2], wbuf3, SDMA_BUF_SIZE/2);
ret = dma_map_sg(NULL, sg, 3, DMA_MEM_TO_MEM);
```

DMA – Direct Memory Access

The same rules described in the single-buffer mapping section apply to scatter/gather.

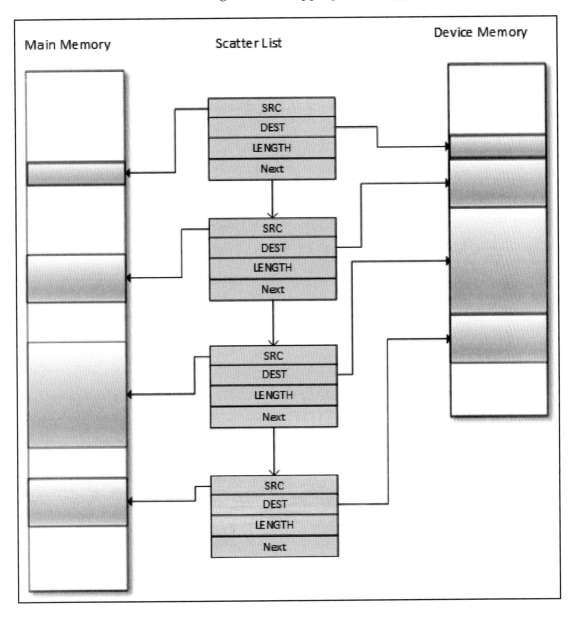

DMA scatter/gather

`dma_map_sg()` and `dma_unmap_sg()` take care of cache coherency. But if one needs to use the same mapping to access (read/write) the data between the DMA transfer, the buffers must be synced between each transfer in an appropriate manner, by either `dma_sync_sg_for_cpu()` if the CPU needs to access the buffers, or `dma_sync_sg_for_device()` if it is the device. Similar functions for single region mapping are `dma_sync_single_for_cpu()` and `dma_sync_single_for_device()`:

```
void dma_sync_sg_for_cpu(struct device *dev,
                    struct scatterlist *sg,
                    int nents,
                    enum dma_data_direction direction);
void dma_sync_sg_for_device(struct device *dev,
                    struct scatterlist *sg, int nents,
                    enum dma_data_direction direction);

void dma_sync_single_for_cpu(struct device *dev, dma_addr_t addr,
                    size_t size,
                    enum dma_data_direction dir)

void dma_sync_single_for_device(struct device *dev,
                    dma_addr_t addr, size_t size,
                    enum dma_data_direction dir)
```

There is no need to call the preceding functions again after the buffer(s) has been unmapped. You can just read the content.

Concept of completion

This section will briefly describe completion and the necessary part of its API that the DMA transfer uses. For a complete description, please feel free to have a look at the kernel documentation at *Documentation/scheduler/completion.txt*. A common pattern in kernel programming involves initiating some activity outside of the current thread, then waiting for that activity to complete.

Completion is a good alternative to `sleep()` when waiting for a buffer to be used. It is suitable for sensing data, which is exactly what the DMA callback does.

Working with completion requires this header:

```
<linux/completion.h>
```

Like other kernel facility data structures, one can create instances of the `struct completion` structure either statically or dynamically:

- Static declaration and initialization looks like this:

    ```
    DECLARE_COMPLETION(my_comp);
    ```

- Dynamic allocation looks like this:

    ```
    struct completion my_comp;
    init_completion(&my_comp);
    ```

When the driver begins some work whose completion must be waited for (a DMA transaction in our case), it just has to pass the completion event to the `wait_for_completion()` function:

```
void wait_for_completion(struct completion *comp);
```

When some other part of the code has decided that the completion has happened (transaction completes), it can wake up anybody (actually the code that needs to access DMA buffer) who is waiting with one of:

```
void complete(struct completion *comp);
void complete_all(struct completion *comp);
```

As one can guess, `complete()` will wake up only one waiting process, while `complete_all()` will wake up every one waiting for that event. Completions are implemented in such a way that they will work properly even if `complete()` is called before `wait_for_completion()`.

Along with code samples used in the next sections, one will have a better understanding of how this works.

DMA engine API

The DMA engine is a generic kernel framework for developing a DMA controller driver. The main goal of DMA is offloading the CPU when it comes to copy memory. One delegates a transaction (I/O data transfers) to the DMA engine by use of channels. A DMA engine, through its driver/API, exposes a set of channels, which can be used by other devices (slaves).

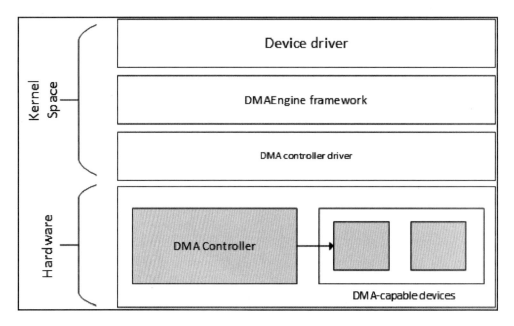

DMA Engine layout

Here we will simply walk through that (slave) API, which is applicable for slave DMA usage only. The mandatory header here is as follows:

```
#include <linux/dmaengine.h>
```

The slave DMA usage is straightforward, and consists of the following steps:

1. Allocate a DMA slave channel.
2. Set slave and controller specific parameters.
3. Get a descriptor for the transaction.
4. Submit the transaction.
5. Issue pending requests and wait for callback notification.

One can see a DMA channel as a highway for I/O data transfer

Allocate a DMA slave channel

One requests a channel using `dma_request_channel()`. Its prototype is as follows:

```
struct dma_chan *dma_request_channel(const dma_cap_mask_t *mask,
                    dma_filter_fn fn, void *fn_param);
```

`mask` is a bitmap mask that represents the capabilities the channel must satisfy. One uses it essentially to specify the transfer types the driver needs to perform:

```
enum dma_transaction_type {
    DMA_MEMCPY,      /* Memory to memory copy */
    DMA_XOR,         /* Memory to memory XOR*/
    DMA_PQ,          /* Memory to memory P+Q computation */
    DMA_XOR_VAL,     /* Memory buffer parity check using XOR */
    DMA_PQ_VAL,      /* Memory buffer parity check using P+Q */
    DMA_INTERRUPT,   /* The device is able to genrrate dummy transfer that
will generate interrupts */
    DMA_SG,          /* Memory to memory scatter gather */
    DMA_PRIVATE,     /* channels are not to be used for global memcpy.
Usually used with DMA_SLAVE */
    DMA_SLAVE,       /* Memory to device transfers */
    DMA_CYCLIC,      /* Device is ableto handle cyclic tranfers */
    DMA_INTERLEAVE,  /* Memoty to memory interleaved transfer */
}
```

The `dma_cap_zero()` and `dma_cap_set()` functions are used to clear the mask and set the capability we need. For example:

```
dma_cap_mask my_dma_cap_mask;
struct dma_chan *chan;
dma_cap_zero(my_dma_cap_mask);
dma_cap_set(DMA_MEMCPY, my_dma_cap_mask); /* Memory to memory copy */
chan = dma_request_channel(my_dma_cap_mask, NULL, NULL);
```

In the preceding excerpt, `dma_filter_fn` is defined as:

```
typedef bool (*dma_filter_fn)(struct dma_chan *chan,
              void *filter_param);
```

If `filter_fn` parameter (which is optional) is `NULL`, `dma_request_channel()` will simply return the first channel that satisfies the capability mask. Otherwise, when the mask parameter is insufficient for specifying the necessary channel, one can use the `filter_fn` routine as a filter for the available channels in the system. The kernel calls the `filter_fn` routine once for each free channel in the system. Upon seeing a suitable channel, `filter_fn` should return `DMA_ACK`, which will tag the given channel to be the return value from `dma_request_channel()`.

A channel allocated through this interface is exclusive to the caller, until `dma_release_channel()` is called:

```
void dma_release_channel(struct dma_chan *chan)
```

Set slave and controller specific parameters

This step introduces a new data structure, `struct dma_slave_config`, which represents the runtime configuration for the DMA slave channel. This allows clients to specify settings, such as the DMA direction, DMA addresses, bus width, DMA burst lengths, and so on, for the peripheral.

```
int dmaengine_slave_config(struct dma_chan *chan,
struct dma_slave_config *config)
```

The `struct dma_slave_config` structure looks like this:

```
/*
 * Please refer to the complete description in
 * include/linux/dmaengine.h
 */
struct dma_slave_config {
    enum dma_transfer_direction direction;
    phys_addr_t src_addr;
    phys_addr_t dst_addr;
    enum dma_slave_buswidth src_addr_width;
    enum dma_slave_buswidth dst_addr_width;
    u32 src_maxburst;
    u32 dst_maxburst;
    [...]
};
```

The following is the meaning of each element in the structure:

- `direction`: This indicates whether the data should go in or out on this slave channel, right now. The possible values are:

    ```
    /* dma transfer mode and direction indicator */
    enum dma_transfer_direction {
        DMA_MEM_TO_MEM, /* Async/Memcpy mode */
        DMA_MEM_TO_DEV, /* From Memory to Device */
        DMA_DEV_TO_MEM, /* From Device to Memory */
        DMA_DEV_TO_DEV, /* From Device to Device */
        [...]
    };
    ```

- `src_addr`: This is the physical address (actually the bus address) of the buffer where the DMA slave data should be read (RX). This element is ignored if the source is memory. `dst_addr` is the physical address (actually the bus address) of the buffer where the DMA slave data should be written (TX), which is ignored if the source is memory. `src_addr_width` is the width in bytes of the source (RX) register where the DMA data should be read. If the source is memory, this may be ignored depending on the architecture. The legal values are 1, 2, 4, or 8. Therefore, `dst_addr_width` is the same as `src_addr_width` but for the destination target (TX).
- Any bus width must be one of the following enumerations:

    ```
    enum dma_slave_buswidth {
            DMA_SLAVE_BUSWIDTH_UNDEFINED = 0,
            DMA_SLAVE_BUSWIDTH_1_BYTE = 1,
            DMA_SLAVE_BUSWIDTH_2_BYTES = 2,
            DMA_SLAVE_BUSWIDTH_3_BYTES = 3,
            DMA_SLAVE_BUSWIDTH_4_BYTES = 4,
            DMA_SLAVE_BUSWIDTH_8_BYTES = 8,
            DMA_SLAVE_BUSWIDTH_16_BYTES = 16,
            DMA_SLAVE_BUSWIDTH_32_BYTES = 32,
            DMA_SLAVE_BUSWIDTH_64_BYTES = 64,
    };
    ```

- `src_maxburs`: This is the maximum number of words (here, consider words as units of the `src_addr_width` member, not in bytes) that can be sent in one burst to the device. Typically, something like half the FIFO depth on I/O peripherals so you do not overflow it. This may or may not be applicable on memory sources. `dst_maxburst` is the same as `src_maxburst` but for the destination target.

For example:

```
struct dma_chan *my_dma_chan;
dma_addr_t dma_src, dma_dst;
struct dma_slave_config my_dma_cfg = {0};

/* No filter callback, neither filter param */
my_dma_chan = dma_request_channel(my_dma_cap_mask, 0, NULL);

/* scr_addr and dst_addr are ignored in this structure for mem to mem copy
*/
my_dma_cfg.direction = DMA_MEM_TO_MEM;
my_dma_cfg.dst_addr_width = DMA_SLAVE_BUSWIDTH_32_BYTES;

dmaengine_slave_config(my_dma_chan, &my_dma_cfg);

char *rx_data, *tx_data;
/* No error check */
rx_data = kzalloc(BUFFER_SIZE, GFP_DMA);
tx_data = kzalloc(BUFFER_SIZE, GFP_DMA);

feed_data(tx_data);

/* get dma addresses */
dma_src_addr = dma_map_single(NULL, tx_data,
BUFFER_SIZE, DMA_MEM_TO_MEM);
dma_dst_addr = dma_map_single(NULL, rx_data,
BUFFER_SIZE, DMA_MEM_TO_MEM);
```

In the preceding excerpt, one calls `dma_request_channel()` function in order to take the owner chip of the DMA channel, on which one calls `dmaengine_slave_config()` to apply its configuration. `dma_map_single()` is called in order to map rx and tx buffers, so that these can be used for purpose of DMA.

Get a descriptor for transaction

If you remember the first step of this section, when one requests a DMA channel, the return value is an instance of the `struct dma_chan` structure. If one looks at its definition in `include/linux/dmaengine.h`, one will notice that it contains a `struct dma_device *device` field, which represents the DMA device (the controller actually) that supplied the channel. The kernel driver of this controller is responsible (it is a rule imposed by the kernel API for DMA controller drivers) for exposing a set of functions to prepare DMA transactions, where each of them correspond to a DMA transaction type (enumerated in step 1). Depending on the transaction type, one has no choice but to choose the dedicated function. Some of these functions are:

- `device_prep_dma_memcpy()`: Prepares a memcpy operation
- `device_prep_dma_sg()`: Prepare a scatter/gather memcpy operation
- `device_prep_dma_xor()`: For a xor operation
- `device_prep_dma_xor_val()`: Prepares a xor validation operation
- `device_prep_dma_pq()`: Prepares a pq operation
- `device_prep_dma_pq_val()`: Prepares a pqzero_sum operation
- `device_prep_dma_memset()`: Prepares a memset operation
- `device_prep_dma_memset_sg()`: For a memset operation over a scatterlist
- `device_prep_slave_sg()`: Prepares a slave DMA operation
- `device_prep_interleaved_dma()`: Transfers an expression in a generic way

Let us have a look at `drivers/dma/imx-sdma.c`, which is the i.MX6 DMA controller (SDMA) driver. Each of these functions returns a pointer to a `struct dma_async_tx_descriptor` structure, which corresponds to the transaction descriptor. With memory-to-memory copy, one will use `device_prep_dma_memcpy`:

```
struct dma_device *dma_dev = my_dma_chan->device;
struct dma_async_tx_descriptor *tx = NULL;

tx = dma_dev->device_prep_dma_memcpy(my_dma_chan, dma_dst_addr,
dma_src_addr, BUFFER_SIZE, 0);

if (!tx) {
    printk(KERN_ERR "%s: Failed to prepare DMA transfer\n",
            __FUNCTION__);
    /* dma_unmap_* the buffer */
}
```

In fact, we should have used `dmaengine_prep_*` DMA engine API. Just note that these functions internally do what we just performed earlier. For example, for memory-to-memory, one could have used the `device_prep_dma_memcpy ()` function:

```
struct dma_async_tx_descriptor *(*device_prep_dma_memcpy)(
        struct dma_chan *chan, dma_addr_t dst, dma_addr_t src,
        size_t len, unsigned long flags)
```

Our sample becomes:

```
struct dma_async_tx_descriptor *tx = NULL;
tx = dma_dev->device_prep_dma_memcpy(my_dma_chan, dma_dst_addr,
dma_src_addr, BUFFER_SIZE, 0);
if (!tx) {
    printk(KERN_ERR "%s: Failed to prepare DMA transfer\n",
            __FUNCTION__);
    /* dma_unmap_* the buffer */
}
```

Please have a look at `include/linux/dmaengine.h`, in the definition of a `struct dma_device` structure, to see how all of these hooks are implemented.

Submit the transaction

To put the transaction in the driver pending queue, one uses `dmaengine_submit()`. Once the descriptor has been prepared and the callback information added, one should place it on the DMA engine drivers pending the queue:

```
dma_cookie_t dmaengine_submit(struct dma_async_tx_descriptor *desc)
```

This function returns a cookie that one can use to check the progress of DMA activity through other DMA engines. `dmaengine_submit()` will not start the DMA operation, it merely adds it to the pending queue. How to start the transaction is discussed in the next step:

```
struct completion transfer_ok;
init_completion(&transfer_ok);
tx->callback = my_dma_callback;

/* Submit our dma transfer */
dma_cookie_t cookie = dmaengine_submit(tx);

if (dma_submit_error(cookie)) {
    printk(KERN_ERR "%s: Failed to start DMA transfer\n", __FUNCTION__);
    /* Handle that */
```

```
[...]
}
```

Issue pending DMA requests and wait for callback notification

Starting the transaction is the last step of the DMA transfer setup. One activates transactions in the pending queue of a channel by calling `dma_async_issue_pending()` on that channel. If the channel is idle then the first transaction in the queue is started and subsequent ones are queued up. On completion of a DMA operation, the next one in the queue is started and a tasklet triggered. This tasklet is in charge of calling the client driver completion callback routine for notification, if set:

```
void dma_async_issue_pending(struct dma_chan *chan);
```

An example would look like this:

```
dma_async_issue_pending(my_dma_chan);
wait_for_completion(&transfer_ok);

dma_unmap_single(my_dma_chan->device->dev, dma_src_addr,
BUFFER_SIZE, DMA_MEM_TO_MEM);
dma_unmap_single(my_dma_chan->device->dev, dma_src_addr,
           BUFFER_SIZE, DMA_MEM_TO_MEM);

/* Process buffer through rx_data and tx_data virtualaddresses. */
```

The `wait_for_completion()` function will block until our DMA callback gets called, which will update (complete) our completion variable in order to resume the previous blocked code. It is a suitable alternative to `while (!done) msleep(SOME_TIME);`.

```
static void my_dma_callback()
{
    complete(transfer_ok);
    return;
}
```

> The DMA engine API function that actually issues pending transactions is `dmaengine_issue_pending(struct dma_chan *chan)`, which is a wrap around `dma_async_issue_pending()`.

Putting it all together – NXP SDMA (i.MX6)

The SDMA engine is a programmable controller in the i.MX6 and each peripheral has its own copy function in this controller. One uses this `enum` to determine their addresses:

```
enum sdma_peripheral_type {
    IMX_DMATYPE_SSI,      /* MCU domain SSI */
    IMX_DMATYPE_SSI_SP,   /* Shared SSI */
    IMX_DMATYPE_MMC,      /* MMC */
    IMX_DMATYPE_SDHC,     /* SDHC */
    IMX_DMATYPE_UART,     /* MCU domain UART */
    IMX_DMATYPE_UART_SP,  /* Shared UART */
    IMX_DMATYPE_FIRI,     /* FIRI */
    IMX_DMATYPE_CSPI,     /* MCU domain CSPI */
    IMX_DMATYPE_CSPI_SP,  /* Shared CSPI */
    IMX_DMATYPE_SIM,      /* SIM */
    IMX_DMATYPE_ATA,      /* ATA */
    IMX_DMATYPE_CCM,      /* CCM */
    IMX_DMATYPE_EXT,      /* External peripheral */
    IMX_DMATYPE_MSHC,     /* Memory Stick Host Controller */
    IMX_DMATYPE_MSHC_SP,  /* Shared Memory Stick Host Controller */
    IMX_DMATYPE_DSP,      /* DSP */
    IMX_DMATYPE_MEMORY,   /* Memory */
    IMX_DMATYPE_FIFO_MEMORY,/* FIFO type Memory */
    IMX_DMATYPE_SPDIF,    /* SPDIF */
    IMX_DMATYPE_IPU_MEMORY, /* IPU Memory */
    IMX_DMATYPE_ASRC,     /* ASRC */
    IMX_DMATYPE_ESAI,     /* ESAI */
    IMX_DMATYPE_SSI_DUAL, /* SSI Dual FIFO */
    IMX_DMATYPE_ASRC_SP,  /* Shared ASRC */
    IMX_DMATYPE_SAI,      /* SAI */
};
```

Despite the generic DMA engine API, any constructor may provide its own custom data structure. This is the case for the `imx_dma_data` structure, which is a private data (used to describe the DMA device type one needs to use) that is to be passed to the `.private` field of the `struct dma_chan` in the filter callback:

```
struct imx_dma_data {
    int dma_request; /* DMA request line */
    int dma_request2; /* secondary DMA request line */
    enum sdma_peripheral_type peripheral_type;
    int priority;
};

enum imx_dma_prio {
    DMA_PRIO_HIGH = 0,
```

```
        DMA_PRIO_MEDIUM = 1,
        DMA_PRIO_LOW = 2
};
```

These structures and enum are all specific to i.MX and are defined in include/linux/platform_data/dma-imx.h. Now, let us write our kernel DMA module. It allocates two buffers (source and destination). Fill the source with predefined data, and perform a transaction in order to copy src into dst. One can improve this module by using data coming from user space (copy_from_user()). This driver is inspired from the one provided in the imx-test package:

```
#include <linux/module.h>
#include <linux/slab.h>     /* for kmalloc */
#include <linux/init.h>
#include <linux/dma-mapping.h>
#include <linux/fs.h>
#include <linux/version.h>
#if (LINUX_VERSION_CODE >= KERNEL_VERSION(3,0,35))
#include <linux/platform_data/dma-imx.h>
#else
#include <mach/dma.h>
#endif

#include <linux/dmaengine.h>
#include <linux/device.h>

#include <linux/io.h>
#include <linux/delay.h>

static int gMajor; /* major number of device */
static struct class *dma_tm_class;
u32 *wbuf;  /* source buffer */
u32 *rbuf;  /* destinationn buffer */

struct dma_chan *dma_m2m_chan;   /* our dma channel */
struct completion dma_m2m_ok;    /* completion variable used in the DMA
callback */
#define SDMA_BUF_SIZE  1024
```

Let us define the filter function. When one requests a DMA channel, the controller driver may perform a lookup in a list of channels (which it has). For fine-grained lookup, one can provide a callback method that will be called on each channel found. It is then up to the callback to choose the suitable channel to use:

```
static bool dma_m2m_filter(struct dma_chan *chan, void *param)
{
    if (!imx_dma_is_general_purpose(chan))
```

```
        return false;
    chan->private = param;
    return true;
}
```

`imx_dma_is_general_purpose` is a special function that checks the controller driver's name. The `open` function will allocate the buffer and request the DMA channel, given our filter function as callback:

```
int sdma_open(struct inode * inode, struct file * filp)
{
    dma_cap_mask_t dma_m2m_mask;
    struct imx_dma_data m2m_dma_data = {0};

    init_completion(&dma_m2m_ok);

    dma_cap_zero(dma_m2m_mask);
    dma_cap_set(DMA_MEMCPY, dma_m2m_mask); /* Set channel capacities */
    m2m_dma_data.peripheral_type = IMX_DMATYPE_MEMORY; /* choose the dma
device type. This is proper to i.MX */
    m2m_dma_data.priority = DMA_PRIO_HIGH;  /* we need high priority */

    dma_m2m_chan = dma_request_channel(dma_m2m_mask, dma_m2m_filter,
&m2m_dma_data);
    if (!dma_m2m_chan) {
        printk("Error opening the SDMA memory to memory channel\n");
        return -EINVAL;
    }

    wbuf = kzalloc(SDMA_BUF_SIZE, GFP_DMA);
    if(!wbuf) {
        printk("error wbuf !!!!!!!!!!!!\n");
        return -1;
    }

    rbuf = kzalloc(SDMA_BUF_SIZE, GFP_DMA);
    if(!rbuf) {
        printk("error rbuf !!!!!!!!!!!!\n");
        return -1;
    }

    return 0;
}
```

The `release` function simply does the reverse of the `open` function; it frees the buffer and releases the DMA channel:

```
int sdma_release(struct inode * inode, struct file * filp)
{
    dma_release_channel(dma_m2m_chan);
    dma_m2m_chan = NULL;
    kfree(wbuf);
    kfree(rbuf);
    return 0;
}
```

In the `read` function, we just compare the source and destination buffer and inform the user about the result.

```
ssize_t sdma_read (struct file *filp, char __user * buf,
size_t count, loff_t * offset)
{
    int i;
    for (i=0; i<SDMA_BUF_SIZE/4; i++) {
        if (*(rbuf+i) != *(wbuf+i)) {
            printk("Single DMA buffer copy falled!,r=%x,w=%x,%d\n",
*(rbuf+i), *(wbuf+i), i);
            return 0;
        }
    }
    printk("buffer copy passed!\n");
    return 0;
}
```

We use completion in order to get notified (woken up) when the transaction has terminated. This callback is called after our transaction has finished and sets our completion variable to the complete state:

```
static void dma_m2m_callback(void *data)
{
    printk("in %s\n",__func__);
    complete(&dma_m2m_ok);
    return ;
}
```

DMA – Direct Memory Access

In the `write` function, we fill our source buffer with the data, perform DMA mapping in order to get physical addresses that correspond to our source and destination buffer, and call `device_prep_dma_memcpy` to get a transaction descriptor. That transaction descriptor is then submitted to the DMA engine with `dmaengine_submit`, which does not perform our transaction yet. It is only after we have called `dma_async_issue_pending` on our DMA channel, that our pending transaction will be processed:

```
ssize_t sdma_write(struct file * filp, const char __user * buf,
                    size_t count, loff_t * offset)
{
    u32 i;
    struct dma_slave_config dma_m2m_config = {0};
    struct dma_async_tx_descriptor *dma_m2m_desc; /* transaction descriptor */

    dma_addr_t dma_src, dma_dst;

    /* No copy_from_user, we just fill the source buffer with predefined data */
    for (i=0; i<SDMA_BUF_SIZE/4; i++) {
        *(wbuf + i) = 0x56565656;
    }

    dma_m2m_config.direction = DMA_MEM_TO_MEM;
    dma_m2m_config.dst_addr_width = DMA_SLAVE_BUSWIDTH_4_BYTES;
    dmaengine_slave_config(dma_m2m_chan, &dma_m2m_config);

    dma_src = dma_map_single(NULL, wbuf, SDMA_BUF_SIZE, DMA_TO_DEVICE);
    dma_dst = dma_map_single(NULL, rbuf, SDMA_BUF_SIZE, DMA_FROM_DEVICE);
    dma_m2m_desc =
dma_m2m_chan->device->device_prep_dma_memcpy(dma_m2m_chan, dma_dst,
dma_src, SDMA_BUF_SIZE,0);
    if (!dma_m2m_desc)
        printk("prep error!!\n");
    dma_m2m_desc->callback = dma_m2m_callback;
    dmaengine_submit(dma_m2m_desc);
    dma_async_issue_pending(dma_m2m_chan);
    wait_for_completion(&dma_m2m_ok);
    dma_unmap_single(NULL, dma_src, SDMA_BUF_SIZE, DMA_TO_DEVICE);
    dma_unmap_single(NULL, dma_dst, SDMA_BUF_SIZE, DMA_FROM_DEVICE);

    return 0;
}

struct file_operations dma_fops = {
    open: sdma_open,
    release: sdma_release,
    read: sdma_read,
```

```
    write: sdma_write,
};
```

The full code is available in the repository of the book: `chapter-12/imx-sdma/imx-sdma-single.c`. There is also a module with which to perform the same task, but using scatter/gather mapping: `chapter-12/imx-sdma/imx-sdma-scatter-gather.c`.

DMA DT binding

DT binding for the DMA channel depends on the DMA controller node, which is SoC dependent, and some parameters (such as DMA cells) may vary from one SoC to another. This example only focuses on the i.MX SDMA controller, which one can find in the kernel source, at *Documentation/devicetree/bindings/dma/fsl-imx-sdma.txt*.

Consumer binding

According to the SDMA event-mapping table, the following code shows the DMA request signals for peripherals in i.MX 6Dual/ 6Quad:

```
uart1: serial@02020000 {
    compatible = "fsl,imx6sx-uart", "fsl,imx21-uart";
    reg = <0x02020000 0x4000>;
    interrupts = <GIC_SPI 26 IRQ_TYPE_LEVEL_HIGH>;
    clocks = <&clks IMX6SX_CLK_UART_IPG>,
             <&clks IMX6SX_CLK_UART_SERIAL>;
    clock-names = "ipg", "per";
    dmas = <&sdma 25 4 0>, <&sdma 26 4 0>;
    dma-names = "rx", "tx";
    status = "disabled";
};
```

The second cells (`25` and `26`) in the DMA property correspond to the DMA request/event ID. Those values come from the SoC manuals (i.MX53 in our case). Please have a look at `https://community.nxp.com/servlet/JiveServlet/download/614186-1-373516/iMX6_Firmware_Guide.pdf` and the Linux reference manual at `https://community.nxp.com/servlet/JiveServlet/download/614186-1-373515/i.MX_Linux_Reference_Manual.pdf`.

The third cell indicates the priority to use. The driver code to request a specified parameter is defined next. One can find the complete code in `drivers/tty/serial/imx.c` in the kernel source tree:

```
static int imx_uart_dma_init(struct imx_port *sport)
{
    struct dma_slave_config slave_config = {};
    struct device *dev = sport->port.dev;
    int ret;

    /* Prepare for RX : */
    sport->dma_chan_rx = dma_request_slave_channel(dev, "rx");
    if (!sport->dma_chan_rx) {
        [...] /* cannot get the DMA channel. handle error */
    }

    slave_config.direction = DMA_DEV_TO_MEM;
    slave_config.src_addr = sport->port.mapbase + URXD0;
    slave_config.src_addr_width = DMA_SLAVE_BUSWIDTH_1_BYTE;
    /* one byte less than the watermark level to enable the aging timer */
    slave_config.src_maxburst = RXTL_DMA - 1;
    ret = dmaengine_slave_config(sport->dma_chan_rx, &slave_config);
    if (ret) {
        [...] /* handle error */
    }

    sport->rx_buf = kzalloc(PAGE_SIZE, GFP_KERNEL);
    if (!sport->rx_buf) {
        [...] /* handle error */
    }

    /* Prepare for TX : */
    sport->dma_chan_tx = dma_request_slave_channel(dev, "tx");
    if (!sport->dma_chan_tx) {
        [...] /* cannot get the DMA channel. handle error */
    }

    slave_config.direction = DMA_MEM_TO_DEV;
    slave_config.dst_addr = sport->port.mapbase + URTX0;
    slave_config.dst_addr_width = DMA_SLAVE_BUSWIDTH_1_BYTE;
    slave_config.dst_maxburst = TXTL_DMA;
    ret = dmaengine_slave_config(sport->dma_chan_tx, &slave_config);
    if (ret) {
        [...] /* handle error */
    }
    [...]
}
```

The magic call here is the `dma_request_slave_channel()` function, which will parse the device node (in the DT) using `of_dma_request_slave_channel()` to gather channel settings, according to the DMA name (refer to the named resource in Chapter 6, *The Concept of Device Tree*).

Summary

DMA is a feature that one finds in many modern CPUs. This chapter gives you the necessary steps to get the most out of this device, using the kernel DMA mapping and DMA engine APIs. After this chapter, I have no doubt you will be able to set up at least a memory-to-memory DMA transfer. One can find further information at *Documentation/dmaengine/*, in the kernel source tree. Therefore, the next chapter deals with an entirely different subject—the Linux device model.

13
Linux Device Model

Until version 2.5, the kernel had no way to describe and manage objects, and the code reusability was not as enhanced as it is now. In other words, there was no device topology nor organization. There was no information on subsystem relationships nor on how the system is put together. Then came the **Linux Device Model** (**LDM**), which introduced:

- The concept of class, to group devices of the same type or devices that expose the same functionalities (for example, mice and keyboards are both input devices).
- Communication with the user space through a virtual filesystem called `sysfs`, in order to let user space manage and enumerate devices and the properties they expose.
- Management of object life cycle, using reference counting (heavily used in managed resources).
- Power management in order to handle the order in which devices should shut down.
- The reusability of the code. Classes and frameworks expose interfaces, behaving like contract that any driver that registers with them must respect.
- LDM brought an **Object Oriented** (**OO**)-like programming style in the kernel.

In this chapter, we will take advantage of LDM and export some properties to the user space through the `sysfs` filesystem.

In this chapter, we will cover the following topics:

- Introducing LDM data structures (Driver, Device, Bus)
- Gathering kernel objects by type
- Dealing with the kernel `sysfs` interface

LDM data structures

The goal is to build a complete DT that will map each physical device present on the system, and introduce their hierarchy. One common and generic structure has been created to represent any object that could be a part of the device model. The upper level of LDM relies on the bus represented in the kernel as an instance of struct bus_type; device driver, represented by a struct device_driver structure, and device, which is the last element represented as an instance of the struct device structure. In this section, we will design a bus driver packt bus, in order to get deep into LDM data structures and mechanisms.

The bus

A bus is a channel link between devices and processors. The hardware entity that manages the bus and exports its protocol to devices is called the bus controller. For example, the USB controller provides USB support. The I2C controller provides I2C bus support. Therefore, the bus controller, being a device on its own, must be registered like any device. It will be the parent of devices that need to sit on the bus. In other words, every device sitting on the bus must have their parent field pointing to the bus device. A bus is represented in the kernel by the struct bus_type structure:

```
struct bus_type {
    const char *name;
    const char *dev_name;
    struct device *dev_root;
    struct device_attribute *dev_attrs; /* use dev_groups instead */
    const struct attribute_group **bus_groups;
    const struct attribute_group **dev_groups;
    const struct attribute_group **drv_groups;

    int (*match)(struct device *dev, struct device_driver *drv);
    int (*probe)(struct device *dev);
    int (*remove)(struct device *dev);
    void (*shutdown)(struct device *dev);

    int (*suspend)(struct device *dev, pm_message_t state);
    int (*resume)(struct device *dev);

    const struct dev_pm_ops *pm;

    struct subsys_private *p;
    struct lock_class_key lock_key;
};
```

The following are the meanings of elements in the structure:

- `match`: This is a callback, called whenever a new device or driver is added to the bus. The callback must be smart enough and should return a nonzero value when there is a match between a device and a driver, both given as parameters. The main purpose of a `match` callback is to allow a bus to determine if a particular device can be handled by a given driver or the other logic, if the given driver supports a given device. Most of the time, the verification is done by a simple string comparison (device and driver name, of table and DT compatible property). For enumerated devices (PCI, USB), the verification is done by comparing the device IDs supported by the driver with the device ID of the given device, without sacrificing bus-specific functionality.
- `probe`: This is a callback when a new device or driver is added to the bus, after the match has occurred. This function is responsible for allocating the specific bus device structure, and call the given driver's `probe` function, which is supposed to manage the device (allocated earlier).
- `remove`: This is called when a device is to removed from the bus.
- `suspend`: This is a method called when a device on the bus needs to be put into sleep mode.
- `resume`: This is called when a device on the bus has to be brought out of sleep mode.
- pm: This is a set of power management operations of the bus, which will call the specific device driver's `pm-ops`.
- `drv_groups`: This is a pointer to a list (array) of `struct attribute_group` elements, each of which has a pointer to a list (array) of `struct attribute` elements. It represents the default attributes of the device drivers on the bus. Attributes passed to this field will be given to every driver registered with the bus. Those attributes can be found in the driver's directory in `/sys/bus/<bus-name>/drivers/<driver-name>`.
- `dev_groups`: This represents the default attributes of the devices on the bus. Attributes passed (through the list/array of the `struct attribute_group` elements) to this field will be given to every device registered with the bus. Those attributes can be found in the device directory in `/sys/bus/<bus-name>/devices/<device-name>`.
- `bus_group`: This holds the set (group) of default attributes added automatically when the bus is registered with the core.

Linux Device Model

Apart from defining a `bus_type`, the bus controller driver must define a bus-specific driver structure that extends the generic `struct device_driver`, and a bus-specific device structure that extends the generic `struct device` structure, both part of the device model core. The bus drivers must also allocate a bus-specific device structure for each physical device discovered when probing, and is responsible for initializing the `bus` and `parent` fields of the device and registering the device with the LDM core. Those fields must point to the bus device and the `bus_type` structures defined in the bus driver. The LDM core uses that to build the device hierarchy and initialize the other fields.

In our example, the following are two helper macros to get the packt device and the packt driver, given a generic `struct device` and `struct driver`:

```
#define to_packt_driver(d) container_of(d, struct packt_driver, driver)
#define to_packt_device(d) container_of(d, struct packt_device, dev)
```

And then comes the structure used to identify a packt device:

```
struct packt_device_id {
    char name[PACKT_NAME_SIZE];
    kernel_ulong_t driver_data;   /* Data private to the driver */
};
```

The following are packt-specific devices and driver structures:

```
/*
 * Bus specific device structure
 * This is what a packt device structure looks like
 */
struct packt_device {
    struct module *owner;
    unsigned char name[30];
    unsigned long price;
    struct device dev;
};

/*
 * Bus specific driver structure
 * This is what a packt driver structure looks like
 * You should provide your device's probe and remove function.
 * may be release too
 */
struct packt_driver {
    int (*probe)(struct packt_device *packt);
    int (*remove)(struct packt_device *packt);
    void (*shutdown)(struct packt_device *packt);
};
```

Each bus internally manages two important lists; the list of devices added and sitting on it, and the list of driver registered with it. Whenever you add/register or remove/unregister a device/driver to/from the bus, the corresponding list is updated with the new entry. The bus driver must provide helper functions to register/unregister device drivers that can handle devices on that bus, as well as helper functions to register/unregister devices sitting on the bus. These helper functions always wrap the generic functions provided by the LDM core, which are driver_register(), device_register(), driver_unregister, and device_unregister().

```c
/*
 * Now let us write and export symbols that people writing
 * drivers for packt devices must use.
 */

int packt_register_driver(struct packt_driver *driver)
{
    driver->driver.bus = &packt_bus_type;
    return driver_register(&driver->driver);
}
EXPORT_SYMBOL(packt_register_driver);

void packt_unregister_driver(struct packt_driver *driver)
{
    driver_unregister(&driver->driver);
}
EXPORT_SYMBOL(packt_unregister_driver);

int packt_device_register(struct packt_device *packt)
{
    return device_register(&packt->dev);
}
EXPORT_SYMBOL(packt_device_register);

void packt_unregister_device(struct packt_device *packt)
{
    device_unregister(&packt->dev);
}
EXPORT_SYMBOL(packt_device_unregister);
```

Linux Device Model

The function used to allocate packt devices is as follows. One must use this to create an instance of any physical device sitting on the bus:

```
/*
 * This function allocate a bus specific device structure
 * One must call packt_device_register to register
 * the device with the bus
 */
struct packt_device * packt_device_alloc(const char *name, int id)
{
    struct packt_device *packt_dev;
    int status;

    packt_dev = kzalloc(sizeof *packt_dev, GFP_KERNEL);
    if (!packt_dev)
            return NULL;

     /* new devices on the bus are son of the bus device */
     strcpy(packt_dev->name, name);
     packt_dev->dev.id = id;
     dev_dbg(&packt_dev->dev,
        "device [%s] registered with packt bus\n", packt_dev->name);

     return packt_dev;

out_err:
    dev_err(&adap->dev, "Failed to register packt client %s\n", packt_dev->name);
    kfree(packt_dev);
    return NULL;
}
EXPORT_SYMBOL_GPL(packt_device_alloc);

int packt_device_register(struct packt_device *packt)
{
    packt->dev.parent = &packt_bus;
    packt->dev.bus = &packt_bus_type;
    return device_register(&packt->dev);
}
EXPORT_SYMBOL(packt_device_register);
```

Bus registration

The bus controller is a device itself, and in 99% of cases buses are platform devices (even buses that offer enumeration). For example, PCI controller is a platform device, so is its respective driver. One must use the `bus_register(struct *bus_type)` function in order to register a bus with the kernel. The packt bus structure looks like this:

```
/*
 * This is our bus structure
 */
struct bus_type packt_bus_type = {
    .name     = "packt",
    .match    = packt_device_match,
    .probe    = packt_device_probe,
    .remove   = packt_device_remove,
    .shutdown = packt_device_shutdown,
};
```

The bus controller is a device itself, it has to be registered with the kernel, and will be used as a parent of the device siting on the bus. This is done in the bus controller's `probe` or `init` function. In the case of the packt bus, the code would be as follows:

```
/*
 * Bus device, the master.
 *
 */
struct device packt_bus = {
    .release = packt_bus_release,
    .parent = NULL, /* Root device, no parent needed */
};

static int __init packt_init(void)
{
    int status;
    status = bus_register(&packt_bus_type);
    if (status < 0)
        goto err0;

    status = class_register(&packt_master_class);
    if (status < 0)
        goto err1;

    /*
     * After this call, the new bus device will appear
     * under /sys/devices in sysfs. Any devices added to this
     * bus will shows up under /sys/devices/packt-0/.
     */
```

```
        device_register(&packt_bus);

    return 0;

err1:
    bus_unregister(&packt_bus_type);
err0:
    return status;
}
```

When a device is registered by the bus controller driver, the parent member of the device must point to the bus controller device and its bus property must point to the bus type to build the physical DT. To register a packt device, one must call `packt_device_register`, given as an argument allocated with `packt_device_alloc`:

```
int packt_device_register(struct packt_device *packt)
{
    packt->dev.parent = &packt_bus;
    packt->dev.bus = &packt_bus_type;
    return device_register(&packt->dev);
}
EXPORT_SYMBOL(packt_device_register);
```

Device driver

A global device hierarchy allows each device in the system to be represented in a common way. This allows the core to easily walk the DT to create such things as properly ordered power management transitions:

```
struct device_driver {
    const char *name;
    struct bus_type *bus;
    struct module *owner;

    const struct of_device_id   *of_match_table;
    const struct acpi_device_id *acpi_match_table;

    int (*probe) (struct device *dev);
    int (*remove) (struct device *dev);
    void (*shutdown) (struct device *dev);
    int (*suspend) (struct device *dev, pm_message_t state);
    int (*resume) (struct device *dev);
    const struct attribute_group **groups;

    const struct dev_pm_ops *pm;
};
```

`struct device_driver` defines a simple set of operations for the core to perform these actions on each device:

- * `name` represents the driver's name. It can be used for matching, by comparing with the device name.
- * `bus` represents the bus the driver sits on. The bus driver must fill this field.
- `module` represents the module owning the driver. In 99% of cases, one should set this field to `THIS_MODULE`.
- `of_match_table` is a pointer to the array of `struct of_device_id`. The `struct of_device_id` structure is used to perform OF match through a special file called DT, passed to the kernel during the boot process:

```
struct of_device_id {
    char compatible[128];
    const void *data;
};
```

- `suspend` and `resume` callbacks provide power management functionality. The `remove` callback is called when the device is physically removed from the system, or when its reference count reaches 0. The `remove` callback is also called during system reboot.
- `probe` is the probe callback that runs when attempting to bind a driver to a device. The bus driver is in charge of calling the device driver's `probe` function.
- `group` is a pointer to a list (array) of `struct attribute_group`, used as a default attribute for the driver. Use this method instead of creating an attribute separately.

Device driver registration

`driver_register()` is the low-level function used to register a device driver with the bus. It adds the driver to the bus's list of drivers. When a device driver is registered with the bus, the core walks through the bus's list of devices and calls the bus's match callback for each device that does not have a driver associated with it in order to find out if there are any devices that the driver can handle.

When a match occurs, the device and the device driver are bound together. The process of associating a device with a device driver is called binding.

Now back to the registration of drivers with our packt bus, one has to use `packt_register_driver(struct packt_driver *driver)`, which is a wrapper around `driver_register()`. The `*driver` parameter must have been filled prior to registering the packt driver. The LDM core provides helper functions for iterating over the list of drivers registered with the bus:

```
int bus_for_each_drv(struct bus_type * bus,
            struct device_driver * start,
            void * data, int (*fn)(struct device_driver *,
            void *));
```

This helper iterates over the bus's list of drivers, and calls the `fn` callback for each driver in the list.

Device

The struct device is the generic data structure used to describe and characterize each device on the system, whether it is physical or not. It contains details about the physical attributes of the device, and provides proper linkage information to build suitable device trees and reference counting:

```
struct device {
    struct device *parent;
    struct kobject kobj;
    const struct device_type *type;
    struct bus_type        *bus;
    struct device_driver *driver;
    void     *platform_data;
    void *driver_data;
    struct device_node     *of_node;
    struct class *class;
    const struct attribute_group **groups;
    void (*release)(struct device *dev);
};
```

- * `parent` represents the device's parent, used to build device tree hierarchy. When registered with a bus, the bus driver is responsible for setting this field with the bus device.
- * `bus` represents the bus the device sits on. The bus driver must fill this field.
- * `type` identifies the device's type.

- `kobj` is the kobject in handle reference counting and device model support.
- `* of_node` is a pointer to the OF (DT) node associated with the device. It is up to the bus driver to set this field.
- `platform_data` is a pointer to the platform data specific to the device. Usually declared in a board-specific file during device provisioning.
- `driver_data` is a pointer to private data for the driver.
- `class` is a pointer to the class that the device belongs to.
- `* group` is a pointer to a list (array) of `struct attribute_group`, used as the default attribute for the device. Use this method instead of creating the attribute separately.
- `release` is a callback called when the device reference count reaches zero. The bus has the responsibility of setting this field. The packt bus driver shows you how to do this.

Device registration

`device_register` is the function provided by the LDM core to register a device with the bus. After this call, the bus list of drivers is iterated over to find the driver that supports this device and then this device is added to the bus's list of devices. `device_register()` internally calls `device_add()`:

```
int device_add(struct device *dev)
{
    [...]
    bus_probe_device(dev);
        if (parent)
                klist_add_tail(&dev->p->knode_parent,
                               &parent->p->klist_children);
    [...]
}
```

The helper function provided by the kernel to iterate over the bus's list of devices is `bus_for_each_dev`:

```
int bus_for_each_dev(struct bus_type * bus,
                     struct device * start, void * data,
                     int (*fn)(struct device *, void *));
```

Whenever a device is added, the core invokes the match method of the bus driver (`bus_type->match`). If the match function says there is a driver for this device, the core will invoke the `probe` function of the bus driver (`bus_type->probe`), given both device and driver as parameters. It is then up to the bus driver to invoke the `probe` method of the device's driver (`driver->probe`). For our packt bus driver, the function used to register a device is `packt_device_register(struct packt_device *packt)`, which internally calls `device_register`, and where the parameter is a packt device allocated with `packt_device_alloc`.

Deep inside LDM

The LDM under the wood relies on three important structures, which are kobject, kobj_type, and kset. Let us see how each of these structures are involved in the device model.

kobject structure

kobject is the core of the device model, running behind the scenes. It brings an OO-like programming style to the kernel, and is mainly used for reference counting and to expose devices hierarchies and relationships between them. kobjects introduce the concept of encapsulation of common object properties such as usage reference counts:

```
struct kobject {
    const char *name;
    struct list_head entry;
    struct kobject *parent;
    struct kset *kset;
    struct kobj_type *ktype;
    struct sysfs_dirent *sd;
    struct kref kref;
    /* Fields out of our interest have been removed */
};
```

- name points to the name of this kobject. One can change this using the `kobject_set_name(struct kobject *kobj, const char *name)` function.
- parent is a pointer to this kobject's parent. It is used to build a hierarchy to describe the relationship between objects.
- sd points to a `struct sysfs_dirent` structure that represents this kobject in sysfs inode inside this structure for sysfs.

- `kref` provides reference counting on the kobject.
- `ktype` describes the object, and `kset` tells us which set (group) of objects this object belongs to.

Each structure that embeds a kobject is embedded and receives the standardized functions that kobjects provide. The embedded kobject will enable the structure to become a part of an object hierarchy.

The `container_of` macro is used to get a pointer on the object to which the kobject belongs. Every kernel device directly or indirectly embeds a kobject property. Prior to be added to the system, the kobject must be allocated using `kobject_create()` function, which will return an empty kobject that one must initialize with `kobj_init()`, given as a parameter the allocated and non-initialized kobject pointer, along with its `kobj_type` pointer:

```
struct kobject *kobject_create(void)
void kobject_init(struct kobject *kobj, struct kobj_type *ktype)
```

The `kobject_add()` function is used to add and link a kobject to the system, at the same time creating its directory according to its hierarchy, along with its default attributes. The reverse function is `kobject_del()`:

```
int kobject_add(struct kobject *kobj, struct kobject *parent,
                const char *fmt, ...);
```

The reverse function of both `kobject_create` and `kobject_add` is `kobject_put`. In the source provided with the book, the excerpt to tie a kobject to the system is:

```
/* Somewhere */
static struct kobject *mykobj;

mykobj = kobject_create();
    if (mykobj) {
        kobject_init(mykobj, &mytype);
        if (kobject_add(mykobj, NULL, "%s", "hello")) {
            err = -1;
            printk("ldm: kobject_add() failed\n");
            kobject_put(mykobj);
            mykobj = NULL;
        }
        err = 0;
    }
```

Linux Device Model

One could have used `kobject_create_and_add`, which internally calls `kobject_create` and `kobject_add`. The following excerpt from `drivers/base/core.c` shows how to use it:

```
static struct kobject * class_kobj   = NULL;
static struct kobject * devices_kobj = NULL;

/* Create /sys/class */
class_kobj = kobject_create_and_add("class", NULL);
if (!class_kobj) {
    return -ENOMEM;
}

/* Create /sys/devices */
devices_kobj = kobject_create_and_add("devices", NULL);

if (!devices_kobj) {
    return -ENOMEM;
}
```

> If a kobject has a NULL parent, then `kobject_add` sets parent to kset. If both are NULL, object becomes a child-member of the top-level sys directory

kobj_type

A `struct kobj_type` structure describes the behavior of kobjects. A `kobj_type` structure describes the type of object that embeds a kobject by means of `ktype` field. Every structure that embeds a kobject needs a corresponding `kobj_type`, which will control what happens when the kobject is created and destroyed, and when attributes are read or written to. Every kobject has a field of the type `struct kobj_type`, which stands for **kernel object type**:

```
struct kobj_type {
    void (*release)(struct kobject *);
    const struct sysfs_ops sysfs_ops;
    struct attribute **default_attrs;
};
```

A `struct kobj_type` structure allows kernel objects to share common operations (`sysfs_ops`), whether those objects are functionally related or not. Fields of that structure are meaningful enough. `release` is a callback called by the `kobject_put()` function whenever your object needs to be freed. You must free memory held by your object here. One can use the `container_of` macro to get a pointer to the object. The `sysfs_ops` field points to sysfs operations, whereas `default_attrs` defines the default attributes associated with this kobject. `sysfs_ops` is a set of callback (sysfs operation) called when a sysfs attribute is accessed. `default_attrs` is a pointer to a list of `struct attribute` elements that will be used as default attributes for each object of this type:

```
struct sysfs_ops {
    ssize_t (*show)(struct kobject *kobj,
                    struct attribute *attr, char *buf);
    ssize_t (*store)(struct kobject *kobj,
                     struct attribute *attr,const char *buf,
                     size_t size);
};
```

`show` is the callback called when one reads an attribute of any kobject which has this `kobj_type`. The buffer size is always `PAGE_SIZE` in length, even if the value to show is a simple `char`. One should set the value of `buf` (using `scnprintf`), and return the size (in bytes) of data actually written into the buffer on success or negative error on failure. `store` is called for write purposes. Its `buf` parameter is at most `PAGE_SIZE` but can be smaller. It returns the size (in bytes) of data actually read from buffer on success or negative error on failure (or if it receives an unwanted value). One can use `get_ktype` to get the `kobj_type` of a given kobject:

```
struct kobj_type *get_ktype(struct kobject *kobj);
```

In the example in the book, our `k_type` variable represents our kobject's type:

```
static struct sysfs_ops s_ops = {
    .show = show,
    .store = store,
};

static struct kobj_type k_type = {
    .sysfs_ops = &s_ops,
    .default_attrs = d_attrs,
};
```

Linux Device Model

Here, the `show` and `store` callbacks are defined as follows:

```
static ssize_t show(struct kobject *kobj, struct attribute *attr, char
*buf)
{
    struct d_attr *da = container_of(attr, struct d_attr, attr);
    printk( "LDM show: called for (%s) attr\n", da->attr.name );
    return scnprintf(buf, PAGE_SIZE,
                    "%s: %d\n", da->attr.name, da->value);
}

static ssize_t store(struct kobject *kobj, struct attribute *attr, const
char *buf, size_t len)
{
    struct d_attr *da = container_of(attr, struct d_attr, attr);
    sscanf(buf, "%d", &da->value);
    printk("LDM store: %s = %d\n", da->attr.name, da->value);

    return sizeof(int);
}
```

ksets

Kernel object sets (**ksets**) mainly group related kernel objects together. ksets are collection of kobjects. In other words, a kset gathers related kobjects into a single place, for example, all block devices:

```
struct kset {
    struct list_head list;
    spinlock_t list_lock;
    struct kobject kobj;
};
```

- `list` is a linked list of all kobjects in the kset
- `list_lock` is a spinlock protecting linked list access
- `kobj` represents the base class for the set

Each registered (added to the system) kset corresponds to a sysfs directory. A kset can be created and added using the `kset_create_and_add()` function, and removed with the `kset_unregister()` function:

```
struct kset * kset_create_and_add(const char *name,
                    const struct kset_uevent_ops *u,
                    struct kobject *parent_kobj);
void kset_unregister (struct kset * k);
```

Adding a kobject to the set is as simple as specifying its kset field to the right kset:

```
static struct kobject foo_kobj, bar_kobj;

example_kset = kset_create_and_add("kset_example", NULL, kernel_kobj);
/*
 * since we have a kset for this kobject,
 * we need to set it before calling the kobject core.
 */
foo_kobj.kset = example_kset;
bar_kobj.kset = example_kset;
retval = kobject_init_and_add(&foo_kobj, &foo_ktype,
                              NULL, "foo_name");
retval = kobject_init_and_add(&bar_kobj, &bar_ktype,
                              NULL, "bar_name");
```

Now in the module `exit` function, after kobject and their attributes have been removed:

```
kset_unregister(example_kset);
```

Attribute

Attributes are sysfs files exported to the user space by kobjects. An attribute represents an object property that can be readable, writable, or both, from the user space. That said, every data structure that embeds a struct kobject can expose either default attributes provided by the kobject itself (if any), or custom ones. In other words, attributes map kernel data to files in sysfs.

An attribute definition looks like this:

```
struct attribute {
        char * name;
        struct module *owner;
        umode_t mode;
};
```

The kernel functions used to add/remove attributes from the filesystem are:

```
int sysfs_create_file(struct kobject * kobj,
                      const struct attribute * attr);
void sysfs_remove_file(struct kobject * kobj,
                       const struct attribute * attr);
```

Let us try to define two properties that we will export, each represented by an attribute:

```
struct d_attr {
    struct attribute attr;
    int value;
};

static struct d_attr foo = {
    .attr.name="foo",
    .attr.mode = 0644,
    .value = 0,
};

static struct d_attr bar = {
    .attr.name="bar",
    .attr.mode = 0644,
    .value = 0,
};
```

To create each enumerated attribute separately, we have to call the following:

```
sysfs_create_file(mykobj, &foo.attr);
sysfs_create_file(mykobj, &bar.attr);
```

A good place to start with attributes is `samples/kobject/kobject-example.c` in the kernel source.

Attributes group

So far, we have seen how to individually add attributes and call (directly or indirectly through a wrapper function such as `device_create_file()`, `class_create_file()`, and so on) `sysfs_create_file()` on each of them. Why bother ourselves with multiple calls if we can do it once? Here is where the attribute group comes in. It relies on the `struct attribute_group` structure:

```
struct attribute_group {
    struct attribute    **attrs;
};
```

Of course, we have removed fields that are not of interest. The `attrs` field is a pointer to NULL terminated list of attributes. Each attribute group must be given a pointer to a list/array of `struct attribute` elements. The group is just a helper wrapper that makes it easier to manage multiple attributes.

The kernel functions used to add/remove group attributes to the filesystem are:

```
int sysfs_create_group(struct kobject *kobj,
                    const struct attribute_group *grp)
void sysfs_remove_group(struct kobject * kobj,
                    const struct attribute_group * grp)
```

The two preceding defined properties can be embedded in a `struct attribute_group`, to make only one call to add both of them to the system:

```
static struct d_attr foo = {
    .attr.name="foo",
    .attr.mode = 0644,
    .value = 0,
};

static struct d_attr bar = {
    .attr.name="bar",
    .attr.mode = 0644,
    .value = 0,
};

/* attrs is a pointer to a list (array) of attributes */
static struct attribute * attrs [] =
{
    &foo.attr,
    &bar.attr,
    NULL,
};
static struct attribute_group my_attr_group = {
    .attrs = attrs,
};
```

The one and only function to call here is this:

```
sysfs_create_group(mykobj, &my_attr_group);
```

It is much better than making a call for each attribute.

Device model and sysfs

Sysfs is a non-persistent virtual filesystem that provides a global view of the system and exposes the kernel object's hierarchy (topology) by means of their kobjects. Each kobjects shows up as a directory, and files in a directory representing kernel variables, exported by the related kobject. These files are called attributes, and can be read or written.

Linux Device Model

If any registered kobject creates a directory in sysfs, where the directory is created depends on the kobject's parent (which is a kobject too). It is natural that directories are created as subdirectories of the kobject's parent. This highlights internal object hierarchies to the user space. Top-level directories in sysfs represent the common ancestors of object hierarchies, that is, the subsystems the objects belong to.

Top-level sysfs directories can be found under the /sys/ directory:

```
/sys$ tree -L 1
├── block
├── bus
├── class
├── dev
├── devices
├── firmware
├── fs
├── hypervisor
├── kernel
├── module
└── power
```

block contains a directory per-block device on the system, each of which contains subdirectories for partitions on the device. bus contains the registered bus on the system. dev contains the registered device nodes in a raw way (no hierarchy), each being a symlink to the real device in the /sys/devices directory. devices gives a view of the topology of devices in the system. firmware shows a system-specific tree of low-level subsystems, such as: ACPI, EFI, OF (DT). fs lists filesystems actually used on the system. kernel holds kernel configuration options and status info. Modules is a list of loaded modules.

Each of these directories corresponds to a kobject, some of which are exported as kernel symbols. These are:

- kernel_kobj which corresponds to /sys/kernel
- power_kobj for /sys/power
- firmware_kobj which is for /sys/firmware, exported in the drivers/base/firmware.c source file
- hypervisor_kobj for /sys/hypervisor, exported in the drivers/base/hypervisor.c
- fs_kobj which corresponds to /sys/fs, exported in the fs/namespace.c file

However, `class/`, `dev/`, `devices/`, are created during the boot by the `devices_init` function in `drivers/base/core.c` in kernel source, `block/` is created in `block/genhd.c`, and `bus/` is created as a kset in `drivers/base/bus.c`.

When a kobject directory is added to sysfs (using `kobject_add`), where it is added depends on the kobject's parent location. If its parent pointer is set, it is added as a subdirectory inside the parent's directory. If the parent pointer is NULL, it is added as a subdirectory inside `kset->kobj`. If neither parent nor kset fields are set, it maps to the root level directory in sysfs (`/sys`).

One can create/remove symbolic links on existing objects (directories), using `sysfs_{create|remove}_link` functions:

```
int sysfs_create_link(struct kobject * kobj,
                      struct kobject * target, char * name);
void sysfs_remove_link(struct kobject * kobj, char * name);
```

This will allow an object to exist in more than one place. The create function will create a symlink named `name` pointing to the `target` kobject sysfs entry. A well know example is devices appearing in both `/sys/bus` and `/sys/devices`. Symbolic links created will be persistent even after `target` removal. You have to know when the `target` is removed, and then remove the corresponding symlink.

Sysfs files and attributes

Now we know that the default set of files are provided via the ktype field in kobjects and ksets, through the `default_attrs` field of `kobj_type`. Default attributes will be sufficient in most of the cases. But sometimes an instance of a ktype may need its own attributes to provide data or functionality not shared by a more general ktype.

Just a recall, the low-level functions used to add/remove new attributes (or group of attributes) on top of default set are:

```
int sysfs_create_file(struct kobject *kobj,
                      const struct attribute *attr);
void sysfs_remove_file(struct kobject *kobj,
                       const struct attribute *attr);
int sysfs_create_group(struct kobject *kobj,
                       const struct attribute_group *grp);
void sysfs_remove_group(struct kobject * kobj,
                        const struct attribute_group * grp);
```

Current interfaces

There are interface layers that currently exist in sysfs. Apart from creating your own ktype or kobject to add your attributes, you can use ones that currently exist: device, driver, bus, and class attributes. Their description are as follows:

Device attributes

Apart from default attributes provided by the kobject embedded in your device structure, you can create custom ones. The structure used for this purpose is `struct device_attribute`, which is nothing but a wrapping around the standard `struct attribute`, and a set of callbacks to show/store the value of the attribute:

```
struct device_attribute {
    struct attribute attr;
    ssize_t (*show)(struct device *dev,
                    struct device_attribute *attr,
                    char *buf);
    ssize_t (*store)(struct device *dev,
                     struct device_attribute *attr,
                     const char *buf, size_t count);
};
```

Their declaration is done through the `DEVICE_ATTR` macro:

```
DEVICE_ATTR(_name, _mode, _show, _store);
```

Whenever you declare a device attribute using `DEVICE_ATTR`, the prefix `dev_attr_` is added to the attribute name. For example, if you declare an attribute with the `_name` parameter set to foo, the attribute will be accessible through the `dev_attr_foo` variable name.

To understand why, let us see how the `DEVICE_ATTR` macro is defined in `include/linux/device.h`:

```
#define DEVICE_ATTR(_name, _mode, _show, _store) \
    struct device_attribute dev_attr_##_name = __ATTR(_name, _mode, _show, _store)
```

Finally, you can add/remove those using the `device_create_file` and `device_remove_file` functions:

```
int device_create_file(struct device *dev,
                       const struct device_attribute * attr);
void device_remove_file(struct device *dev,
                        const struct device_attribute * attr);
```

The following sample is a demonstration of how to put it all together:

```
static ssize_t foo_show(struct device *child,
    struct device_attribute *attr, char *buf)
{
    return sprintf(buf, "%d\n", foo_value);
}

static ssize_t bar_show(struct device *child,
        struct device_attribute *attr, char *buf)
{
    return sprintf(buf, "%d\n", bar_value);
}
```

Here are the static declarations of the attribute:

```
static DEVICE_ATTR(foo, 0644, foo_show, NULL);
static DEVICE_ATTR(bar, 0644, bar_show, NULL);
```

The following code shows how to actually create files on the system:

```
if ( device_create_file(dev, &dev_attr_foo) != 0 )
    /* handle error */
if ( device_create_file(dev, &dev_attr_bar) != 0 )
    /* handle error*/
```

For cleanup, the attribute removal is done in the remove function as follows:

```
device_remove_file(wm->dev, &dev_attr_foo);
device_remove_file(wm->dev, &dev_attr_bar);
```

You may wonder how and why we used to define the same set of store/show callbacks for all attributes of the same kobject/ktype, and now, we use a custom one for each attribute. The first reason is because, the device subsystem defines its own attribute structure, which wraps the standard one, and secondly, instead of showing/storing the value of the attribute, it uses the container_of macro to extract the struct device_attribute giving a generic struct attribute, and then executes the show/store callback depending on the user action. The following is the excerpt from drivers/base/core.c, showing sysfs_ops of the device kobject:

```
static ssize_t dev_attr_show(struct kobject *kobj,
                struct attribute *attr,
                char *buf)
{
    struct device_attribute *dev_attr = to_dev_attr(attr);
    struct device *dev = kobj_to_dev(kobj);
    ssize_t ret = -EIO;
```

```c
        if (dev_attr->show)
                ret = dev_attr->show(dev, dev_attr, buf);
        if (ret >= (ssize_t)PAGE_SIZE) {
                print_symbol("dev_attr_show: %s returned bad count\n",
                             (unsigned long)dev_attr->show);
        }
        return ret;
}

static ssize_t dev_attr_store(struct kobject *kobj, struct attribute *attr,
                              const char *buf, size_t count)
{
        struct device_attribute *dev_attr = to_dev_attr(attr);
        struct device *dev = kobj_to_dev(kobj);
        ssize_t ret = -EIO;

        if (dev_attr->store)
                ret = dev_attr->store(dev, dev_attr, buf, count);
        return ret;
}

static const struct sysfs_ops dev_sysfs_ops = {
   .show  = dev_attr_show,
   .store = dev_attr_store,
};
```

The principle is the same for bus (in `drivers/base/bus.c`), driver (in `drivers/base/bus.c`), and class (in `drivers/base/class.c`) attributes. They use the `container_of` macro to extract their specific attribute structure, and then call the show/store callback embedded in it.

Bus attributes

It relies on the `struct bus_attribute` structure:

```c
struct bus_attribute {
   struct attribute attr;
   ssize_t (*show)(struct bus_type *, char * buf);
   ssize_t (*store)(struct bus_type *, const char * buf, size_t count);
};
```

Bus attributes are declared using the BUS_ATTR macro:

```
BUS_ATTR(_name, _mode, _show, _store)
```

Any bus attribute declared using BUS_ATTR will have the prefix bus_attr_ added to the attribute variable name:

```
#define BUS_ATTR(_name, _mode, _show, _store)        \
struct bus_attribute bus_attr_##_name = __ATTR(_name, _mode, _show, _store)
```

They are created/removed using bus_{create|remove}_file functions:

```
int bus_create_file(struct bus_type *, struct bus_attribute *);
void bus_remove_file(struct bus_type *, struct bus_attribute *);
```

Device drivers attributes

The structure used is struct driver_attribute:

```
struct driver_attribute {
        struct attribute attr;
        ssize_t (*show)(struct device_driver *, char * buf);
        ssize_t (*store)(struct device_driver *, const char * buf,
                         size_t count);
};
```

The declaration relies on the DRIVER_ATTR macro, which will prefix the attribute variable name with driver_attr_:

```
DRIVER_ATTR(_name, _mode, _show, _store)
```

The macro definition is:

```
#define DRIVER_ATTR(_name, _mode, _show, _store) \
struct driver_attribute driver_attr_##_name = __ATTR(_name, _mode, _show, _store)
```

Creation/removal relies on driver_{create|remove}_file functions:

```
int driver_create_file(struct device_driver *,
                       const struct driver_attribute *);
void driver_remove_file(struct device_driver *,
                        const struct driver_attribute *);
```

Class attributes

The `struct class_attribute` is the base structure:

```
struct class_attribute {
        struct attribute         attr;
        ssize_t (*show)(struct device_driver *, char * buf);
        ssize_t (*store)(struct device_driver *, const char * buf,
                         size_t count);
};
```

The declaration of a class attribute relies on `CLASS_ATTR`:

```
CLASS_ATTR(_name, _mode, _show, _store)
```

As the macro's definition shows, any class attribute declared with `CLASS_ATTR` will have the prefix `class_attr_` added to the attribute variable name:

```
#define CLASS_ATTR(_name, _mode, _show, _store) \
struct class_attribute class_attr_##_name = __ATTR(_name, _mode, _show, _store)
```

Finally, file creation and removal is done with the `class_{create|remove}_file` functions:

```
int class_create_file(struct class *class,
        const struct class_attribute *attr);

void class_remove_file(struct class *class,
        const struct class_attribute *attr);
```

Notice that `device_create_file()`, `bus_create_file()`, `driver_create_file()`, and `class_create_file()` all make an internal call to `sysfs_create_file()`. As they all are kernel objects, they have a `kobject` embedded into their structure. That `kobject` is then passed as a parameter to `sysfs_create_file`, as you can see as follows:

```
int device_create_file(struct device *dev,
                const struct device_attribute *attr)
{
    [...]
    error = sysfs_create_file(&dev->kobj, &attr->attr);
    [...]
}

int class_create_file(struct class *cls,
                const struct class_attribute *attr)
```

```
{
    [...]
    error =
        sysfs_create_file(&cls->p->class_subsys.kobj,
                        &attr->attr);
    return error;
}

int bus_create_file(struct bus_type *bus,
                struct bus_attribute *attr)
{
    [...]
    error =
        sysfs_create_file(&bus->p->subsys.kobj,
                        &attr->attr);
    [...]
}
```

Allow sysfs attribute files to be pollable

Here we will see how not to make CPU wasting polling to sense sysfs attributes data availability. The idea is to use the `poll` or `select` system calls to wait for the attribute's content to change. The patch to make sysfs attributes pollable was created by **Neil Brown** and **Greg Kroah-Hartman**. The kobject manager (the driver which has access to the kobject) must support notification to allow `poll` or `select` to return (be released) when the content changes. The magic function that does the trick comes from the kernel side, and is `sysfs_notify()`:

```
void sysfs_notify(struct kobject *kobj, const char *dir,
                const char *attr)
```

If the `dir` parameter is non-NULL, it is used to find a subdirectory, which contains the attribute (presumably created by `sysfs_create_group`). This has a cost of one int per attribute, one `wait_queuehead` per kobject, one int per open file.

`poll` will return `POLLERR|POLLPRI`, and `select` will return the fd whether it is waiting for read, write, or exceptions. The blocking poll is from the user's side. `sysfs_notify()` should be called only after you adjust your kernel attribute value.

Think of the `poll()` (or `select()`) code as a **subscriber** to notice a change in an attribute of interest, and `sysfs_notify()` as a **publisher**, notifying subscribers of any changes.

The following is an excerpt of code provided with the book, which is the store function of an attribute:

```
static ssize_t store(struct kobject *kobj, struct attribute *attr,
                    const char *buf, size_t len)
{
    struct d_attr *da = container_of(attr, struct d_attr, attr);

    sscanf(buf, "%d", &da->value);
    printk("sysfs_foo store %s = %d\n", a->attr.name, a->value);

    if (strcmp(a->attr.name, "foo") == 0){
        foo.value = a->value;
        sysfs_notify(mykobj, NULL, "foo");
    }
    else if(strcmp(a->attr.name, "bar") == 0){
        bar.value = a->value;
        sysfs_notify(mykobj, NULL, "bar");
    }
    return sizeof(int);
}
```

The code from the user space must behave like this in order to sense the data change:

1. Open the file attributes.
2. Make a dummy read of all the contents.
3. Call poll requesting POLLERR|POLLPRI (select/exceptfds works too).
4. When `poll` (or `select`) returns (which indicates that a value has changed), read the content of files whose data changed.
5. Close the files and go to the top of the loop.

When in doubt of a sysfs attribute being pollable, set a suitable timeout value. The user space example is provided with the book sample.

Summary

Now you are familiar with the LDM concept and with its data structures (bus, class, device drivers, and devices), including low-level data structures, which are `kobject`, `kset`, `kobj_types`, and attributes (or group of those), how objects are represented within the kernel (hence sysfs and devices topology) is not a secret anymore. You will be able to create an attribute (or group), exposing your device or driver feature through sysfs. If the previous topic seems clear to you, we will move to the next `chapter 14`, *Pin Control and GPIO Subsystem*, which heavily uses the power of `sysfs`.

14
Pin Control and GPIO Subsystem

Most embedded Linux driver and kernel engineers write using GPIOs or play with pins multiplexing. By pins, I mean outgoing line of component. SoC does multiplex pins, meaning that a pin may have several functions, for example, `MX6QDL_PAD_SD3_DAT1` in `arch/arm/boot/dts/imx6dl-pinfunc.h` can be either an SD3 data line 1, UART1's cts/rts, Flexcan2's Rx, or normal GPIO.

The mechanism by which one choses the mode a pin should work on is called pin muxing. The system responsible for is called the pin controller. In the second part of the chapter, we will discuss the **General Purpose Input Output** (**GPIO**), which is a special function (mode) in which a pin can operate.

In this chapter, we will:

- Walk through the pin control subsystem, and see how one can declare their nodes in DT
- Explore both legacy integer-based GPIO interfaces, as well as the new descriptor-based interface API
- Deal with GPIO mapped to IRQ
- Handle sysfs interfaces dedicated to GPIOs

Pin control subsystem

The **Pin control** (**pinctrl**) subsystem allows managing pin muxing. In the DT, devices that need pins to be multiplexed in a certain way must declare the pin control configuration they need.

Pin Control and GPIO Subsystem

The pinctrl subsystem provides:

- Pin multiplexing, which allows for reusing the same pin for different purposes, such as one pin being a UART TX pin, GPIO line, or HSI data line. Multiplexing can affect groups of pins or individual pins.
- Pin configuration, applying electronic properties of pins such as pull-up, pull-down, driver strength, debounce period, and so on.

The purpose of this book is limited to using functions exported by the pin controller driver, and does not not how to write a pin controller driver.

Pinctrl and the device tree

The pinctrl is nothing but a way to gather pins (not only GPIO), and pass them to the driver. The pin controller driver is responsible for parsing pin descriptions in the DT and applying their configuration in the chip. The driver usually needs a set of two nested nodes to describe group of pins configurations. The first node describes the function of the group (what purpose the group will be used for), the second holds the pins configuration.

How pin groups are assigned in the DT heavily depends on the platform, and thus the pin controller driver. Every pin control state is given an integer ID starting at 0 and contiguous. One can use a name property, which will be mapped on top of IDs, so that the same name always points to the same ID.

Each client device's own binding determines the set of states that must be defined in its DT node, and whether to define the set of state IDs that must be provided, or whether to define the set of state names that must be provided. In any case, a pin configuration node can be assigned to a device by means of two properties:

- `pinctrl-<ID>`: This allows for giving the list of pinctrl configurations needed for a certain state of the device. It is a list of phandles, each of which points to a pin configuration node. These referenced pin configuration nodes must be child nodes of the pin controller that they configure. Multiple entries may exist in this list so that multiple pin controllers may be configured, or so that a state may be built from multiple nodes for a single pin controller, each contributing part of the overall configuration.
- `pinctrl-name`: This allows for giving a name to each state in a list. List entry 0 defines the name for integer state ID 0, list entry 1 for state ID 1, and so on. The state ID 0 is commonly given the name *default*. The list of standardized states can be found in `include/linux/pinctrl/pinctrl-state.h`.

- The following is an excerpt of DT, showing some device nodes, along with their pin control nodes:

```
usdhc@0219c000 { /* uSDHC4 */
    non-removable;
    vmmc-supply = <&reg_3p3v>;
    status = "okay";
    pinctrl-names = "default";
    pinctrl-0 = <&pinctrl_usdhc4_1>;
};

gpio-keys {
    compatible = "gpio-keys";
    pinctrl-names = "default";
    pinctrl-0 = <&pinctrl_io_foo &pinctrl_io_bar>;
};

iomuxc@020e0000 {
    compatible = "fsl,imx6q-iomuxc";
    reg = <0x020e0000 0x4000>;

    /* shared pinctrl settings */
    usdhc4 { /* first node describing the function */
        pinctrl_usdhc4_1: usdhc4grp-1 { /* second node */
            fsl,pins = <
                MX6QDL_PAD_SD4_CMD__SD4_CMD     0x17059
                MX6QDL_PAD_SD4_CLK__SD4_CLK     0x10059
                MX6QDL_PAD_SD4_DAT0__SD4_DATA0  0x17059
                MX6QDL_PAD_SD4_DAT1__SD4_DATA1  0x17059
                MX6QDL_PAD_SD4_DAT2__SD4_DATA2  0x17059
                MX6QDL_PAD_SD4_DAT3__SD4_DATA3  0x17059
                MX6QDL_PAD_SD4_DAT4__SD4_DATA4  0x17059
                MX6QDL_PAD_SD4_DAT5__SD4_DATA5  0x17059
                MX6QDL_PAD_SD4_DAT6__SD4_DATA6  0x17059
                MX6QDL_PAD_SD4_DAT7__SD4_DATA7  0x17059
            >;
        };
    };
    [...]
    uart3 {
        pinctrl_uart3_1: uart3grp-1 {
            fsl,pins = <
                MX6QDL_PAD_EIM_D24__UART3_TX_DATA 0x1b0b1
                MX6QDL_PAD_EIM_D25__UART3_RX_DATA 0x1b0b1
            >;
        };
    };
    // GPIOs (Inputs)
```

Pin Control and GPIO Subsystem

```
        gpios {
            pinctrl_io_foo: pinctrl_io_foo {
                fsl,pins = <
                    MX6QDL_PAD_DISP0_DAT15__GPIO5_IO09   0x1f059
                    MX6QDL_PAD_DISP0_DAT13__GPIO5_IO07   0x1f059
                >;
            };
            pinctrl_io_bar: pinctrl_io_bar {
                fsl,pins = <
                    MX6QDL_PAD_DISP0_DAT11__GPIO5_IO05   0x1f059
                    MX6QDL_PAD_DISP0_DAT9__GPIO4_IO30    0x1f059
                    MX6QDL_PAD_DISP0_DAT7__GPIO4_IO28    0x1f059
                    MX6QDL_PAD_DISP0_DAT5__GPIO4_IO26    0x1f059
                >;
            };
        };
    };
```

In the preceding example, a pin configuration is given in the form `<PIN_FUNCTION>` `<PIN_SETTING>`. For example:

```
MX6QDL_PAD_DISP0_DAT15__GPIO5_IO09    0x80000000
```

`MX6QDL_PAD_DISP0_DAT15__GPIO5_IO09` represents the pin function, which is GPIO in this case, and `0x80000000` represents the pin settings.

For this line,

```
MX6QDL_PAD_EIM_D25__UART3_RX_DATA  0x1b0b1
```

`MX6QDL_PAD_EIM_D25__UART3_RX_DATA` represents the pin function, which is the RX line of UART3, and `0x1b0b1` represent is settings.

The pin function is a macro whose value is meaningful for pin controller driver only. These are generally defined in header files located in `arch/<arch>/boot/dts/`. If one uses a UDOO quad, for example, which has an i.MX6 quad core (ARM), the pin function header would be `arch/arm/boot/dts/imx6q-pinfunc.h`. The following is the macro corresponding to the fifth line of the GPIO5 controller:

```
#define MX6QDL_PAD_DISP0_DAT11__GPIO5_IO05   0x19c 0x4b0 0x000 0x5 0x0
```

`<PIN_SETTING>` can be used to set up things like pull-ups, pull-downs, keepers, drive strength, and so on. How it should be specified depends on the pin controller binding, and the meaning of its value depends on the SoC data sheet, generally in the IOMUX section. On i.MX6 IOMUXC, only lower than 17 bits are used for this purpose.

Pin Control and GPIO Subsystem

These preceding nodes are called from the corresponding driver-specific node. Moreover, these pins are configured during corresponding driver initialization. Prior to selecting a pin group state, one must get the pin control first using the `pinctrl_get()` function, call `pinctrl_lookup_state()` in order to check whether the requested state exist or not, and finally `pinctrl_select_state()` to apply the state.

The following is a sample that shows how to get a pincontrol and apply its default configuration:

```
struct pinctrl *p;
struct pinctrl_state *s;
int ret;

p = pinctrl_get(dev);
if (IS_ERR(p))
    return p;

s = pinctrl_lookup_state(p, name);
if (IS_ERR(s)) {
    devm_pinctrl_put(p);
    return ERR_PTR(PTR_ERR(s));
}

ret = pinctrl_select_state(p, s);
if (ret < 0) {
    devm_pinctrl_put(p);
    return ERR_PTR(ret);
}
```

One usually performs such steps during driver initialization. The suitable place for this code could be within the `probe()` function.

> `pinctrl_select_state()` internally calls `pinmux_enable_setting()`, which in turn calls the `pin_request()` on each pin in the pin control node.

A pin control can be released with the `pinctrl_put()` function. One can use the resource-managed version of the API. That said, one can use `pinctrl_get_select()`, given the name of the state to select, in order to configure pinmux. The function is defined in `include/linux/pinctrl/consumer.h` as follows:

```
static struct pinctrl *pinctrl_get_select(struct device *dev,
                        const char *name)
```

where `*name` is the state name as written in `pinctrl-name` property. If the name of the state is `default`, one can just call `pinctr_get_select_default()` function, which is a wrapper around `pinctl_get_select()`:

```
static struct pinctrl * pinctrl_get_select_default(
                            struct device *dev)
{
    return pinctrl_get_select(dev, PINCTRL_STATE_DEFAULT);
}
```

Let us see a real example in a board-specific dts file (`am335x-evm.dts`):

```
dcan1: d_can@481d0000 {
    status = "okay";
    pinctrl-names = "default";
    pinctrl-0 = <&d_can1_pins>;
};
```

And in the corresponding driver:

```
pinctrl = devm_pinctrl_get_select_default(&pdev->dev);
if (IS_ERR(pinctrl))
    dev_warn(&pdev->dev,"pins are not configured from the driver\n");
```

The pin control core will automatically claim the `default` pinctrl state for us when the device is probed. If one defines an `init` state, the pinctrl core will automatically set pinctrl to this state before the `probe()` function, and then switch to the `default` state after `probe()` (unless the driver explicitly changed states already).

The GPIO subsystem

From the hardware point of view, a GPIO is a functionality, a mode in which a pin can operate. From a software point of view, a GPIO is nothing but a digital line, which can operate as an input or output, and can have only two values: (1 for high or 0 for low). Kernel GPIO subsystems provide every function you can imagine to set up and handle GPIO line from within your driver:

- Prior to using a GPIO from within the driver, one should claim it to the kernel. This is a way to take the ownership of the GPIO, preventing other drivers from accessing the same GPIO. After taking the ownership of the GPIO, one can:
  - Set the direction

- Toggle its output state (driving line high or low) if used as output
- Set the debounce-interval and read the state, if used as input. For GPIO lines mapped to IRQ, one can define at what edge/level the interrupt should be triggered, and register a handler that will be run whenever the interrupt occurs.

There are actually two different ways to deal with GPIO in the kernel, as follows:

- The legacy and depreciated integer-based interface, where GPIOs are represented by integer
- The new and recommended descriptor-based interface, where a GPIO is represented and described by an opaque structure, with a dedicated API

The integer-based GPIO interface: legacy

The integer-based interface is the most well-known. The GPIO is identified by an integer, which is used for every operation that needs to be performed on the GPIO. The following is the header that contains legacy GPIO access functions:

```
#include <linux/gpio.h>
```

There are well known functions to handle GPIO in kernel.

Claiming and configuring the GPIO

One can allocate and take the ownership of a GPIO using the `gpio_request()` function:

```
static int  gpio_request(unsigned gpio, const char *label)
```

`gpio` represents the GPIO number we are interested in, and `label` is the label used by the kernel for the GPIO in sysfs, as we can see in `/sys/kernel/debug/gpio`. You have to check the value returned, where 0 mean success, and negative error code on error. Once done with the GPIO, it should be set free with the `gpio_free()` function:

```
void gpio_free(unsigned int gpio)
```

If in doubt, one can use `gpio_is_valid()` function to check whether this GPIO number is valid on the system prior to allocate it:

```
static bool gpio_is_valid(int number)
```

Once we own the GPIO, we can change its direction, depending on the need, and whether it should be an input or output, using the `gpio_direction_input()` or `gpio_direction_output()` functions:

```
static int  gpio_direction_input(unsigned gpio)
static int  gpio_direction_output(unsigned gpio, int value)
```

`gpio` is the GPIO number we need to set the direction. There is a second parameter when it comes to configuring the GPIO as output: `value`, which is the state the GPIO should be in once the output direction is effective. Here again, the return value is zero or a negative error number. These functions are internally mapped on top of lower level callback functions exposed by the driver of the GPIO controller that provides the GPIO we use. In the next `Chapter 15`, *GPIO Controller Drivers - gpio_chip*, dealing with GPIO controller drivers, we will see that a GPIO controller, through its `struct gpio_chip` structure, must expose a generic set of callback functions to use its GPIOs.

Some GPIO controllers offer the possibility to change the GPIO debounce-interval (this is only useful when the GPIO line is configured as input). This feature is platform-dependent. One can use `int gpio_set_debounce()` to achieve that:

```
static  int  gpio_set_debounce(unsigned gpio, unsigned debounce)
```

where `debounce` is the debounce time in ms.

All the preceding functions should be called in a context that may sleep. It is a good practice to claim and configure GPIOs from within the driver's `probe` function.

Accessing the GPIO – getting/setting the value

You should pay attention when accessing GPIO. In an atomic context, especially in an interrupt handler, one has to be sure the GPIO controller callback functions will not sleep. A well-designed controller driver should be able to inform other drivers (actually clients) whether call to its methods may sleep or not. This can be checked with `gpio_cansleep()` function.

None of the functions used to access GPIO return an error code. That is why you should pay attention and check return values during GPIO allocation and configuration.

In atomic context

There are GPIO controllers that can be accessed and managed through simple memory read/write operations. These are generally embedded in the SoC, and do not need to sleep. `gpio_cansleep()` will always return `false` for those controllers. For such GPIOs, you can get/set their value from within an IRQ handler, using the well-known `gpio_get_value()` or `gpio_set_value()`, depending on the GPIO line being configured as input or output :

```
static int  gpio_get_value(unsigned gpio)
void gpio_set_value(unsigned int gpio, int value);
```

`gpio_get_value()` should be used when the GPIO is configured as input (using `gpio_direction_input()`), and return the actual value (state) of the GPIO. On the other hand, `gpio_set_value()` will affect the value of the GPIO, which should have been configured as an output using `gpio_direction_output()`. For both function, `value` can be considered as `Boolean`, where zero means low, and non-zero value mean high.

In a non-atomic context (that may sleep)

On the other hand, there are GPIO controllers wired on buses such as SPI and I2C. Since functions accessing those buses may lead to sleep, the `gpio_cansleep()` function should always return `true` (it is up to the GPIO controller to take of returning true). In this case, you should not access those GPIOs from within the IRQ handled, at least not in the top half (the hard IRQ). Moreover, the accessors you have to use as your general-purpose access should be suffixed with `_cansleep`.

```
static int gpio_get_value_cansleep(unsigned gpio);
void gpio_set_value_cansleep(unsigned gpio, int value);
```

They behave exactly like accessors without the `_cansleep()` name suffix, with the only difference being that they prevent the kernel from printing warnings when the GPIOs are accessed.

GPIOs mapped to IRQ

Input GPIOs can often be used as IRQ signals. Such IRQs can be edge-triggered or level-triggered. The configuration depends on your needs. The GPIO controller is responsible for providing the mapping between the GPIO and its IRQ. One can use `goio_to_irq()` to map a given GPIO number to its IRQ number:

```
int gpio_to_irq(unsigned gpio);
```

The return value is the IRQ number, on which one can call `request_irq()` (or the threaded version `request_threaded_irq()`) in order to register a handler for this IRQ:

```
static irqreturn_t my_interrupt_handler(int irq, void *dev_id)
{
    [...]
    return IRQ_HANDLED;
}

[...]
int gpio_int = of_get_gpio(np, 0);
int irq_num = gpio_to_irq(gpio_int);
int error = devm_request_threaded_irq(&client->dev, irq_num,
                    NULL, my_interrupt_handler,
                    IRQF_TRIGGER_RISING | IRQF_ONESHOT,
                    input_dev->name, my_data_struct);
if (error) {
    dev_err(&client->dev, "irq %d requested failed, %d\n",
        client->irq, error);
    return error;
}
```

Putting it all together

The following code is a summary putting into practice all the concepts discussed regarding integer-based interfaces. This driver manages four GPIOs: two buttons (btn1 and btn2), and two LEDs (green and red). Btn1 is mapped to an IRQ, and whenever its state changes to LOW, the state of btn2 is applied to LEDs. For example, if the state of btn1 goes LOW while btn2 is high, GREEN and RED led will be driven to HIGH:

```
#include <linux/init.h>
#include <linux/module.h>
#include <linux/kernel.h>
#include <linux/gpio.h>          /* For Legacy integer based GPIO */
#include <linux/interrupt.h>     /* For IRQ */

static unsigned int GPIO_LED_RED = 49;
static unsigned int GPIO_BTN1 = 115;
static unsigned int GPIO_BTN2 = 116;
static unsigned int GPIO_LED_GREEN = 120;
static unsigned int irq;

static irq_handler_t btn1_pushed_irq_handler(unsigned int irq,
                    void *dev_id, struct pt_regs *regs)
{
    int state;
```

```c
    /* read BTN2 value and change the led state */
    state = gpio_get_value(GPIO_BTN2);
    gpio_set_value(GPIO_LED_RED, state);
    gpio_set_value(GPIO_LED_GREEN, state);

    pr_info("GPIO_BTN1 interrupt: Interrupt! GPIO_BTN2 state is %d\n",
state);
    return IRQ_HANDLED;
}

static int __init helloworld_init(void)
{
    int retval;

    /*
     * One could have checked whether the GPIO is valid on the controller
or not,
     * using gpio_is_valid() function.
     * Ex:
     *  if (!gpio_is_valid(GPIO_LED_RED)) {
     *      pr_infor("Invalid Red LED\n");
     *      return -ENODEV;
     *  }
     */
    gpio_request(GPIO_LED_GREEN, "green-led");
    gpio_request(GPIO_LED_RED, "red-led");
    gpio_request(GPIO_BTN1, "button-1");
    gpio_request(GPIO_BTN2, "button-2");

    /*
     * Configure Button GPIOs as input
     *
     * After this, one can call gpio_set_debounce()
     * only if the controller has the feature
     *
     * For example, to debounce a button with a delay of 200ms
     *  gpio_set_debounce(GPIO_BTN1, 200);
     */
    gpio_direction_input(GPIO_BTN1);
    gpio_direction_input(GPIO_BTN2);

    /*
     * Set LED GPIOs as output, with their initial values set to 0
     */
    gpio_direction_output(GPIO_LED_RED, 0);
    gpio_direction_output(GPIO_LED_GREEN, 0);

    irq = gpio_to_irq(GPIO_BTN1);
```

```c
        retval = request_threaded_irq(irq, NULL,\
                        btn1_pushed_irq_handler, \
                        IRQF_TRIGGER_LOW | IRQF_ONESHOT, \
                        "device-name", NULL);

    pr_info("Hello world!\n");
    return 0;
}

static void __exit hellowolrd_exit(void)
{
    free_irq(irq, NULL);
    gpio_free(GPIO_LED_RED);
    gpio_free(GPIO_LED_GREEN);
    gpio_free(GPIO_BTN1);
    gpio_free(GPIO_BTN2);

    pr_info("End of the world\n");
}

module_init(hellowolrd_init);
module_exit(hellowolrd_exit);

MODULE_AUTHOR("John Madieu <john.madieu@gmail.com>");
MODULE_LICENSE("GPL");
```

The descriptor-based GPIO interface: the new and recommended way

With the new descriptor-based GPIO interface, a GPIO is characterized by a coherent `struct gpio_desc` structure:

```c
struct gpio_desc {
    struct gpio_chip  *chip;
    unsigned long flags;
    const char *label;
};
```

One should use the following header to be able to use the new interface:

```c
#include <linux/gpio/consumer.h>
```

With the descriptor-based interface, prior to allocating and taking the ownership of GPIOs, those GPIOs must have been mapped somewhere. By mapped, I mean they should be assigned to your device, whereas with the legacy integer-based interface, you just have to fetch a number anywhere and request it as a GPIO. Actually, there are three kinds of mapping in the kernel:

- **Platform data mapping**: The mapping is done in the board file.
- **Device tree**: The mapping is done in DT style, the same as discussed in the preceding sections. This is the mapping we will discuss in this book.
- **Advanced Configuration and Power Interface mapping** (ACPI): The mapping is done in ACPI style. Generally used on x86-based systems.

GPIO descriptor mapping - the device tree

GPIO descriptor mappings are defined in the consumer device's node. The property that contains a GPIO descriptor mapping must be named `<name>-gpios` or `<name>-gpio`, where `<name>` is meaningful enough to describe the function for which those GPIOs will be used.

One should always suffix the property name with either `-gpio` or `-gpios` because every descriptor-based interface function relies on the `gpio_suffixes[]` variable, defined in `drivers/gpio/gpiolib.h` and shown as follows:

```
/* gpio suffixes used for ACPI and device tree lookup */
static const char * const gpio_suffixes[] = { "gpios", "gpio" };
```

Let us see how by having a look at the function used to look for GPIO descriptors mappings in devices in DT:

```
static struct gpio_desc *of_find_gpio(struct device *dev,
                        const char *con_id,
                        unsigned int idx,
                        enum gpio_lookup_flags *flags)
{
    char prop_name[32]; /* 32 is max size of property name */
    enum of_gpio_flags of_flags;
    struct gpio_desc *desc;
    unsigned int i;

    for (i = 0; i < ARRAY_SIZE(gpio_suffixes); i++) {
        if (con_id)
            snprintf(prop_name, sizeof(prop_name), "%s-%s",
                con_id,
                gpio_suffixes[i]);
```

Pin Control and GPIO Subsystem

```
                else
                        snprintf(prop_name, sizeof(prop_name), "%s",
                                gpio_suffixes[i]);

                desc = of_get_named_gpiod_flags(dev->of_node,
                                                prop_name, idx,
                                        &of_flags);
                if (!IS_ERR(desc) || (PTR_ERR(desc) == -EPROBE_DEFER))
                        break;
        }

        if (IS_ERR(desc))
                return desc;

        if (of_flags & OF_GPIO_ACTIVE_LOW)
                *flags |= GPIO_ACTIVE_LOW;

        return desc;
}
```

Now, let us consider the following node, which is an excerpt of `Documentation/gpio/board.txt`:

```
foo_device {
    compatible = "acme,foo";
    [...]
    led-gpios = <&gpio 15 GPIO_ACTIVE_HIGH>, /* red */
                <&gpio 16 GPIO_ACTIVE_HIGH>, /* green */
                <&gpio 17 GPIO_ACTIVE_HIGH>; /* blue */

    power-gpios = <&gpio 1 GPIO_ACTIVE_LOW>;
    reset-gpios = <&gpio 1 GPIO_ACTIVE_LOW>;
};
```

This is what a mapping should look like, with meaningful name.

Allocating and using GPIO

One can use either `gpiog_get()` or `gpiod_get_index()` to allocate a GPIO descriptor:

```
struct gpio_desc *gpiod_get_index(struct device *dev,
                                  const char *con_id,
                                  unsigned int idx,
                                  enum gpiod_flags flags)
struct gpio_desc *gpiod_get(struct device *dev,
                            const char *con_id,
                            enum gpiod_flags flags)
```

On error, these functions will return -ENOENT if no GPIO with the given function is assigned, or another error on which one can use the IS_ERR() macro. The first function returns the GPIO descriptor structure that corresponds to the GPIO at a given index, whereas the second function returns the GPIO at index 0 (useful for one-GPIO mapping). dev is the device to which the GPIO descriptor will belong. It is your device. con_id is the function within the GPIO consumer. It corresponds to the <name> prefix of the property name in the DT. idx is the index (starting from 0) of the GPIO for which one needs a descriptor. flags is an optional parameter that determines the GPIO initialization flags, to configure direction and/or output value. It is an instance of enum gpiod_flags, defined in include/linux/gpio/consumer.h:

```
enum gpiod_flags {
    GPIOD_ASIS = 0,
    GPIOD_IN = GPIOD_FLAGS_BIT_DIR_SET,
    GPIOD_OUT_LOW = GPIOD_FLAGS_BIT_DIR_SET |
                    GPIOD_FLAGS_BIT_DIR_OUT,
    GPIOD_OUT_HIGH = GPIOD_FLAGS_BIT_DIR_SET |
                     GPIOD_FLAGS_BIT_DIR_OUT |
                     GPIOD_FLAGS_BIT_DIR_VAL,
};
```

Now let us allocate GPIO descriptors for mappings defined in the preceding DT:

```
struct gpio_desc *red, *green, *blue, *power;

red = gpiod_get_index(dev, "led", 0, GPIOD_OUT_HIGH);
green = gpiod_get_index(dev, "led", 1, GPIOD_OUT_HIGH);
blue = gpiod_get_index(dev, "led", 2, GPIOD_OUT_HIGH);

power = gpiod_get(dev, "power", GPIOD_OUT_HIGH);
```

The LED GPIOs will be active-high, while the power GPIO will be active-low (that is, gpiod_is_active_low(power) will be true). The reverse operation of allocation is done with the gpiod_put() function:

```
gpiod_put(struct gpio_desc *desc);
```

Let us see how one can release red and blue GPIO LEDs:

```
gpiod_put(blue);
gpiod_put(red);
```

Before we go further, keep in mind that apart from the `gpiod_get()`/`gpiod_get_index()` and `gpio_put()` functions, which completely differ from `gpio_request()` and `gpio_free()`, one can perform API translation from integer-based interfaces to descriptor-based ones just by changing the `gpio_` prefix into `gpiod_`.

That said, to change direction, one should use the `gpiod_direction_input()` and `gpiod_direction_output()` functions:

```
int gpiod_direction_input(struct gpio_desc *desc);
int gpiod_direction_output(struct gpio_desc *desc, int value);
```

`value` is the state to apply to the GPIO once the direction is set to output. If the GPIO controller has this feature, one can set the debounce timeout of a given GPIO using its descriptor:

```
int gpiod_set_debounce(struct gpio_desc *desc, unsigned debounce);
```

In order to access a GPIO given its descriptor, the same attention must be paid as with the integer-based interface. In other words, one should take care whether one is in an atomic (cannot sleep) or non-atomic context, and then use the appropriate function:

```
int gpiod_cansleep(const struct gpio_desc *desc);

/* Value get/set from sleeping context */
int gpiod_get_value_cansleep(const struct gpio_desc *desc);
void gpiod_set_value_cansleep(struct gpio_desc *desc, int value);

/* Value get/set from non-sleeping context */
int gpiod_get_value(const struct gpio_desc *desc);
void gpiod_set_value(struct gpio_desc *desc, int value);
```

For a GPIO descriptor mapped to IRQ, one can use `gpiod_to_irq()` in order to get the IRQ number that corresponds to the given GPIO descriptor, which can be used with the `request_irq()` function:

```
int gpiod_to_irq(const struct gpio_desc *desc);
```

At any given time in the code, one can switch from the descriptor-based interface to the legacy integer-based interface and vice versa, using the `desc_to_gpio()` or `gpio_to_desc()` functions:

```
/* Convert between the old gpio_ and new gpiod_ interfaces */
struct gpio_desc *gpio_to_desc(unsigned gpio);
int desc_to_gpio(const struct gpio_desc *desc);
```

Putting it all together

The driver bellows summarizes the concepts introduced in descriptor-based interfaces. The principle is the same, as are the GPIOs:

```
#include <linux/init.h>
#include <linux/module.h>
#include <linux/kernel.h>
#include <linux/platform_device.h>    /* For platform devices */
#include <linux/gpio/consumer.h>      /* For GPIO Descriptor */
#include <linux/interrupt.h>          /* For IRQ */
#include <linux/of.h>                 /* For DT*/

/*
 * Let us consider the below mapping in device tree:
 *
 *     foo_device {
 *         compatible = "packt,gpio-descriptor-sample";
 *         led-gpios = <&gpio2 15 GPIO_ACTIVE_HIGH>, // red
 *                     <&gpio2 16 GPIO_ACTIVE_HIGH>, // green
 *
 *         btn1-gpios = <&gpio2 1 GPIO_ACTIVE_LOW>;
 *         btn2-gpios = <&gpio2 31 GPIO_ACTIVE_LOW>;
 *     };
 */

static struct gpio_desc *red, *green, *btn1, *btn2;
static unsigned int irq;

static irq_handler_t btn1_pushed_irq_handler(unsigned int irq,
                          void *dev_id, struct pt_regs *regs)
{
    int state;

    /* read the button value and change the led state */
    state = gpiod_get_value(btn2);
    gpiod_set_value(red, state);
    gpiod_set_value(green, state);

    pr_info("btn1 interrupt: Interrupt! btn2 state is %d)\n",
            state);
    return IRQ_HANDLED;
}

static const struct of_device_id gpiod_dt_ids[] = {
    { .compatible = "packt,gpio-descriptor-sample", },
    { /* sentinel */ }
};
```

Pin Control and GPIO Subsystem

```c
static int my_pdrv_probe (struct platform_device *pdev)
{
    int retval;
    struct device *dev = &pdev->dev;

    /*
     * We use gpiod_get/gpiod_get_index() along with the flags
     * in order to configure the GPIO direction and an initial
     * value in a single function call.
     *
     * One could have used:
     *   red = gpiod_get_index(dev, "led", 0);
     *   gpiod_direction_output(red, 0);
     */
    red = gpiod_get_index(dev, "led", 0, GPIOD_OUT_LOW);
    green = gpiod_get_index(dev, "led", 1, GPIOD_OUT_LOW);

    /*
     * Configure GPIO Buttons as input
     *
     * After this, one can call gpiod_set_debounce()
     * only if the controller has the feature
     * For example, to debounce  a button with a delay of 200ms
     *   gpiod_set_debounce(btn1, 200);
     */
    btn1 = gpiod_get(dev, "led", 0, GPIOD_IN);
    btn2 = gpiod_get(dev, "led", 1, GPIOD_IN);

    irq = gpiod_to_irq(btn1);
    retval = request_threaded_irq(irq, NULL, \
                        btn1_pushed_irq_handler, \
                        IRQF_TRIGGER_LOW | IRQF_ONESHOT, \
                        "gpio-descriptor-sample", NULL);
    pr_info("Hello! device probed!\n");
    return 0;
}

static void my_pdrv_remove(struct platform_device *pdev)
{
    free_irq(irq, NULL);
    gpiod_put(red);
    gpiod_put(green);
    gpiod_put(btn1);
    gpiod_put(btn2);
    pr_info("good bye reader!\n");
}

static struct platform_driver mypdrv = {
```

```
        .probe    = my_pdrv_probe,
        .remove   = my_pdrv_remove,
        .driver   = {
            .name          = "gpio_descriptor_sample",
            .of_match_table = of_match_ptr(gpiod_dt_ids),
            .owner         = THIS_MODULE,
        },
    };
    module_platform_driver(mypdrv);
    MODULE_AUTHOR("John Madieu <john.madieu@gmail.com>");
    MODULE_LICENSE("GPL");
```

The GPIO interface and the device tree

Whatever interface one needs to use GPIO for, how to specify GPIOs depends on the controller providing them, especially regarding its #gpio-cells property, which determines the number of cells used for a GPIO specifier. A GPIO specifier contains at least the controller phandle, and one or more argument, where the number of arguments on #gpio-cells property of the controller that provides the GPIO. The first cell is generally the GPIO offset number on the controller, and the second represents the GPIO flags.

GPIO properties should be named [<name>-]gpios], with <name> being the purpose of this GPIO for the device. Keep in mind this rule is a must for descriptor-based interfaces, and becomes <name>-gpios (note the absence of square brackets, meaning that the <name> prefix is mandatory):

```
gpio1: gpio1 {
    gpio-controller;
    #gpio-cells = <2>;
};
gpio2: gpio2 {
    gpio-controller;
    #gpio-cells = <1>;
};
[...]

cs-gpios = <&gpio1 17 0>,
           <&gpio2 2>;
           <0>, /* holes are permitted, means no GPIO 2 */
           <&gpio1 17 0>;

reset-gpios = <&gpio1 30 0>;
cd-gpios = <&gpio2 10>;
```

In the preceding sample, CS GPIOs contain both controller-1 and controller-2 GPIOs. If one does not need to specify a GPIO at a given index in the list, one can use <0>. The reset GPIO has two cells (two arguments after the controller phandle), whereas CD GPIO has only one cell. You can see how meaningful the names are that I gave to my GPIO specifier.

The legacy integer-based interface and device tree

This interface relies on the following header:

```
#include <linux/of_gpio.h>
```

There are two functions you should remember when you need to support DT from within your driver using legacy integer-based interfaces; these are of_get_named_gpio() and of_get_named_gpio_count():

```
int of_get_named_gpio(struct device_node *np,
                      const char *propname, int index)
int of_get_named_gpio_count(struct device_node *np,
                            const char* propname)
```

Given a device node, the former returns the GPIO number of the property *propname at index position. The second just returns the number of GPIOs specified in the property:

```
int n_gpios = of_get_named_gpio_count(dev.of_node,
                                      "cs-gpios"); /* return 4 */
int second_gpio = of_get_named_gpio(dev.of_node, "cs-gpio", 1);
int rst_gpio = of_get_named_gpio("reset-gpio", 0);
gpio_request(second_gpio, "my-gpio);
```

There are drivers still supporting the old specifier, where GPIO properties are named [<name>-gpio] or gpios. In that case, one should use unnamed API versions, by means of of_get_gpio() and of_gpio_count():

```
int of_gpio_count(struct device_node *np)
int of_get_gpio(struct device_node *np, int index)
```

The DT node would look like:

```
my_node@addr {
    compatible = "[...]";

    gpios = <&gpio1 2 0>, /* INT */
            <&gpio1 5 0>; /* RST */
    [...]
};
```

The code in the driver would look like this:

```
struct device_node *np = dev->of_node;

if (!np)
    return ERR_PTR(-ENOENT);

int n_gpios = of_gpio_count(); /* Will return 2 */
int gpio_int = of_get_gpio(np, 0);
if (!gpio_is_valid(gpio_int)) {
    dev_err(dev, "failed to get interrupt gpio\n");
    return ERR_PTR(-EINVAL);
}

gpio_rst = of_get_gpio(np, 1);
if (!gpio_is_valid(pdata->gpio_rst)) {
    dev_err(dev, "failed to get reset gpio\n");
    return ERR_PTR(-EINVAL);
}
```

One can summarize this by rewriting the first driver (the one for integer-based interfaces), in order to comply with the platform drivers structure, and use DT API:

```
#include <linux/init.h>
#include <linux/module.h>
#include <linux/kernel.h>
#include <linux/platform_device.h>    /* For platform devices */
#include <linux/interrupt.h>           /* For IRQ */
#include <linux/gpio.h>         /* For Legacy integer based GPIO */
#include <linux/of_gpio.h>      /* For of_gpio* functions */
#include <linux/of.h>           /* For DT*/

/*
 * Let us consider the following node
 *
 *     foo_device {
 *         compatible = "packt,gpio-legacy-sample";
 *         led-gpios = <&gpio2 15 GPIO_ACTIVE_HIGH>, // red
 *                     <&gpio2 16 GPIO_ACTIVE_HIGH>, // green
 *
 *         btn1-gpios = <&gpio2 1 GPIO_ACTIVE_LOW>;
 *         btn2-gpios = <&gpio2 1 GPIO_ACTIVE_LOW>;
 *     };
 */

static unsigned int gpio_red, gpio_green, gpio_btn1, gpio_btn2;
static unsigned int irq;
```

Pin Control and GPIO Subsystem

```c
static irq_handler_t btn1_pushed_irq_handler(unsigned int irq, void *dev_id,
                                    struct pt_regs *regs)
{
    /* The content of this function remains unchanged */
    [...]
}

static const struct of_device_id gpio_dt_ids[] = {
    { .compatible = "packt,gpio-legacy-sample", },
    { /* sentinel */ }
};

static int my_pdrv_probe (struct platform_device *pdev)
{
    int retval;
    struct device_node *np = &pdev->dev.of_node;

    if (!np)
        return ERR_PTR(-ENOENT);

    gpio_red = of_get_named_gpio(np, "led", 0);
    gpio_green = of_get_named_gpio(np, "led", 1);
    gpio_btn1 = of_get_named_gpio(np, "btn1", 0);
    gpio_btn2 = of_get_named_gpio(np, "btn2", 0);

    gpio_request(gpio_green, "green-led");
    gpio_request(gpio_red, "red-led");
    gpio_request(gpio_btn1, "button-1");
    gpio_request(gpio_btn2, "button-2");

    /* Code to configure GPIO and request IRQ remains unchanged */
    [...]
    return 0;
}

static void my_pdrv_remove(struct platform_device *pdev)
{
    /* The content of this function remains unchanged */
    [...]
}

static struct platform_driver mypdrv = {
    .probe   = my_pdrv_probe,
    .remove  = my_pdrv_remove,
    .driver  = {
        .name    = "gpio_legacy_sample",
```

```
        .of_match_table = of_match_ptr(gpio_dt_ids),
        .owner    = THIS_MODULE,
    },
};
module_platform_driver(mypdrv);

MODULE_AUTHOR("John Madieu <john.madieu@gmail.com>");
MODULE_LICENSE("GPL");
```

GPIO mapping to IRQ in the device tree

One can easily map GPIO to IRQ in the device tree. Two properties are used to specify an interrupt:

- `interrupt-parent`: This is the GPIO controller for GPIO
- `interrupts`: This is the interrupts specifier list

This applies to legacy and descriptor-based interface. The IRQ specifier depends on the `#interrupt-cell` property of the GPIO controller providing this GPIO. `#interrupt-cell` determine the number of cells used when specifying the interrupt. Generally, the first cell represents the GPIO number to map to an IRQ and the second cell represents what level/edge should trigger the interrupt. In any case, interrupt specifier always depends on its parent (the one which has the interrupt-controller set), so refer to its binding documentation in the kernel source:

```
gpio4: gpio4 {
    gpio-controller;
    #gpio-cells = <2>;
    interrupt-controller;
    #interrupt-cells = <2>;
};

my_label: node@0 {
    reg = <0>;
    spi-max-frequency = <1000000>;
    interrupt-parent = <&gpio4>;
    interrupts = <29 IRQ_TYPE_LEVEL_LOW>;
};
```

There are two solutions for obtaining the corresponding IRQ:

1. **Your device sits on a known bus (I2C or SPI)**: The IRQ mapping will be done for you, and made available either through the `struct i2c_client` or `struct spi_device` structure given to your `probe()` function (by means of `i2c_client.irq` or `spi_device.irq`).
2. **Your device sits on the pseudo-platform bus**: The `probe()` function will be given a `struct platform_device`, on which you can call `platform_get_irq()`:

```
int platform_get_irq(struct platform_device *dev, unsigned int num);
```

Feel free to have a look at Chapter 6, *The Concept of Device Tree*.

GPIO and sysfs

The sysfs GPIO interface lets people manage and control GPIOs through sets or files. It is located under `/sys/class/gpio`. The device model is heavily used here, and there are three kinds of entries available:

- `/sys/class/gpio/`: This is where everything begins. This directory contains two special files, `export` and `unexport`:
    - `export`: This allow us to ask the kernel to export control of a given GPIO to user space by writing its number to this file. Example: `echo 21 > export` will create a GPIO21 node for GPIO #21, if that's not requested by kernel code.
    - `unexport`: This reverses the effect of exporting to user space. Example: `echo 21 > unexport` will remove any GPIO21 node exported using the export file.

- `/sys/class/gpio/gpioN/`: This directory corresponds to the GPIO number N (where N is global to the system, not relative to the chip), exported either using the `export` file, or from within the kernel. For example: `/sys/class/gpio/gpio42/` (for GPIO #42) with the following read/write attributes:
    - The `direction` file is used to get/set GPIO direction. Allowed values are either `in` or `out` strings. This value may normally be written. Writing as out defaults to initializing the value as low. To ensure glitch-free operation, low and high values may be written to configure the GPIO as an output with that initial value. This attribute will not exist if the kernel code has exported this GPIO, disabling direction (see the `gpiod_export()` or `gpio_export()` function).
    - The `value` attribute lets us get/set the state of the GPIO line, depending on the direction, input or output. If the GPIO is configured as an output, any non-zero value written will be treated as HIGH state. If configured as an output, writing 0 will set the output low, whereas 1 will set the output high. If the pin can be configured as an interrupt-generating lines and if it has been configured to generate, one can call the `poll(2)` system call on that file and `poll(2)` will return whenever the interrupt was triggered. Using `poll(2)` wil require setting the events POLLPRI and POLLERR. If one uses `select(2)` instead, one should set the file descriptor in exceptfds. After `poll(2)` returns, either `lseek(2)` to the beginning of the sysfs file and read the new value or close the file and re-open it to read the value. It is the same principle as we discussed regarding the pollable sysfs attribute.
    - `edge` determines the signal edge that will let the `poll()` or `select()` function return. Allowed values are `none`, `rising`, `falling`, or `both`. This file is readable/writable, and exists only if the pin can be configured as an interrupt-generating input pin.
    - `active_low` reads as either 0 (false) or 1 (true). Writing any non-zero value will invert the *value* attribute for both reading and writing. Existing and subsequent `poll(2)` support configurations through the edge attribute for rising and falling edges will follow this setting. The relevant function from kernel to set this value is `gpio_sysf_set_active_low()`.

Exporting a GPIO from kernel code

Apart from using /sys/class/gpio/export file to export a GPIO to user space, one can use functions like gpio_export (for legacy interface) or gpioD_export (the new interface) from the kernel code in order to explicitly manage export of GPIOs which have already been requested using gpio_request() or gpiod_get():

```
int gpio_export(unsigned gpio, bool direction_may_change);

int gpiod_export(struct gpio_desc *desc, bool direction_may_change);
```

The direction_may_change parameter decides if one can change the signal direction from input to output and vice versa. The reverse operations from kernel are gpio_unexport() or gpiod_unexport():

```
void gpio_unexport(unsigned gpio); /* Integer-based interface */
void gpiod_unexport(struct gpio_desc *desc) /* Descriptor-based */
```

Once exported, one can use gpio_export_link() (or gpiod_export_link() for descriptor-based interfaces) in order to create symbolic links from elsewhere in sysfs, which will point to the GPIO sysfs node. Drivers can use this to provide the interface under their own device in sysfs with a descriptive name:

```
int gpio_export_link(struct device *dev, const char *name,
                     unsigned gpio)
int gpiod_export_link(struct device *dev, const char *name,
                      struct gpio_desc *desc)
```

One could use this in the probe() function for descriptor-based interfaces as follows:

```
static struct gpio_desc *red, *green, *btn1, *btn2;

static int my_pdrv_probe (struct platform_device *pdev)
{
    [...]
    red = gpiod_get_index(dev, "led", 0, GPIOD_OUT_LOW);
    green = gpiod_get_index(dev, "led", 1, GPIOD_OUT_LOW);
    gpiod_export(&pdev->dev, "Green_LED", green);
    gpiod_export(&pdev->dev, "Red_LED", red);

        [...]
    return 0;
}
```

For integer-based interfaces, the code would look like this:

```
static int my_pdrv_probe (struct platform_device *pdev)
{
    [...]

    gpio_red = of_get_named_gpio(np, "led", 0);
    gpio_green = of_get_named_gpio(np, "led", 1);
    [...]

    int gpio_export_link(&pdev->dev, "Green_LED", gpio_green)
    int gpio_export_link(&pdev->dev, "Red_LED", gpio_red)
    return 0;
}
```

Summary

Dealing with a GPIO from within the kernel is an easy task, as shown in this chapter. Both legacy and new interfaces are discussed, giving the possibility to choose the one that fits your needs, in order to write enhanced GPIO drivers. You'll be able to handle IRQs mapped to GPIOs. The next chapter will deal with the chip that provides and exposes GPIO lines, known as the GPIO controller.

15
GPIO Controller Drivers – gpio_chip

In the previous chapter, we dealt with GPIO lines. Those lines are exposed to the system by means of a special device called the GPIO controller. This chapter will explain step by step how to write drivers for such devices, thus covering the following topics:

- GPIO controller driver architecture and data structures
- Sysfs interface for GPIO controllers
- GPIO controllers representation in DT

Driver architecture and data structures

Drivers for such devices should provide:

- Methods to establish GPIO direction (input and output).
- Methods used to access GPIO values (get and set).
- Methods to map a given GPIO to IRQ and return the associated number.
- Flag saying whether calls to its methods may sleep, this is very important.
- Optional `debugfs dump` method (showing extra state like pullup config).
- Optional numbers called base number, from which GPIO numbering should start. It will be automatically assigned if omitted.

In the kernel, a GPIO controller is represented as an instance of `struct gpio_chip`, defined in `linux/gpio/driver.h`:

```
struct gpio_chip {
  const char *label;
  struct device *dev;
  struct module *owner;

  int (*request)(struct gpio_chip *chip, unsigned offset);
  void (*free)(struct gpio_chip *chip, unsigned offset);
  int (*get_direction)(struct gpio_chip *chip, unsigned offset);
  int (*direction_input)(struct gpio_chip *chip, unsigned offset);
  int (*direction_output)(struct gpio_chip *chip, unsigned offset,
          int value);
  int (*get)(struct gpio_chip *chip,unsigned offset);
  void (*set)(struct gpio_chip *chip, unsigned offset, int value);
  void (*set_multiple)(struct gpio_chip *chip, unsigned long *mask,
          unsigned long *bits);
  int (*set_debounce)(struct gpio_chip *chip, unsigned offset,
          unsigned debounce);

  int (*to_irq)(struct gpio_chip *chip, unsigned offset);

  int base;
  u16 ngpio;
  const char *const *names;
  bool can_sleep;
  bool irq_not_threaded;
  bool exported;

#ifdef CONFIG_GPIOLIB_IRQCHIP
  /*
   * With CONFIG_GPIOLIB_IRQCHIP we get an irqchip
   * inside the gpiolib to handle IRQs for most practical cases.
   */
  struct irq_chip *irqchip;
  struct irq_domain *irqdomain;
  unsigned int irq_base;
  irq_flow_handler_t  irq_handler;
  unsigned int irq_default_type;
#endif

#if defined(CONFIG_OF_GPIO)
  /*
   * If CONFIG_OF is enabled, then all GPIO controllers described in the
   * device tree automatically may have an OF translation
   */
  struct device_node *of_node;
```

```
        int of_gpio_n_cells;
        int (*of_xlate)(struct gpio_chip *gc,
            const struct of_phandle_args *gpiospec, u32 *flags);
}
```

The following is the meaning of each element in the structure:

- `request` is an optional hook for chip-specific activation. If provided, it is executed prior to allocating GPIO whenever one calls `gpio_request()` or `gpiod_get()`.
- `free` is an optional hook for chip-specific deactivation. If provided, it is executed before the GPIO is deallocated whenever one calls `gpiod_put()` or `gpio_free()`.
- `get_direction` is executed whenever one needs to know the direction of the GPIO `offset`. Return value should be 0 to mean out, and 1 to mean in, (same as `GPIOF_DIR_XXX`), or negative error.
- `direction_input` configures the signal `offset` as input, or returns error.
- `get` returns value of GPIO `offset`; for output signals, this returns either the value actually sensed, or zero.
- `set` assigns output value to GPIO `offset`.
- `set_multiple` is called when one needs to assign output values for multiple signals defined by `mask`. If not provided, the kernel will install a generic hook that will walk through `mask` bits and execute `chip->set(i)` on each bit set.

Please see the following which shows how one can implement this function:

```
        static void gpio_chip_set_multiple(struct gpio_chip *chip,
            unsigned long *mask, unsigned long *bits)
{
  if (chip->set_multiple) {
    chip->set_multiple(chip, mask, bits);
  } else {
    unsigned int i;

    /* set outputs if the corresponding mask bit is set */
    for_each_set_bit(i, mask, chip->ngpio)
      chip->set(chip, i, test_bit(i, bits));
  }
}
```

- `set_debounce` if supported by the controller, this hook is an optional callback provided to set the debounce time for specified GPIO.
- `to_irq` is an optional hook to provide GPIO to IRQ mapping. This is called whenever one wants to execute the `gpio_to_irq()` or `gpiod_to_irq()` function. This implementation may not sleep.
- `base` identifies the first GPIO number handled by this chip; or, if negative during registration, the kernel will automatically (dynamically) assign one.
- `ngpio` is the number of GPIOs this controller provides, starts from `base`, to (base + ngpio - 1).
- `names`, if set, must be an array of strings to use as alternative names for the GPIOs in this chip. The array must be `ngpio` sized, and any GPIO that does not need an alias may have its entry set to NULL in the array.
- `can_sleep` is a Boolean flag to be set if `get()`/`set()` method may sleep. It is the case for GPIO controller (also known as expander) sitting on a bus, such as I2C or SPI, whose accesses may lead to sleep. This implies that if the chip supports IRQs, these IRQs need to be threaded as the chip access may sleep when, for example, reading out the IRQ status registers. For GPIO controller mapped to memory (part of SoC), this can be set to false.
- `irq_not_threaded` is a Boolean flag and must be set if `can_sleep` is set, but the IRQs don't need to be threaded.

Each chip exposes a number of signals, identified in method calls by offset values in the range 0 (ngpio - 1). When those signals are referenced through calls like `gpio_get_value(gpio)`, the offset is calculated by subtracting base from the GPIO number.

After every callback has been defined and other fields set, one should call `gpiochip_add()` on the configured `struct gpio_chip` structure in order to register the controller with the kernel. When it comes to unregister, use `gpiochip_remove()`. That is all. You can see how easy it is to write your own GPIO controller driver. In the book sources repository, you will find a working GPIO controller driver, for MCP23016 I2C I/O expander from microchip, whose data sheet is available at
http://ww1.microchip.com/downloads/en/DeviceDoc/20090C.pdf.

To write such drivers, you should include:

```
#include <linux/gpio.h>
```

The following is an excerpt from the driver we have written for our controller, just to show you how easy the task, of writing a GPIO controller driver is:

```
#define GPIO_NUM 16
struct mcp23016 {
  struct i2c_client *client;
  struct gpio_chip chip;
};

static int mcp23016_probe(struct i2c_client *client,
        const struct i2c_device_id *id)
{
  struct mcp23016 *mcp;
  if (!i2c_check_functionality(client->adapter,
      I2C_FUNC_SMBUS_BYTE_DATA))
    return -EIO;

  mcp = devm_kzalloc(&client->dev, sizeof(*mcp), GFP_KERNEL);
  if (!mcp)
    return -ENOMEM;

  mcp->chip.label = client->name;
  mcp->chip.base = -1;
  mcp->chip.dev = &client->dev;
  mcp->chip.owner = THIS_MODULE;
  mcp->chip.ngpio = GPIO_NUM; /* 16 */
  mcp->chip.can_sleep = 1; /* may not be accessed from actomic context */
  mcp->chip.get = mcp23016_get_value;
  mcp->chip.set = mcp23016_set_value;
  mcp->chip.direction_output = mcp23016_direction_output;
  mcp->chip.direction_input = mcp23016_direction_input;
  mcp->client = client;
  i2c_set_clientdata(client, mcp);

  return gpiochip_add(&mcp->chip);
}
```

To request a self-owned GPIO from within the controller driver, one should not use gpio_request(). A GPIO driver can use the following functions instead to request and free descriptors without being pinned to the kernel forever:

```
struct gpio_desc *gpiochip_request_own_desc(struct gpio_desc *desc, const char *label)
void gpiochip_free_own_desc(struct gpio_desc *desc)
```

Descriptors requested with gpiochip_request_own_desc() must be released with gpiochip_free_own_desc().

Pin controller guideline

Depending on the controller you write the driver for, you may need to implement some pin control operation to handle pin multiplexing, configuration, and so on:

- For a pin controller that can only do simple GPIO, a simple `struct gpio_chip` will be sufficient to handle it. There is no need to set up a `struct pinctrl_desc` structure, just write the GPIO controller driver as it.
- If the controller can generate interrupts on top of the GPIO functionality, a `struct irq_chip` must be set up and registered to the IRQ subsystem.
- For a controller having pin multiplexing, advanced pin driver strength, complex biasing, you should set up the following three interfaces :
    - `struct gpio_chip`, discussed earlier in this chapter
    - `struct irq_chip`, discussed in the next chapter (*Chapter 16, Advanced IRQ Management*)
    - `struct pinctrl_desc`, not discussed in the book, but well explained in the kernel documentation in *Documentation/pinctrl.txt*

Sysfs interface for GPIO controller

On successful `gpiochip_add()`, a directory entry with a path like `/sys/class/gpio/gpiochipX/` will be created, where X is the GPIO controller base (controller providing GPIOs starting at #X), having the following attributes:

- `base`, whose value is same as X, and which corresponds to `gpio_chip.base` (if assigned statically), and being the first GPIO managed by this chip.
- `label`, which is provided for diagnostics (not always unique).
- `ngpio`, which tells how many GPIOs this controller provides (N to N + ngpio - 1). This is the same as defined in `gpio_chip.ngpios`.

All of the preceding attributes are read-only.

GPIO controllers and DT

Every GPIO controller declared in the DT must have the Boolean property `gpio-controller` set. Some controllers provide IRQ mapped to the GPIO. In that case, the property `interrupt-cells` should be set too and usually one uses 2, but it depends on the need. The first cell is the pin number, and the second represents the interrupt flag.

`gpio-cells` should be set to identify how many cells are used to describe a GPIO specifier. One usually uses <2>, the first cell to identify the GPIO number, and the second for flags. Actually, most of the nonmemory mapped GPIO controllers do not use the flags:

```
expander_1: mcp23016@27 {
    compatible = "microchip,mcp23016";
    interrupt-controller;
    gpio-controller;
    #gpio-cells = <2>;
    interrupt-parent = <&gpio6>;
    interrupts = <31 IRQ_TYPE_LEVEL_LOW>;
    reg = <0x27>;
    #interrupt-cells=<2>;
};
```

The preceding sample is the node of our GPIO-controller device, and the complete device driver is provided with the sources of the book.

Summary

This chapter is much more than a basis to write the driver for a GPIO controller that you may encounter. It explains the main structure to describe such devices. The next chapter deals with advanced IRQ management, in which we will see how to manage an interrupt controller and thus add such functionality in the driver of the MCP23016 expander from microchip.

16
Advanced IRQ Management

Linux is a system on which devices notify the kernel about particular events by means of IRQs. The CPU exposes IRQ lines, shared or not, and used by connected devices, so that when a device needs the CPU it sends a request to the CPU. When the CPU gets this request it stops its actual job and saves its context, in order to serve the request issued by the device. After serving the device, its state is restored back to exactly where it stopped when the interruption occurred. There are so many IRQ lines, that another device is responsible for them to the CPU. That device is the interrupt controller:

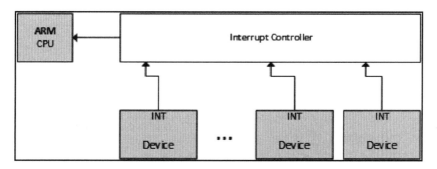

Interrupt controller and IRQ lines

Advanced IRQ Management

Not only can devices raise interrupts, some processor operations can do that too. There are two different kinds of interrupts:

1. Synchronous interrupts called **exceptions**, produced by the CPU while processing instructions. These are **non-maskable interrupts** (**NMI**), and result from a critical malfunction such as hardware failure. They are always processed by the CPU.
2. Asynchronous interrupts called **interrupts**, are issued by other hardware devices. These are normal and maskable interrupts. It is what we will discuss in the next sections of this chapter. Therefore, let us go a bit deeper into exceptions:

Exceptions are consequences of programming errors handled by the kernel, which sends a signal to the program and tries to recover from the error. These are classified in two categories, enumerated below:

- **Processor-detected exceptions**: Those the CPU generates in response to an anomalous condition, and it is divided into three groups:
    - Faults, which can generally be corrected (bogus instruction).
    - Traps, which occur in user process (invalid memory access, division by zero), are also a mechanism to switch to kernel mode in response to a system call. If the kernel code does cause a trap, it immediately panics.
    - Aborts, the serious errors.
- **Programmed exception**: These are requested by the programmer, handled like a trap.

The following array lists unmaskable interrupts (for more details refer to `http://wiki.osdev.org/Exceptions`):

Interrupt number	Description
0	Divide by zero error
1	Debug exception
2	NMI interrupt
3	Breakpoint
4	INTO detected overflow
5	BOUND range exceeded
6	Invalid opcode

7	Coprocessor (device) not available
8	Double fault
9	Coprocessor segment overrun
10	Invalid task state segment
11	Segment not present
12	Stack fault
13	General protection fault
14	Page fault
15	Reserved
16	Coprocessor error
17 - 31	Reserved
32 - 255	Maskable interrupts

NMIs are enough to cover the whole exception list. Back to maskable interrupts, their number depends on the number of devices connected, and how they actually share those IRQ lines. Sometimes, they are not enough and some of them need multiplexing. The commonly used method is by means of a GPIO controller, which also acts as an interrupt controller. In this chapter, we will deal with the API that the kernel offers to manage IRQ and the ways by which multiplexing can be done, and get deeper in interrupt controller driver writing.

That said, in this chapter the following topics will be covered:

- Interrupt controllers and interrupt multiplexing
- Advanced peripheral IRQs management
- Interrupt requests and propagations (chained or nested)
- GPIOLIB irqchip API
- Handling interrupt controllers from DT

Multiplexing interrupts and interrupt controllers

Having a single interrupt from the CPU is usually not enough. Most systems have tens and hundreds of them. Now comes interrupt controller, allowing them to be multiplexed. Very often architecture or platform-specific offers specific facilities, such as:

- Masking/unmasking individual interrupts
- Setting priorities
- SMP affinity
- Exotic things like wake-up interrupts

IRQ management and interrupt controller drivers both rely on the IRQ domain, its turn built on top of the following structures:

- `struct irq_chip`: This structure implements a set of methods describing how to drive the interrupt controller, and which are directly called by core IRQ code.
- `struct irqdomain` structure, which provides:
    - A pointer to the firmware node for a given interrupt controller (fwnode)
    - A method to convert a firmware description of an IRQ into an ID local to this interrupt controller (hwirq)
    - A way to retrieve the Linux view of an IRQ from hwirq
- `struct irq_desc`: This structure is the Linux's view of an interrupt, containing all the core stuff, and one to one mapping to the Linux interrupt number
- `struct irq_action`: This structure Linux uses to describe an IRQ handler
- `struct irq_data`: This is embedded in the `struct irq_desc` structure, and contains:
    - The data that is relevant to the `irq_chip` managing this interrupt
    - Both the Linux IRQ number and the hwirq
    - A pointer to the `irq_chip`

Almost every `irq_chip` call is given an `irq_data` as a parameter, from which you can obtain the corresponding `irq_desc`.

All the preceding structures are part of the IRQ domain API. An interrupt controller is represented in the kernel by an instance of `struct irq_chip` structure, which describes the actual hardware device, and some methods used by the IRQ core:

```
struct irq_chip {
    struct device *parent_device;
    const char    *name;
    void (*irq_enable)(struct irq_data *data);
    void (*irq_disable)(struct irq_data *data);

    void (*irq_ack)(struct irq_data *data);
    void (*irq_mask)(struct irq_data *data);
    void (*irq_unmask)(struct irq_data *data);
    void (*irq_eoi)(struct irq_data *data);

    int (*irq_set_affinity)(struct irq_data *data, const struct cpumask *dest, bool force);
    int (*irq_retrigger)(struct irq_data *data);
    int (*irq_set_type)(struct irq_data *data, unsigned int flow_type);
    int (*irq_set_wake)(struct irq_data *data, unsigned int on);

    void (*irq_bus_lock)(struct irq_data *data);
    void (*irq_bus_sync_unlock)(struct irq_data *data);

    int (*irq_get_irqchip_state)(struct irq_data *data, enum irqchip_irq_state which, bool *state);
    int (*irq_set_irqchip_state)(struct irq_data *data, enum irqchip_irq_state which, bool state);

    unsigned long flags;
};
```

The following is the meaning of elements in the structure:

- `parent_device`: This is a pointer to the parent of this irqchip.
- `name`: This is the name for `/proc/interrupts` file.
- `irq_enable`: This hook enables the interrupt, and its default value is `chip->unmask` if NULL.
- `irq_disable`: This disables the interrupt.
- * `irq_ack`: This is the start of a new interrupt. Some controllers do not need this. Linux calls this function as soon as an interrupt is raised, far before it is serviced. Some implementations map this function to `chip->disable()`, so that another interrupt request on the line will not cause another interrupt until after the current interrupt request has been serviced.

- `irq_mask`: This is the hook that masks an interrupt source in the hardware, so that it cannot be raised anymore.
- `irq_unmask`: This hook unmasks an interrupt source.
- `irq_eoi`: eoi stands for **end of interrupt**. Linux invokes this hook right after an IRQ servicing completes. One uses this function to reconfigure the controller as necessary in order to receive another interrupt request on that line. Some implementations map this function to `chip->enable()` to reverse operations done in `chip->ack()`.
- `irq_set_affinity`: This sets the CPU affinity only on SMP machines. In SMP environments, this function sets the CPU on which the interrupt will be serviced. This function isn't used in single processor machines.
- `irq_retrigger`: This retriggers the interrupt in the hardware, which resends an IRQ to the CPU.
- `irq_set_type`: This sets the flow type (IRQ_TYPE_LEVEL/ and so on) of an IRQ.
- `irq_set_wake`: This enables/disables power-management wake-on of an IRQ.
- `irq_bus_lock`: This functions to lock access to slow bus (I2C) chips. Locking a mutex here is sufficient.
- `irq_bus_sync_unlock`: This functions to sync and unlock slow bus (I2C) chips. Unlock the mutex previously locked.
- `irq_get_irqchip_state` and `irq_set_irqchip_state`: These respectively return or set the internal state of an interrupt.

Each interrupt controller is given a domain, which is for the controller what an address space is for a process (see Chapter 11, *Kernel Memory Management*). The interrupt controller domain is described in the kernel as an instance of `struct irq_domain` structure. It manages mappings between hardware IRQ and Linux IRQ (that is, virtual IRQ). It is the hardware interrupt number translation object:

```
struct irq_domain {
   const char *name;
   const struct irq_domain_ops *ops;
   void *host_data;
   unsigned int flags;

   /* Optional data */
   struct fwnode_handle *fwnode;
   [...]
};
```

- `name` is the name of the interrupt domain.
- `ops` is a pointer to the irq_domain methods.
- `host_data` is private data pointer for use by the owner. Not touched by the irqdomain core code.
- `flags` is host per `irq_domain` flags.
- `fwnode` is optional. It is a pointer to DT nodes associated with the `irq_domain`. Used when decoding DT interrupt specifiers.

An interrupt controller driver creates and registers an `irq_domain` by calling one of the `irq_domain_add_<mapping_method>()` functions, where `<mapping_method>` is the method by which hwirq should be mapped to Linux IRQ. These are:

1. `irq_domain_add_linear()`: This uses a fixed size table indexed by the hwirq number. When a hwirq is mapped, an `irq_desc` is allocated for the hwirq, and the IRQ number is stored in the table. This linear mapping is suitable for fixed and small numbers of hwirq (~ < 256). The inconvenience of this mapping is the table size, being as large as the largest possible hwirq number. Therefore, IRQ number lookup time is fixed, and `irq_desc` are allocated for in-use IRQs only. The majority of drivers should use the linear map. This function has the following prototype:

   ```
   struct irq_domain *irq_domain_add_linear(struct device_node
   *of_node,
                               unsigned int size,
                               const struct irq_domain_ops *ops,
                               void *host_data)
   ```

2. `irq_domain_add_tree()`: This is where the `irq_domain` maintains the mapping between Linux IRQs and hwirq numbers in a radix tree. When a hwirq is mapped, an `irq_desc` is allocated and the hwirq is used as the lookup key for the radix tree. The tree map is a good choice if the hwirq number can be very large since it does not need to allocate a table as large as the largest hwirq number. The disadvantage is that hwirq to IRQ number lookup is dependent on how many entries are in the table. Very few drivers should need this mapping. It has the following prototype:

   ```
   struct irq_domain *irq_domain_add_tree(struct device_node *of_node,
                               const struct irq_domain_ops *ops,
                               void *host_data)
   ```

Advanced IRQ Management

3. `irq_domain_add_nomap()`: You will probably never use this method. Nonetheless, its entire description is available in *Documentation/IRQ-domain.txt*, in the kernel source tree. Its prototype is:

```
struct irq_domain *irq_domain_add_nomap(struct device_node
*of_node,
                    unsigned int max_irq,
                    const struct irq_domain_ops *ops,
                    void *host_data)
```

`of_node` is a pointer to interrupt controller's DT node. `size` represents the number of interrupts in the domain. `ops` represents map/unmap domain callbacks, and `host_data` is the controller's private data pointer.

Since the IRQ domain began empty at creation time (no mapping), you should use `irq_create_mapping()` function in order to create mapping and assign it to the domain. In the next section, will we decide the right place in the code to create mappings:

```
unsigned int irq_create_mapping(struct irq_domain *domain,
                    irq_hw_number_t hwirq)
```

- domain : This is the domain to which this hardware interrupt belongs, or NULL for default domain
- Hwirq: This is the hardware IRQ number in that domain space

When writing driver for GPIO controllers that are also interrupt controllers, `irq_create_mapping()` is called from within `gpio_chip.to_irq()` callback function, like:

```
return irq_create_mapping(gpiochip->irq_domain, offset);
```

Other people prefer creating the mapping in advance for each hwirq inside the `probe` function like:

```
for (j = 0; j < gpiochip->chip.ngpio; j++) {
    irq = irq_create_mapping(
            gpiochip ->irq_domain, j);
}
```

hwirq is the GPIO offset from the gpiochip.

If a mapping for the hwirq doesn't already exist, the function will allocate a new Linux `irq_desc` structure, associate it with the hwirq, and call the `irq_domain_ops.map()` (by means of the `irq_domain_associate()` function) callback so that the driver can perform any required hardware setup:

```
struct irq_domain_ops {
    int (*map)(struct irq_domain *d, unsigned int virq, irq_hw_number_t hw);
    void (*unmap)(struct irq_domain *d, unsigned int virq);
    int (*xlate)(struct irq_domain *d, struct device_node *node,
            const u32 *intspec, unsigned int intsize,
            unsigned long *out_hwirq, unsigned int *out_type);
};
```

- `.map()`: This creates or updates a mapping between a **virtual irq** (**virq**) number and a hwirq number. This is called only once for a given mapping. It generally maps the virq with a given handler using `irq_set_chip_and_handler*`, so that, calling `generic_handle_irq()` or `handle_nested_irq` will trigger the right handler. The magic here is called the `irq_set_chip_and_handler()` function:

    ```
    void irq_set_chip_and_handler(unsigned int irq,
            struct irq_chip *chip, irq_flow_handler_t handle)
    ```

where:

- `irq`: This is the Linux IRQ given as parameter to `map()` function.
- `chip`: This is your `irq_chip`. Some controllers are quite dumb and need almost nothing in their `irq_chip` structure. In this case, you should pass `dummy_irq_chip` defined in `kernel/irq/dummychip.c`, which is a kernel `irq_chip` structure defined for such controllers.

- `handle`: This determines the wrapper function that will call the real handler register using `request_irq()`. Its value depends on the IRQ being edge or level-triggered. In either case, `handle` should be set to `handle_edge_irq`, or `handle_level_irq`. Both are kernel helper functions that do some trick before and after calling the real IRQ handler. An example is shown as follows:

    ```
    static int pcf857x_irq_domain_map(struct irq_domain  *domain,
                        unsigned int irq, irq_hw_number_t hw)
    {
        struct pcf857x *gpio = domain->host_data;

        irq_set_chip_and_handler(irq, &dummy_irq_chip,handle_level_irq);
    #ifdef CONFIG_ARM
        set_irq_flags(irq, IRQF_VALID);
    #else
        irq_set_noprobe(irq);
    #endif
        gpio->irq_mapped |= (1 << hw);

        return 0;
    }
    ```

- `xlate`: Given a DT node and interrupt specifier, this hook decodes the hardware IRQ number and Linux IRQ type value. Depending on the `#interrupt-cells` specified in your DT controller node, the kernel provides a generics translation function:
    - `irq_domain_xlate_twocell()`: Generic translation function is for direct two cell binding. DT IRQ specifier which works with two cell bindings where the cell values map directly to the hwirq number and Linux irq flags.
    - `irq_domain_xlate_onecell()`: Generic xlate for direct one cell bindings.
    - `irq_domain_xlate_onetwocell()`: Generic xlate for one or two cell bindings.

An example of domain operation is given as follows:

```
static struct irq_domain_ops mcp23016_irq_domain_ops = {
    .map   = mcp23016_irq_domain_map,
    .xlate = irq_domain_xlate_twocell,
};
```

Advanced IRQ Management

When an interrupt is received, `irq_find_mapping()` function should be used to find the Linux IRQ number from the hwirq number. Of course, the mapping must exist prior to being returned. A Linux IRQ number is always tied to a `struct irq_desc` structure, which is the structure by which Linux describes an IRQ:

```
struct irq_desc {
    struct irq_common_data irq_common_data;
    struct irq_data irq_data;
    unsigned int __percpu *kstat_irqs;
    irq_flow_handler_t handle_irq;
    struct irqaction *action;
    unsigned int irqs_unhandled;
    raw_spinlock_t lock;
    struct cpumask *percpu_enabled;
    atomic_t threads_active;
    wait_queue_head_t wait_for_threads;
#ifdef CONFIG_PM_SLEEP
    unsigned int nr_actions;
    unsigned int no_suspend_depth;
    unsigned int  force_resume_depth;
#endif
#ifdef CONFIG_PROC_FS
    struct proc_dir_entry *dir;
#endif
    int parent_irq;
    struct module *owner;
    const char *name;
};
```

Some fields that are not described here are internal, and are used by the IRQ core:

- `irq_common_data` is a per IRQ and chip data passed down to chip functions
- `kstat_irqs` is per CPU IRQ statistics since boot
- `handle_irq` is high level IRQ events handler
- `action` represents the list of the IRQ action for this descriptor
- `irqs_unhandled` is the stats field for spurious unhandled interrupts
- `lock` represents locking for SMP
- `threads_active` is the number of IRQ action threads currently running for this descriptor
- `wait_for_threads` represents the wait queue for `sync_irq` to wait for threaded handlers
- `nr_actions` is the number of installed actions on this descriptor

Advanced IRQ Management

- `no_suspend_depth` and `force_resume_depth` represents the number of `irqactions` on a IRQ descriptor with `IRQF_NO_SUSPEND` or `IRQF_FORCE_RESUME` flags set
- `dir` represents `/proc/irq/` procfs entry
- `name` names the flow handler, visible in `/proc/interrupts` output

The `irq_desc.action` field is a list of `irqaction` structures, each of which records the address of an interrupt handler for the associated interrupt source. Each call to the kernel's `request_irq()` function (or the threaded version o) creates an add one `struct irqaction` structure to the end of the list. For example, for a shared interrupt, this field will contain as many IRQ actions as there are handlers registered;

```
struct irqaction {
    irq_handler_t handler;
    void *dev_id;
    void __percpu *percpu_dev_id;
    struct irqaction *next;
    irq_handler_t thread_fn;
    struct task_struct *thread;
    unsigned int irq;
    unsigned int flags;
    unsigned long thread_flags;
    unsigned long thread_mask;
    const char *name;
    struct proc_dir_entry *dir;
};
```

- `handler` is the non-threaded (hard) interrupt handler function
- `name` is the device's name
- `dev_id` is a cookie to identify the device
- `percpu_dev_id` is a cookie to identify the device
- `next` is a pointer to the next IRQ action for shared interrupts
- `irq` is the Linux interrupt number
- `flags` represent the IRQ's flags (see `IRQF_*`)
- `thread_fn` is the threaded interrupt handler function for threaded interrupts
- `thread` is a pointer to the thread structure in case of threaded interrupts
- `thread_flags` represents the flags related to thread
- `thread_mask` is a bitmask for keeping track of thread activity
- `dir` points to the `/proc/irq/NN/<name>/` entry

Interrupt handlers referenced by the `irqaction.handler` field are simply functions associated with the handling of interrupts from particular external devices, and they have minimal knowledge (if any) of the means by which those interrupt requests are delivered to the host microprocessor. They are not microprocessor-level interrupt service routines, and therefore do not exit through RTE or similar interrupt-related opcodes. This makes interrupt-driven device drivers largely portable across different microprocessor architectures

The following is the definition of important fields of the `struct irq_data` structure, which is a per IRQ chip data passed down to chip functions:

```
struct irq_data {
    [...]
    unsigned int irq;
    unsigned long hwirq;
    struct irq_common_data *common;
    struct irq_chip *chip;
    struct irq_domain *domain;
    void *chip_data;
};
```

- `irq` is the interrupt number (Linux IRQ)
- `hwirq` is the hardware interrupt number, local to the `irq_data.domain` interrupt domain
- `common` points to data shared by all irqchips
- `chip` represents the low level interrupt controller hardware access
- `domain` represents the interrupt translation domain, responsible for mapping between the hwirq number and the Linux irq number
- `chip_data` is a platform-specific per-chip private data for the chip methods, to allow shared chip implementations

Advanced peripheral IRQs management

In Chapter 3, *Kernel Facilities and Helper Functions*, we introduced peripheral IRQs, using `request_irq()` and `request_threaded_irq()`. With `request_irq()`, one registers a handler (top half) that will be executed in atomic context, from which one can schedule a bottom half using one of a differing mechanism discussed in that same chapter. On the other hand, with `request_thread_irq()`, one can provide top and bottom halves to the function, so that the former will be run as hardirq handler, which may decide to raise the second and threaded handler, which will be run in a kernel thread.

Advanced IRQ Management

The problem with those approaches is that sometimes, drivers requesting an IRQ do not know about the nature of the interrupt that provides this IRQ line, especially when the interrupt controller is a discrete chip (typically a GPIO expander connected over SPI or I2C buses). Now comes `request_any_context_irq()`, function with which drivers requesting an IRQ know whether the handler will run in a thread context or not, and call `request_threaded_irq()` or `request_irq()` accordingly. It means that whether the IRQ associated to our device comes from an interrupt controller that may not sleep (memory mapped one) or from one that can sleep (behind I2C/SPI bus), there will be no need to change the code. Its prototype is the following:

```
int request_any_context_irq ( unsigned int irq, irq_handler_t handler,
            unsigned long flags,  const char * name,  void * dev_id);
```

The following is the meaning of each parameter in the function:

- `irq` represents the interrupt line to allocate.
- `handler` is the function to be called when the IRQ occurs. Depending on the context, this function may run as hardirq or may be threaded.
- `flags` represents the interrupt type flags. It is the same as those in `request_irq()`.
- `name` will be used for debug purposes to name the interrupt in `/proc/interrupts`.
- `dev_id` is a cookie passed back to the handler function.

`request_any_context_irq()` means that one can either get a hardirq or a treaded one. It works like the usual `request_irq()`, except that it checks whether the IRQ level is configured as nested or not, and calls the right backend. In other words, it selects either a hardIRQ or threaded handling method depending on the context. This function returns a negative value on failure. On success, it returns either `IRQC_IS_HARDIRQ` or `IRQC_IS_NESTED`. The following is a use case:

```
static irqreturn_t packt_btn_interrupt(int irq, void *dev_id)
{
    struct btn_data *priv = dev_id;

    input_report_key(priv->i_dev, BTN_0,
                    gpiod_get_value(priv->btn_gpiod) & 1);
    input_sync(priv->i_dev);
    return IRQ_HANDLED;
}

static int btn_probe(struct platform_device *pdev)
{
```

Advanced IRQ Management

```
    struct gpio_desc *gpiod;
    int ret, irq;

    [...]
    gpiod = gpiod_get(&pdev->dev, "button", GPIOD_IN);
    if (IS_ERR(gpiod))
        return -ENODEV;

    priv->irq = gpiod_to_irq(priv->btn_gpiod);
    priv->btn_gpiod = gpiod;

    [...]

    ret = request_any_context_irq(priv->irq,
                packt_btn_interrupt,
                (IRQF_TRIGGER_FALLING | IRQF_TRIGGER_RISING),
                "packt-input-button", priv);
    if (ret < 0) {
        dev_err(&pdev->dev,
            "Unable to acquire interrupt for GPIO line\n");
        goto err_btn;
    }

    return ret;
}
```

The preceding code is an excerpt of the driver sample of an input device driver. Actually, it is the one used in the next chapter. The advantage using `request_any_context_irq()` is that, one does not need to care about what can be done in the IRQ handler, as the context in which the handler will run depends on the interrupt controller that provides the IRQ line. In our example, if the GPIO below to a controller seating on an I2C or SPI bus, the handler will be threaded. Otherwise, the handler will run in hardirq.

Interrupt request and propagation

Let us consider the following figure, which represents a chained IRQ flow

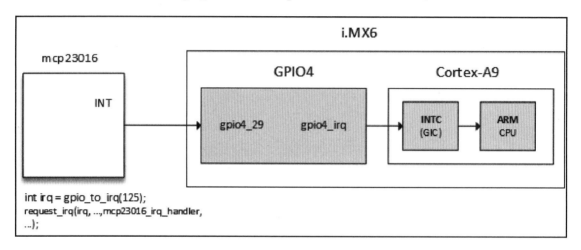

Interrupt requests are always performed on Linux IRQ (not hwirq). The general function to request IRQ on Linux is `request_threaded_irq()` or `request_irq()`, which internally calls the former:

```
int request_threaded_irq(unsigned int irq, irq_handler_t handler,
               irq_handler_t thread_fn, unsigned long irqflags,
               const char *devname, void *dev_id)
```

When called, the function extracts the `struct irq_desc` associated with the IRQ using the `irq_to_desc()` macro. It then allocates a new `struct irqaction` structure and sets it up, filling parameters such as handler, flags, and so on.

```
action->handler = handler;
action->thread_fn = thread_fn;
action->flags = irqflags;
action->name = devname;
action->dev_id = dev_id;
```

That same function finally inserts/registers the descriptor in the proper IRQ list by invoking `__setup_irq()` (by means of `setup_irq()`) function, defined in `kernel/irq/manage.c`.

Now, when an IRQ is raised, the kernel executes a few assembler codes in order to save the current state, and jumps to the arch specific handler, `handle_arch_irq`, which is set with the `handle_irq` field of `struct machine_desc` of our platform in the `setup_arch()` function, in `arch/arm/kernel/setup.c`:

```
handle_arch_irq = mdesc->handle_irq
```

For SoCs that use ARM GIC, `handle_irq` callback is set with `gic_handle_irq`, in either `drivers/irqchip/irq-gic.c`, or `drivers/irqchip/irq-gic-v3.c`:

```
set_handle_irq(gic_handle_irq);
```

`gic_handle_irq()` calls `handle_domain_irq()`, which executes `generic_handle_irq()`, its turn calling `generic_handle_irq_desc()` that ends by calling `desc->handle_irq()`. Have a look at `include/linux/irqdesc.h` for the last call and `arch/arm/kernel/irq.c` for other function calls. `handle_irq` is the actual call for the flow handler, which we registered as `mcp23016_irq_handler`.

`gic_hande_irq()` is a GIC interrupt handler. `generic_handle_irq()` will execute the handler of the SoC's GPIO4 IRQ, which will look for GPIOs pin responsible for the interrupt, and call `generic_handle_irq_desc()`, and so on. Now that you are familiar with interrupt propagation, let us switch to a practical example by writing our own interrupt controller.

Chaining IRQ

This section describes how interrupt handlers of a parent, call the child's interrupt handlers, in turn calling their child's interrupt handlers, and so on. The kernel offers two approaches on how to call interrupt handlers for child devices in the IRQ handler of the parent (interrupt controller) device. These are the chained and nested methods:

Chained interrupts

This approach is used for SoC's internal GPIO controller, which are memory mapped and whose access does not sleep. Chained means that those interrupts are just chains of function calls (for example, SoC's GPIO module interrupt handler is being called from GIC interrupt handler, just as a function call). `generic_handle_irq()` is used for interrupts chaining child IRQ handlers and are called inside of the parent hwirq handler. Even from within the child interrupt handlers, we still are in atomic context (HW interrupt). One cannot call functions that may sleep.

Nested interrupts

This method is used by controllers that sit on slow buses, like I2C (for example, GPIO expander), and whose access may sleep (I2C functions may sleep). Nested means those interrupts handlers that do not run in the HW context (they are not really hwirq, they are not in atomic context), but they are threaded instead, and can be preempted (or interrupted by another interrupt). `handle_nested_irq()` is used for creating nested interrupt child IRQs. Handlers are being called inside of the new thread created by the `handle_nested_irq()` function; we need them to be run in process context, so that we can call sleeping bus functions (like I2C functions that may sleep).

Case study – GPIO and IRQ chip

Let us consider the following figure that ties an interrupt controller device to another one, which we will use to describe interrupt multiplexing:

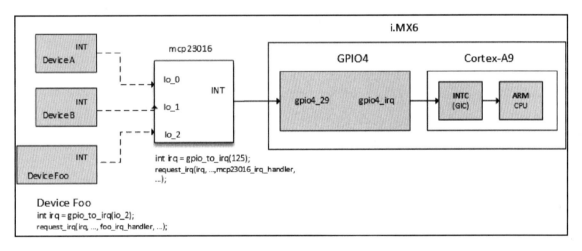

mcp23016 IRQ flow

Consider that you have configured `io_1` and `io_2` as interrupts. Even if interrupt happens on `io_1` or `io_2`, the same interrupt line will be triggered to the interrupt controller. Now the GPIO driver has to figure out reading the interrupt status register of the GPIO to find which interrupt (`io_1` or `io_2`) has really fired. Therefore, in this case a single interrupt line is a multiplex for 16 GPIO interrupts.

Now let us mangle the original driver of the mcp23016 written in Chapter 15, *GPIO Controller Drivers – gpio_chip* in order to support IRQ domain API first, which will let it act as an interrupt controller as well. The second part will introduce the new and recommended gpiolib irqchip API. This will be used as a step-by-step guide to write the interrupt controller driver, at least for the GPIO controller:

Legacy GPIO and IRQ chip

1. The first step, allocate a `struct irq_domain` to our gpiochip that will store the mapping between hwirq and virq. The linear mapping is suitable for us. We do that in the `probe` function. That domain will hold the number of IRQ our drivers wish to provide. For example, for an I/O expander, the number of IRQs could be the number of GPIOs the expander provides:

    ```
    my_gpiochip->irq_domain = irq_domain_add_linear(
    client->dev.of_node,
                my_gpiochip->chip.ngpio, &mcp23016_irq_domain_ops,
    NULL);
    ```

    `host_data` parameter is NULL. Therefore, you can pass whatever data structure you need. Prior to allocating the domain, our domain ops structure should be defined:

    ```
    static struct irq_domain_ops mcp23016_irq_domain_ops = {
        .map   = mcp23016_irq_domain_map,
        .xlate = irq_domain_xlate_twocell,
    };
    ```

    And prior to filling our IRQ domain ops structure, we must define at least the `.map()` callback:

    ```
    static int mcp23016_irq_domain_map(
                struct irq_domain *domain,
                unsigned int virq, irq_hw_number_t hw)
    {
       irq_set_chip_and_handler(virq,
                &dummy_irq_chip, /* Dumb irqchip */
                handle_level_irq); /* Level trigerred irq */
       return 0;
    }
    ```

Our controller is not smart enough. There is then no need to set up an `irq_chip`. We will use the one provided by the kernel for this kind of chip: `dummy_irq_chip`. Some controllers are smart enough and need an `irq_chip` to be set up. Have a look in `drivers/gpio/gpio-mcp23s08.c`.

The next ops callback is `.xlate`. Here again, we use a helper provided by the kernel. `irq_domain_xlate_twocell` is a helper able to parse an interrupt specifier with two cells. We can add this `interrupt-cells = <2>;` in our controller DT node.

2. The next step is to fill the domain with IRQ mappings, using `irq_create_mapping()` function. In our driver, will do it in the `gpiochip.to_irq` callback, so that whenever someone will call `gpio{d}_to_irq()` on the GPIO, the mapping will be returned if it exists, or it will be created if it doesn't:

```
static int mcp23016_to_irq(struct gpio_chip *chip,
                             unsigned offset)
{
    return irq_create_mapping(chip->irq_domain, offset);
}
```

We could have done that for each GPIO in the `probe` function, and just call `irq_find_mapping()` in `.to_irq` function.

3. Now still in the `probe` function, we need to register our controller's IRQ handler, which in turn is responsible for calling the right handler that raised the interrupt on its pins:

```
devm_request_threaded_irq(client->irq, NULL,
                  mcp23016_irq, irqflags,
                  dev_name(chip->parent), mcp);
```

The function `mcp23016` should have been defined prior to registering the IRQ:

```
static irqreturn_t mcp23016_irq(int irq, void *data)
{
    struct mcp23016 *mcp = data;
    unsigned int child_irq, i;
    /* Do some stuff */
    [...]
    for (i = 0; i < mcp->chip.ngpio; i++) {
        if (gpio_value_changed_and_raised_irq(i)) {
            child_irq =
                irq_find_mapping(mcp->chip.irqdomain, i);
```

```
                handle_nested_irq(child_irq);
        }
    }

    return IRQ_HANDLED;
}
```

`handle_nested_irq()` already descried in the preceding section will create a dedicated thread for each handler registered.

New gpiolib irqchip API

Almost every GPIO controller driver was using IRQ domain for the same purpose. Instead of each of them rolling their own irqdomain handling and so on, kernel developers decided to move that code to gpiolib framework, by means of `GPIOLIB_IRQCHIP` Kconfig symbol, in order to harmonize the development and avoid redundant code.

That portion of code helps in handling management of GPIO irqchips and the associated `irq_domain` and resource allocation callbacks, as well as their setup, using the reduced set of helper functions. These are `gpiochip_irqchip_add()` and `gpiochip_set_chained_irqchip()`.

`gpiochip_irqchip_add()`: This adds an irqchip to a gpiochip. What this function does:

- Sets `gpiochip.to_irq` field to `gpiochip_to_irq`, which is an IRQ callback that just returns `irq_find_mapping(chip->irqdomain, offset)`;
- Allocates an irq_domain to the gpiochip using `irq_domain_add_simple()` function, passing a kernel IRQ core `irq_domain_ops` called `gpiochip_domain_ops` and defined in `drivers/gpio/gpiolib.c`
- Create mapping from 0 to `gpiochip.ngpio` using `irq_create_mapping()` function

Its prototype is as follows:

```
int gpiochip_irqchip_add(struct gpio_chip *gpiochip,
            struct irq_chip *irqchip,
            unsigned int first_irq,
            irq_flow_handler_t handler,
            unsigned int type)
```

Where `gpiochip` is our GPIO chip, the one to add the irqchip to, `irqchip` is the irqchip to add to the gpiochip. `first_irq` if not dynamically assigned, is the base (first) IRQ to allocate gpiochip IRQs from. `handler` is the IRQ handler to use (often a predefined IRQ core function), and `type` is the default type for IRQs on this irqchip, pass `IRQ_TYPE_NONE` to have the core avoid setting up any default type in the hardware.

This function will handle two celled simple IRQs (because it sets `irq_domain_ops.xlate` to `irq_domain_xlate_twocell`) and assumes all the pins on the gpiochip can generate a unique IRQ.

```
static const struct irq_domain_ops gpiochip_domain_ops = {
    .map   = gpiochip_irq_map,
    .unmap = gpiochip_irq_unmap,
    /* Virtually all GPIO irqchips are twocell:ed */
    .xlate = irq_domain_xlate_twocell,
};
```

`gpiochip_set_chained_irqchip()`: This function sets a chained irqchip to a `gpio_chip` from a parent IRQ and passes a pointer to the `struct gpio_chip` as handler data:

```
void gpiochip_set_chained_irqchip(struct gpio_chip *gpiochip,
                    struct irq_chip *irqchip, int parent_irq,
                    irq_flow_handler_t parent_handler)
```

`parent_irq` is the IRQ number to which this chip is connected. In case of our mcp23016 as shown in the figure in the section *Case study-GPIO and IRQ chip*, it corresponds to the IRQ of `gpio4_29` line. In other words, it is the parent IRQ number for this chained irqchip. `parent_handler` is the parent interrupt handler for the accumulated IRQ coming out of the gpiochip. If the interrupt is nested rather than cascaded (chained), pass `NULL` in this handler argument.

With this new API, the only code to add to our `probe` function is:

```
/* Do we have an interrupt line? Enable the irqchip */
if (client->irq) {
    status = gpiochip_irqchip_add(&gpio->chip, &dummy_irq_chip,
                        0, handle_level_irq, IRQ_TYPE_NONE);
    if (status) {
        dev_err(&client->dev, "cannot add irqchip\n");
        goto fail_irq;
    }

    status = devm_request_threaded_irq(&client->dev, client->irq,
                        NULL, mcp23016_irq, IRQF_ONESHOT |
                        IRQF_TRIGGER_FALLING | IRQF_SHARED,
```

```
                    dev_name(&client->dev), gpio);
    if (status)
        goto fail_irq;

    gpiochip_set_chained_irqchip(&gpio->chip,
                    &dummy_irq_chip, client->irq, NULL);
}
```

IRQ core does everything for us. No need to define even the `gpiochip.to_irq` function, since the API already sets it. Our example uses the IRQ core `dummy_irq_chip`, but one could have defined its own as well. Since the v4.10 version of the kernel, two other functions have been added: these are `gpiochip_irqchip_add_nested()` and `gpiochip_set_nested_irqchip()`. Have a look at *Documentation/gpio/driver.txt* for more details. A driver that uses this API in the same kernel version is `drivers/gpio/gpio-mcp23s08.c`.

Interrupt controller and DT

Now we will declare our controller in the DT. If you remember in Chapter 6: *The Concept of Device Tree*, every interrupt controller must have the Boolean property interrupt-controller set. The second mandatory Boolean property is `gpio-controller`, since it is a GPIO controller too. We need to define how many cells are needed for an interrupt specifier for our device. Since we have set the `irq_domain_ops.xlate` field to `irq_domain_xlate_twocell`, `#interrupt-cells` should be 2:

```
expander: mcp23016@20 {
    compatible = "microchip,mcp23016";
    reg = <0x20>;
    interrupt-controller;
    #interrupt-cells = <2>;
    gpio-controller;
    #gpio-cells = <2>;
    interrupt-parent = <&gpio4>;
    interrupts = <29 IRQ_TYPE_EDGE_FALLING>;
};
```

`interrupt-parent` and `interrupts` properties are describing interrupt line connection.

Finally, let us say that we have a driver for mcp23016 and drivers for two other devices: `foo_device` and `bar_device`, all running in the CPU of course. In the `foo_device` driver, one wants to request interrupt for events when `foo_device` changes level on the `io_2` pin of mcp23016. The `bar_device` driver requires `io_8` and `io_12` respectively for reset and power GPIOs. Let us declare this in the DT:

```
foo_device: foo_device@1c {
    reg = <0x1c>;
    interrupt-parent = <&expander>;
    interrupts = <2 IRQ_TYPE_EDGE_RISING>;
};

bar_device {
    reset-gpios = <&expander 8 GPIO_ACTIVE_HIGH>;
    power-gpios = <&expander 12 GPIO_ACTIVE_HIGH>;
    /* Other properties do here */
};
```

Summary

Now IRQ multiplexing has no more secrets for you. We discussed the most important elements of IRQ management under Linux systems, the IRQ domain API. You have the basics to develop interrupt controller drivers, as well as managing their binding from within the DT. IRQ propagation has been discussed in order to understand what happens from the request to the handling. This chapter will help you to understand the interrupt driven part of the next chapter, which deals with input device drivers.

17
Input Devices Drivers

Input devices are devices by which one can interact with the system. Such devices are buttons, keyboards, touchscreens, mouse, and so on. They work by sending events, caught and broadcasted over the system by the input core. This chapter will explain each structure used by input core to handle input devices. That being said, we will see how one can manage events from the user space.

In this chapter, we will cover following topics:

- Input core data structures
- Allocating and registering input devices, and well as polled devices family
- Generating and reporting events to the input core
- Input device from user space
- Writing a driver example

Input device structures

First of all, the main file to include in order to interface with the input subsystem is `linux/input.h`:

```
#include <linux/input.h>
```

Whatever type of input device it is, whatever type of event it sends, an input device is represented in the kernel as an instance of `struct input_dev`:

```
struct input_dev {
    const char *name;
    const char *phys;

    unsigned long evbit[BITS_TO_LONGS(EV_CNT)];
```

Input Devices Drivers

```
    unsigned long keybit[BITS_TO_LONGS(KEY_CNT)];
    unsigned long relbit[BITS_TO_LONGS(REL_CNT)];
    unsigned long absbit[BITS_TO_LONGS(ABS_CNT)];
    unsigned long mscbit[BITS_TO_LONGS(MSC_CNT)];

    unsigned int repeat_key;

    int rep[REP_CNT];
    struct input_absinfo *absinfo;
    unsigned long key[BITS_TO_LONGS(KEY_CNT)];

    int (*open)(struct input_dev *dev);
    void (*close)(struct input_dev *dev);

    unsigned int users;
    struct device dev;

    unsigned int num_vals;
    unsigned int max_vals;
    struct input_value *vals;

    bool devres_managed;
};
```

The meaning of the fields are as follows:

- `name` represents the name of the device.
- `phys` is the physical path to the device in the system hierarchy.
- `evbit` is a bitmap of types of events supported by the device. Some types of areas are as follows:
    - `EV_KEY` for devices supporting sending key events (keyboards, button, and so on)
    - `EV_REL` for device supporting sending relative positions (mouse, digitizers, and so on)
    - `EV_ABS` for device supporting sending absolute positions(joystick)

    The list of events is available in the kernel source in the `include/linux/input-event-codes.h` file. One uses the `set_bit()` macro in order to set the appropriate bit depending on our input device capabilities. Of course a device can support more than one type of event. For example, a mouse will set both `EV_KEY` and `EV_REL`.

```
    set_bit(EV_KEY, my_input_dev->evbit);
    set_bit(EV_REL, my_input_dev->evbit);
```

- `keybit` is for an `EV_KEY` type enabled device, a bitmap of keys/buttons that this device exposes. For example, `BTN_0`, `KEY_A`, `KEY_B`, and so on. The complete list of keys/buttons is in the `include/linux/input-event-codes.h` file.
- `relbit` is for an `EV_REL` type enabled device, a bitmap of relative axes for the device. For example, `REL_X`, `REL_Y`, `REL_Z`, `REL_RX`, and so on. Have a look at `include/linux/input-event-codes.h` for the complete list.
- `absbit` is for an `EV_ABS` type enabled device, bitmap of absolute axes for the device. For example, `ABS_Y`, `ABS_X`, and so on. Have a look at the same previous file for the complete list.
- `mscbit` is for `EV_MSC` type enabled device, a bitmap of miscellaneous events supported by the device.
- `repeat_key` stores the key code of the last key pressed; used to implement software auto repeat.
- `rep`, current values for auto repeat parameters (delay, rate).
- `absinfo` is an array of `&struct input_absinfo` elements holding information about absolute axes (current value, min, max, flat, fuzz, resolution). You should use the `input_set_abs_params()` function in order to set those values.

```
void input_set_abs_params(struct input_dev *dev, unsigned int axis,
                          int min, int max, int fuzz, int flat)
```

- `min` and `max` specify lower and upper bound values. `fuzz` indicates the expected noise on the specified channel of the specified input device. The following is an example in which we set each channel's bound only:

```
#define ABSMAX_ACC_VAL 0x01FF
#define ABSMIN_ACC_VAL -(ABSMAX_ACC_VAL)
[...]
set_bit(EV_ABS, idev->evbit);
input_set_abs_params(idev, ABS_X, ABSMIN_ACC_VAL,
                     ABSMAX_ACC_VAL, 0, 0);
input_set_abs_params(idev, ABS_Y, ABSMIN_ACC_VAL,
                     ABSMAX_ACC_VAL, 0, 0);
input_set_abs_params(idev, ABS_Z, ABSMIN_ACC_VAL,
                     ABSMAX_ACC_VAL, 0, 0);
```

- `key` reflects the current state of the device's keys/buttons.
- `open` is a method called when the very first user calls `input_open_device()`. Use this method to prepare the device, such as interrupt request, polling thread start, and so on.

- `close` is called when the very last user calls `input_close_device()`. Here you can stop polling (which consumes lot of resource).
- `users` stores the number of users (input handlers) that opened this device. It is used by `input_open_device()` and `input_close_device()` to make sure that `dev->open()` is only called when the first user opens the device and `dev->close()` is called when the very last user closes the device.
- `dev` is the struct device associated with this device (for device model).
- `num_vals` is the number of values queued in the current frame.
- `max_vals` is the maximum number of values queued in a frame.
- `Vals` is the array of values queued in the current frame.
- `devres_managed` indicates that devices are managed with `devres` framework and needs not be explicitly unregistered or freed.

Allocating and registering an input device

Prior to the registering and sending the event with an input device, it should be allocated with the `input_allocate_device()` function. In order to free the previously allocated memory for a non-registered input device, `input_free_device()` function should be used. If the device has already been registered, `input_unregister_device()` should be used instead. Like every function where memory allocation is needed, we can use a resource-managed version of functions:

```
struct input_dev *input_allocate_device(void)
struct input_dev *devm_input_allocate_device(struct device *dev)

void input_free_device(struct input_dev *dev)
static void devm_input_device_unregister(struct device *dev,
                                         void *res)
int input_register_device(struct input_dev *dev)
void input_unregister_device(struct input_dev *dev)
```

Device allocation may sleep and therefore must not be called in the atomic context or with a spinlock held.

The following is an excerpt of the `probe` function of an input device siting on the I2C bus:

```
struct input_dev *idev;
int error;

idev = input_allocate_device();
if (!idev)
    return -ENOMEM;

idev->name = BMA150_DRIVER;
idev->phys = BMA150_DRIVER "/input0";
idev->id.bustype = BUS_I2C;
idev->dev.parent = &client->dev;

set_bit(EV_ABS, idev->evbit);
input_set_abs_params(idev, ABS_X, ABSMIN_ACC_VAL,
                    ABSMAX_ACC_VAL, 0, 0);
input_set_abs_params(idev, ABS_Y, ABSMIN_ACC_VAL,
                    ABSMAX_ACC_VAL, 0, 0);
input_set_abs_params(idev, ABS_Z, ABSMIN_ACC_VAL,
                    ABSMAX_ACC_VAL, 0, 0);

error = input_register_device(idev);
if (error) {
    input_free_device(idev);
    return error;
}

error = request_threaded_irq(client->irq,
            NULL, my_irq_thread,
            IRQF_TRIGGER_RISING | IRQF_ONESHOT,
            BMA150_DRIVER, NULL);
if (error) {
    dev_err(&client->dev, "irq request failed %d, error %d\n",
            client->irq, error);
    input_unregister_device(bma150->input);
    goto err_free_mem;
}
```

Polled input device sub-class

Polled input devices is a special type of input device, which relies on polling to sense device state changes, whereas the generic input device type relies on IRQ to sense change and send events to the input core.

A polled input device is described in the kernel as an instance of `struct input_polled_dev` structure, which is a wrapper around the generic `struct input_dev` structure:

```
struct input_polled_dev {
    void *private;

    void (*open)(struct input_polled_dev *dev);
    void (*close)(struct input_polled_dev *dev);
    void (*poll)(struct input_polled_dev *dev);
    unsigned int poll_interval; /* msec */
    unsigned int poll_interval_max; /* msec */
    unsigned int poll_interval_min; /* msec */

    struct input_dev *input;

    bool devres_managed;
};
```

The following are the meanings of elements in this structure:

- `private` is the driver's private data.
- `open` is an optional method that prepares a device for polling (enabled the device and maybe flushes device state).
- `close` is an optional method that is called when the device is no longer being polled. It is used to put devices into the low power mode.
- `poll` is a mandatory method called whenever the device needs to be polled. It is called at the frequency of `poll_interval`.
- `poll_interval` is the frequency at which the `poll()` method should be called. Defaults to 500 msec unless overridden when registering the device.
- `poll_interval_max` specifies the upper bound for the poll interval. Defaults to the initial value of `poll_interval`.
- `poll_interval_min` specifies the lower bound for the poll interval. Defaults to 0.
- `input` is the input device around which the polled device is built. It must be properly initialized by the driver (ID, name, bits). Polled input device just provides an interface to use polling instead of IRQ, to sense device state change.

Allocating/freeing the `struct input_polled_dev` structure is done using `input_allocate_polled_device()` and `input_free_polled_device()`. You should take care of initializing mandatory fields of the `struct input_dev` embedded in it. Polling interval should be set too, otherwise, it defaults to 500 msec. One can use resource manage version too. Both prototypes are as follows:

```
struct input_polled_dev *devm_input_allocate_polled_device(struct
device *dev)
struct input_polled_dev *input_allocate_polled_device(void)
void input_free_polled_device(struct input_polled_dev *dev)
```

For resource-managed devices, the field `input_dev->devres_managed` will be set to true by the input core.

After allocation and proper fields initialization, the polled input device can be registered using `input_register_polled_device()`, which returns 0 on success. The reverse operation (unregister) is done with the `input_unregister_polled_device()` function:

```
int  input_register_polled_device(struct input_polled_dev *dev)
void input_unregister_polled_device(struct input_polled_dev *dev)
```

A typical example of the `probe()` function for such a device looks as follows:

```
static int button_probe(struct platform_device *pdev)
{
    struct my_struct *ms;
    struct input_dev *input_dev;
    int retval;
    ms = devm_kzalloc(&pdev->dev, sizeof(*ms), GFP_KERNEL);
    if (!ms)
        return -ENOMEM;
    ms->poll_dev = input_allocate_polled_device();
    if (!ms->poll_dev) {
        kfree(ms);
        return -ENOMEM;
    }

    /* This gpio is not mapped to IRQ */
    ms->reset_btn_desc = gpiod_get(dev, "reset", GPIOD_IN);

    ms->poll_dev->private = ms ;
    ms->poll_dev->poll = my_btn_poll;
    ms->poll_dev->poll_interval = 200; /* Poll every 200ms */
    ms->poll_dev->open = my_btn_open; /* consist */
```

Input Devices Drivers

```c
    input_dev = ms->poll_dev->input;
    input_dev->name = "System Reset Btn";

    /* The gpio belong to an expander sitting on I2C */
    input_dev->id.bustype = BUS_I2C;
    input_dev->dev.parent = &pdev->dev;

    /* Declare the events generated by this driver */
    set_bit(EV_KEY, input_dev->evbit);
    set_bit(BTN_0, input_dev->keybit); /* buttons */

    retval = input_register_polled_device(mcp->poll_dev);
    if (retval) {
        dev_err(&pdev->dev, "Failed to register input device\n");
        input_free_polled_device(ms->poll_dev);
        kfree(ms);
    }
    return retval;
}
```

The following is how our `struct my_struct` structure looks:

```c
struct my_struct {
    struct gpio_desc *reset_btn_desc;
    struct input_polled_dev *poll_dev;
}
```

And following is how the `open` function looks:

```c
static void my_btn_open(struct input_polled_dev *poll_dev)
{
    struct my_strut *ms = poll_dev->private;
    dev_dbg(&ms->poll_dev->input->dev, "reset open()\n");
}
```

The `open` method is used to prepare resources needed by the device. We do not really need this method for this example.

Generating and reporting an input event

Device allocation and registration are essential, but they are not the main goal of an input device driver, which is designed to report even to the input core. Depending on the type of event your device can support, the kernel provides appropriate APIs to report them to the core.

Given an `EV_XXX` capable device, the corresponding report function would be `input_report_xxx()`. The following table shows a mapping between the most important event types and their report functions:

Event type	Report function	Code example
EV_KEY	input_report_key()	input_report_key(poll_dev->input, BTN_0, gpiod_get_value(ms-> reset_btn_desc) & 1);
EV_REL	input_report_rel()	input_report_rel(nunchuk->input, REL_X, (nunchuk->report.joy_x - 128)/10);
EV_ABS	input_report_abs()	input_report_abs(bma150->input, ABS_X, x_value); input_report_abs(bma150->input, ABS_Y, y_value); input_report_abs(bma150->input, ABS_Z, z_value);

Their respective prototypes are as follows:

```
void input_report_abs(struct input_dev *dev,
                unsigned int code, int value)
void input_report_key(struct input_dev *dev,
                unsigned int code, int value)
void input_report_rel(struct input_dev *dev,
                unsigned int code, int value)
```

Input Devices Drivers

The list of available report functions can be found in `include/linux/input.h` in the kernel source file. They all have the same skeleton:

- `dev` is the input device responsible for the event.
- `code` represents the event code, for example, `REL_X` or `KEY_BACKSPACE`. The complete list is in `include/linux/input-event-codes.h`.
- `value` is the value the event carries. For `EV_REL` event type, it carries the relative change. For a `EV_ABS` (joysticks and so on.) event type, it contains an absolute new value. For `EV_KEY` event type, it should be set to 0 for key release, 1 for key press, and 2 for auto repeat.

After all changes have been reported, the driver should call `input_sync()` on the input device, in order to indicate that this event is complete. The input subsystem will collect these into a single packet and send it through `/dev/input/event<X>`, which is the character device representing our `struct input_dev` on the system, and where `<X>` is the interface number assigned to the driver by the input core:

```
void input_sync(struct input_dev *dev)
```

Let us see an example, which is an excerpt of the `bma150` digital acceleration sensors drivers in `drivers/input/misc/bma150.c`:

```
static void threaded_report_xyz(struct bma150_data *bma150)
{
  u8 data[BMA150_XYZ_DATA_SIZE];
  s16 x, y, z;
  s32 ret;

  ret = i2c_smbus_read_i2c_block_data(bma150->client,
      BMA150_ACC_X_LSB_REG, BMA150_XYZ_DATA_SIZE, data);
  if (ret != BMA150_XYZ_DATA_SIZE)
    return;

  x = ((0xc0 & data[0]) >> 6) | (data[1] << 2);
  y = ((0xc0 & data[2]) >> 6) | (data[3] << 2);
  z = ((0xc0 & data[4]) >> 6) | (data[5] << 2);

  /* sign extension */
  x = (s16) (x << 6) >> 6;
  y = (s16) (y << 6) >> 6;
  z = (s16) (z << 6) >> 6;

  input_report_abs(bma150->input, ABS_X, x);
  input_report_abs(bma150->input, ABS_Y, y);
  input_report_abs(bma150->input, ABS_Z, z);
```

```
    /* Indicate this event is complete */
    input_sync(bma150->input);
}
```

In the preceding sample, `input_sync()` tells the core to consider the three reports as the same event. It makes sense since the position has three axes (X, Y, Z) and we do not want X, Y, or Z to be reported separately.

The best place to report the event is inside the `poll` function for a polled device, or inside the IRQ routine (threaded part or not) for an IRQ enabled device. If you perform some operations that may sleep, you should process your report inside the threaded part of the IRQ handled:

```
static void my_btn_poll(struct input_polled_dev *poll_dev)
{
    struct my_struct *ms = poll_dev->private;
    struct i2c_client *client = mcp->client;
    input_report_key(poll_dev->input, BTN_0,
                    gpiod_get_value(ms->reset_btn_desc) & 1);
    input_sync(poll_dev->input);
}
```

User space interface

Every registered input device is represented by a `/dev/input/event<X>` char device, from which we can read the event from the user space. An application reading this file will receive event packets in the `struct input_event` format:

```
struct input_event {
  struct timeval time;
  __u16 type;
  __u16 code;
  __s32 value;
}
```

Let us see the meaning of each element in the structure:

- `time` is the timestamp, it returns the time at which the event happened.
- `type` is the event type. For example, `EV_KEY` for a key press or release, `EV_REL` for relative moment, or `EV_ABS` for an absolute one. More types are defined in `include/linux/input-event-codes.h`.
- `code` is the event code, for example: `REL_X` or `KEY_BACKSPACE`, again a complete list is in `include/linux/input-event-codes.h`.

- **value** is the value that the event carries. For `EV_REL` event type, it carries the relative change. For an `EV_ABS` (joysticks and so on) event type, it contains the absolute new value. For `EV_KEY` event type, it should be set to `0` for key release, `1` for keypress and `2` for auto repeat.

A user space application can use blocking and non-blocking reads, but also `poll()` or `select()` system calls in order to get notified of events after opening this device. Following is an example with `select()` system call, with the complete source code provided in the book source repository:

```c
#include <unistd.h>
#include <fcntl.h>
#include <stdio.h>
#include <stdlib.h>
#include <linux/input.h>
#include <sys/select.h>

#define INPUT_DEVICE "/dev/input/event1"

int main(int argc, char **argv)
{
    int fd;
    struct input_event event;
    ssize_t bytesRead;

    int ret;
    fd_set readfds;

    fd = open(INPUT_DEVICE, O_RDONLY);
    /* Let's open our input device */
    if(fd < 0){
        fprintf(stderr, "Error opening %s for reading", INPUT_DEVICE);
        exit(EXIT_FAILURE);
    }

    while(1){
        /* Wait on fd for input */
        FD_ZERO(&readfds);
        FD_SET(fd, &readfds);

        ret = select(fd + 1, &readfds, NULL, NULL, NULL);
        if (ret == -1) {
            fprintf(stderr, "select call on %s: an error ocurred",
                    INPUT_DEVICE);
            break;
        }
```

```c
        else if (!ret) { /* If we have decided to use timeout */
            fprintf(stderr, "select on %s: TIMEOUT", INPUT_DEVICE);
            break;
        }
        /* File descriptor is now ready */
        if (FD_ISSET(fd, &readfds)) {
            bytesRead = read(fd, &event,
                            sizeof(struct input_event));
            if(bytesRead == -1)
                /* Process read input error*/
            if(bytesRead != sizeof(struct input_event))
                /* Read value is not an input even */
            /*
             * We could have done a switch/case if we had
             * many codes to look for
             */
            if(event.code == BTN_0) {
                /* it concerns our button */
                if(event.value == 0){
                    /* Process Release */
                    [...]
                }
                else if(event.value == 1){
                    /* Process KeyPress */
                    [...]
                }
            }
        }
    }
    close(fd);
    return EXIT_SUCCESS;
}
```

Putting it all together

So far, we have described structures used when writing drivers for input devices, and how they can be managed from the user space.

1. Allocate a new input device, according to its type, polled or not, using `input_allocate_polled_device()` or `input_allocate_device()`.
2. Fill in the mandatory fields or not (if necessary):

    - Specify type of event the device supports by using `set_bit()` helper macro on the `input_dev.evbit` field

- Depending on event type, `EV_REL`, `EV_ABS`, `EV_KEY` or other, indicate code this device can report using either `input_dev.relbit, input_dev.absbit, input_dev.keybit,` or other.
- Specify `input_dev.dev` in order to set up a proper device tree
- Fill `abs_info` if necessary
- For polled device, indicate at which interval the `poll()` function should be called:

3. Write your `open()` function if necessary, in which you should prepare and set up resource used by the device. This function is called only once. In this function, setup GPIO, request interrupt if needed, initialize the device.
4. Write your `close()` function, in which you will release and deallocate what you have done in the `open()` function. For example, free GPIO, IRQ, put device to power saving mode.
5. Pass either your `open()` or `close()` function (or both) to `input_dev.open` and `input_dev.close` fields.
6. Register your device using `input_register_polled_device()` if polled, or `input_register_device()` if not.
7. In your IRQ function (threaded or not) or in your `poll()` function, gather and report events depending on their types, using either `input_report_key()`, `input_report_rel(), input_report_abs()` or other, and then, call `input_sync()` on the input device to indicate the end of frame (the report is complete).

The usual way is to use classic input devices if no IRQ is provided, or else fall back to polled device:

```
if(client->irq > 0){
    /* Use generic input device */
} else {
    /* Use polled device */
}
```

To see how to manage such devices from the user space, please refer to the example provided within the source of the book.

Driver examples

One can summarize thing in the two following drivers. The first one is a polled input device, based on a GPIO non-mapped to IRQ. The polled input core will poll the GPIO to sense any change. This driver is configured to send 0 key code. Each GPIO state corresponds either to key press or key release:

```c
#include <linux/kernel.h>
#include <linux/module.h>
#include <linux/slab.h>
#include <linux/of.h>                   /* For DT*/
#include <linux/platform_device.h>      /* For platform devices */
#include <linux/gpio/consumer.h>        /* For GPIO Descriptor interface */
#include <linux/input.h>
#include <linux/input-polldev.h>

struct poll_btn_data {
    struct gpio_desc *btn_gpiod;
    struct input_polled_dev *poll_dev;
};

static void polled_btn_open(struct input_polled_dev *poll_dev)
{
    /* struct poll_btn_data *priv = poll_dev->private; */
    pr_info("polled device opened()\n");
}

static void polled_btn_close(struct input_polled_dev *poll_dev)
{
    /* struct poll_btn_data *priv = poll_dev->private; */
    pr_info("polled device closed()\n");
}

static void polled_btn_poll(struct input_polled_dev *poll_dev)
{
    struct poll_btn_data *priv = poll_dev->private;

    input_report_key(poll_dev->input, BTN_0,
gpiod_get_value(priv->btn_gpiod) & 1);
    input_sync(poll_dev->input);
}

static const struct of_device_id btn_dt_ids[] = {
    { .compatible = "packt,input-polled-button", },
    { /* sentinel */ }
};
```

Input Devices Drivers

```c
static int polled_btn_probe(struct platform_device *pdev)
{
    struct poll_btn_data *priv;
    struct input_polled_dev *poll_dev;
    struct input_dev *input_dev;
    int ret;

    priv = devm_kzalloc(&pdev->dev, sizeof(*priv), GFP_KERNEL);
    if (!priv)
        return -ENOMEM;

    poll_dev = input_allocate_polled_device();
    if (!poll_dev){
        devm_kfree(&pdev->dev, priv);
        return -ENOMEM;
    }

    /* We assume this GPIO is active high */
    priv->btn_gpiod = gpiod_get(&pdev->dev, "button", GPIOD_IN);
    poll_dev->private = priv;
    poll_dev->poll_interval = 200; /* Poll every 200ms */
    poll_dev->poll = polled_btn_poll;
    poll_dev->open = polled_btn_open;
    poll_dev->close = polled_btn_close;
    priv->poll_dev = poll_dev;

    input_dev = poll_dev->input;
    input_dev->name = "Packt input polled Btn";
    input_dev->dev.parent = &pdev->dev;

    /* Declare the events generated by this driver */
    set_bit(EV_KEY, input_dev->evbit);
    set_bit(BTN_0, input_dev->keybit); /* buttons */

    ret = input_register_polled_device(priv->poll_dev);
    if (ret) {
        pr_err("Failed to register input polled device\n");
        input_free_polled_device(poll_dev);
        devm_kfree(&pdev->dev, priv);
        return ret;
    }

    platform_set_drvdata(pdev, priv);
    return 0;
}

static int polled_btn_remove(struct platform_device *pdev)
{
```

```
        struct poll_btn_data *priv = platform_get_drvdata(pdev);
        input_unregister_polled_device(priv->poll_dev);
         input_free_polled_device(priv->poll_dev);
         gpiod_put(priv->btn_gpiod);
        return 0;
}

static struct platform_driver mypdrv = {
    .probe      = polled_btn_probe,
    .remove     = polled_btn_remove,
    .driver     = {
        .name          = "input-polled-button",
        .of_match_table = of_match_ptr(btn_dt_ids),
        .owner         = THIS_MODULE,
    },
};
module_platform_driver(mypdrv);

MODULE_LICENSE("GPL");
MODULE_AUTHOR("John Madieu <john.madieu@gmail.com>");
MODULE_DESCRIPTION("Polled input device");
```

This second driver sends events to the input core according to the IRQ on which the button's GPIO is mapped. When using IRQ to sense key press or release, it is a good practice to trig the interrupt on edge change:

```
#include <linux/kernel.h>
#include <linux/module.h>
#include <linux/slab.h>
#include <linux/of.h>                    /* For DT*/
#include <linux/platform_device.h>       /* For platform devices */
#include <linux/gpio/consumer.h>         /* For GPIO Descriptor interface */
#include <linux/input.h>
#include <linux/interrupt.h>

struct btn_data {
   struct gpio_desc *btn_gpiod;
   struct input_dev *i_dev;
   struct platform_device *pdev;
   int irq;
};

static int btn_open(struct input_dev *i_dev)
{
    pr_info("input device opened()\n");
    return 0;
}
```

Input Devices Drivers

```c
static void btn_close(struct input_dev *i_dev)
{
    pr_info("input device closed()\n");
}

static irqreturn_t packt_btn_interrupt(int irq, void *dev_id)
{
    struct btn_data *priv = dev_id;

    input_report_key(priv->i_dev, BTN_0, gpiod_get_value(priv->btn_gpiod) & 1);
    input_sync(priv->i_dev);
    return IRQ_HANDLED;
}

static const struct of_device_id btn_dt_ids[] = {
    { .compatible = "packt,input-button", },
    { /* sentinel */ }
};

static int btn_probe(struct platform_device *pdev)
{
    struct btn_data *priv;
    struct gpio_desc *gpiod;
    struct input_dev *i_dev;
    int ret;

    priv = devm_kzalloc(&pdev->dev, sizeof(*priv), GFP_KERNEL);
    if (!priv)
        return -ENOMEM;

    i_dev = input_allocate_device();
    if (!i_dev)
        return -ENOMEM;

    i_dev->open = btn_open;
    i_dev->close = btn_close;
    i_dev->name = "Packt Btn";
    i_dev->dev.parent = &pdev->dev;
    priv->i_dev = i_dev;
    priv->pdev = pdev;

    /* Declare the events generated by this driver */
    set_bit(EV_KEY, i_dev->evbit);
    set_bit(BTN_0, i_dev->keybit); /* buttons */

    /* We assume this GPIO is active high */
    gpiod = gpiod_get(&pdev->dev, "button", GPIOD_IN);
```

```c
    if (IS_ERR(gpiod))
        return -ENODEV;

    priv->irq = gpiod_to_irq(priv->btn_gpiod);
    priv->btn_gpiod = gpiod;

    ret = input_register_device(priv->i_dev);
    if (ret) {
        pr_err("Failed to register input device\n");
        goto err_input;
    }

    ret = request_any_context_irq(priv->irq,
                    packt_btn_interrupt,
                    (IRQF_TRIGGER_FALLING | IRQF_TRIGGER_RISING),
                    "packt-input-button", priv);
    if (ret < 0) {
        dev_err(&pdev->dev,
            "Unable to acquire interrupt for GPIO line\n");
        goto err_btn;
    }

    platform_set_drvdata(pdev, priv);
    return 0;

err_btn:
    gpiod_put(priv->btn_gpiod);
err_input:
    printk("will call input_free_device\n");
    input_free_device(i_dev);
    printk("will call devm_kfree\n");
    return ret;
}

static int btn_remove(struct platform_device *pdev)
{
    struct btn_data *priv;
    priv = platform_get_drvdata(pdev);
    input_unregister_device(priv->i_dev);
    input_free_device(priv->i_dev);
    free_irq(priv->irq, priv);
    gpiod_put(priv->btn_gpiod);
    return 0;
}

static struct platform_driver mypdrv = {
    .probe      = btn_probe,
    .remove     = btn_remove,
```

```
        .driver      = {
        .name        = "input-button",
        .of_match_table = of_match_ptr(btn_dt_ids),
        .owner       = THIS_MODULE,
        },
};
module_platform_driver(mypdrv);

MODULE_LICENSE("GPL");
MODULE_AUTHOR("John Madieu <john.madieu@gmail.com>");
MODULE_DESCRIPTION("Input device (IRQ based)");
```

For both examples, when a device matches the module, a node will be created in /dev/input directory. The node corresponds to event0 in our example. One can use udevadm tool in order to display information about the device:

```
# udevadm info /dev/input/event0
P: /devices/platform/input-button.0/input/input0/event0
N: input/event0
S: input/by-path/platform-input-button.0-event
E: DEVLINKS=/dev/input/by-path/platform-input-button.0-event
E: DEVNAME=/dev/input/event0
E: DEVPATH=/devices/platform/input-button.0/input/input0/event0
E: ID_INPUT=1
E: ID_PATH=platform-input-button.0
E: ID_PATH_TAG=platform-input-button_0
E: MAJOR=13
E: MINOR=64
E: SUBSYSTEM=input
E: USEC_INITIALIZED=74842430
```

The tool that actually allows us to print the event key to the screen is evtest, given the path of the input device:

```
# evtest /dev/input/event0
input device opened()
Input driver version is 1.0.1
Input device ID: bus 0x0 vendor 0x0 product 0x0 version 0x0
Input device name: "Packt Btn"
Supported events:
Event type 0 (EV_SYN)
Event type 1 (EV_KEY)
Event code 256 (BTN_0)
```

Since the second module is based on IRQ, one can easily check if the IRQ request succeeded, and how many time it has been fired:

```
$ cat /proc/interrupts | grep packt
160:   0  0  0  0  gpio-mxc  0  packt-input-button
```

Finally, one can successively push/release the button, and check whether the GPIO's state changed or not:

```
$ cat /sys/kernel/debug/gpio | grep button
gpio-193 (button-gpio ) in hi
$ cat /sys/kernel/debug/gpio | grep button
gpio-193 (button-gpio ) in lo
```

Summary

This chapter described the whole input framework, and highlighted the difference between polled and interrupt driven input devices. By the end of this chapter, you have the necessary knowledge to write a driver for any input driver, whatever its type, and whatever input event it supports. The user space interface was discussed too, with a sample provided. The next chapter discusses another important framework, the RTC, which is the key element of time management in PC as well as embedded devices.

18
RTC Drivers

Real Time Clock (**RTC**) are devices used to tracks absolute time in nonvolatile memory, which may be internal to the processor, or externally connected through the I2C or SPI bus.

One may use an RTC to do the following:

- Read and set the absolute clock, and generate interrupts during clock updates
- Generate periodic interrupts
- Set alarms

RTCs and the system clock have different purposes. The former is a hardware clock that maintains absolute time and date in a nonvolatile manner, whereas the last is a software clock maintained by the kernel and used to implement the `gettimeofday(2)` and `time(2)` system calls, as well as setting timestamps on files, and so on. The system clock reports seconds and microseconds from a start point, defined to be the POSIX epoch: `1970-01-01 00:00:00 +0000 (UTC)`.

In this chapter, we will cover the following topics:

- Introducing RTC framework API
- Describing such driver's architecture, along with a dummy driver example
- Dealing with alarms
- Managing RTC devices from user space, either through the sysfs interface, or using the hwclock tool

RTC framework data structures

There are three main data structures used by the RTC framework on Linux systems. They are `strcut rtc_time`, `struct rtc_device`, and `struct rtc_class_ops` structures. The former is an opaque structure that represents a given date and time; the second structure represents the physical RTC device; and the last one represents a set of operations exposed by the driver and used by the RTC core to read/update a device's date/time/alarm.

The only header needed to pull RTC functions from within your driver is :

```
#include <linux/rtc.h>
```

The same file contains all of the three structures enumerated in the preceding section:

```
struct rtc_time {
    int tm_sec;   /* seconds after the minute */
    int tm_min;   /* minutes after the hour - [0, 59] */
    int tm_hour;  /* hours since midnight - [0, 23] */
    int tm_mday;  /* day of the month - [1, 31] */
    int tm_mon;   /* months since January - [0, 11] */
    int tm_year;  /* years since 1900 */
    int tm_wday;  /* days since Sunday - [0, 6] */
    int tm_yday;  /* days since January 1 - [0, 365] */
    int tm_isdst; /* Daylight saving time flag */
};
```

This structure is similar to the `struct tm` in `<time.h>`, used to pass time. The next structure is `struct rtc_device`, which represent the chip in the kernel:

```
struct rtc_device {
    struct device dev;
    struct module *owner;

    int id;
    char name[RTC_DEVICE_NAME_SIZE];

    const struct rtc_class_ops *ops;
    struct mutex ops_lock;

    struct cdev char_dev;
    unsigned long flags;

    unsigned long irq_data;
    spinlock_t irq_lock;
    wait_queue_head_t irq_queue;

    struct rtc_task *irq_task;
```

```
    spinlock_t irq_task_lock;
    int irq_freq;
    int max_user_freq;

    struct work_struct irqwork;
};
```

The following are the meanings of the elements of the structure:

- dev: This is the device structure.
- owner: This is the module that owns this RTC device. Using THIS_MODULE will be enough.
- id: This is the global index given to the RTC device by the kernel /dev/rtc<id>.
- name: This is the name given to the RTC device.
- ops: This is a set of operations (like read/set time/alarm) exposed by this RTC device to be managed by the core or from user space.
- ops_lock: This is a mutex used internally by the kernel to protect ops functions call.
- cdev: This is the char device associated to this RTC, /dev/rtc<id>.

The next important structure is struct rtc_class_ops, which is a set of functions used as callback to perform standard and limited on the RTC device. It is the communication interface between top-layer and bottom-layer RTC drivers:

```
struct rtc_class_ops {
    int (*open)(struct device *);
    void (*release)(struct device *);
    int (*ioctl)(struct device *, unsigned int, unsigned long);
    int (*read_time)(struct device *, struct rtc_time *);
    int (*set_time)(struct device *, struct rtc_time *);
    int (*read_alarm)(struct device *, struct rtc_wkalrm *);
    int (*set_alarm)(struct device *, struct rtc_wkalrm *);
    int (*read_callback)(struct device *, int data);
    int (*alarm_irq_enable)(struct device *, unsigned int enabled);
};
```

All of the hooks in the preceding code are given a struct device structure as parameter, which is the same as the one embedded in the struct rtc_device structure. This means that from within these hooks, one can access the RTC device itself at any given time, using the to_rtc_device() macro, which is built on top of the container_of() macro.

```
#define to_rtc_device(d) container_of(d, struct rtc_device, dev)
```

The `open()`, `release()`, and `read_callback()` hooks are internally called by the kernel when the `open()`, `close()`, or `read()` functions are called on the device from user space.

`read_time()` is a driver function that reads the time from the device and fills the `struct rtc_time` output argument. This function should return 0 on success, or else the negative error code.

`set_time()` is a driver function that updates the device's time according to the `struct rtc_time` structure given as the input parameter. Return parameter's remarks are the same as the `read_time` function.

If your device supports an alarm feature, `read_alarm()` and `set_alarm()` should be provided by the driver to read/set the alarm on the device. The `struct rtc_wkalrm` will be described later in the chapter. `alarm_irq_enable()` should be provided too, to enable the alarm.

RTC API

An RTC device is represented in the kernel as an instance of the `struct rtc_device` structure. Unlike other kernel framework devices registrations(where the device is given as parameter to the registering function), the RTC device is built by the core, and registered first before the `rtc_device` structure gets returned to the driver. The device is built and registered with the kernel using the `rtc_device_register()` function:

```
struct rtc_device *rtc_device_register(const char *name,
                        struct device *dev,
                        const struct rtc_class_ops *ops,
                        struct module *owner)
```

One can see the meaning of each parameter of the functions, as follows:

- `name`: This is your RTC device name. It could be the chip's name, for example: ds1343.
- `dev`: This is the parent device, used for device model purposes. For chips sitting on I2C or SPI buses, for example, dev could be set with `spi_device.dev`, or `i2c_client.dev`.
- `ops`: This is your RTC ops, filled according to the features the RTC has, or those your driver can support.
- `owner`: This is the module to which this RTC device belongs. In most cases, `THIS_MODULE` is enough.

The registration should be performed in the `probe` function, and obviously, one can use the resource-managed version of this function:

```
struct rtc_device *devm_rtc_device_register(struct device *dev,
                        const char *name,
                        const struct rtc_class_ops *ops,
                        struct module *owner)
```

Both functions return a pointer on a `struct rtc_device` structure built by the kernel on success, or a pointer error on which you should use `IS_ERR` and `PTR_ERR` macros.

Associated reverse operations are `rtc_device_unregister()` and `devm_rtc_device_unregister()`:

```
void rtc_device_unregister(struct rtc_device *rtc)
void devm_rtc_device_unregister(struct device *dev,
                    struct rtc_device *rtc)
```

Reading and setting time

The driver is responsible for providing functions to read and set the device's time. These are the least an RTC driver can provide. When it comes to reading, the read callback function is given a pointer to an allocated/zeroed `struct rtc_time` structure, which the driver has to fill. Therefore, RTCs almost always store/restitute time in **Binary Coded Decimal (BCD)**, where each quartet (series of 4 bits) represents a number between 0 and 9 (rather than between 0 and 15). The kernel provides two macros, `bcd2bin()` and `bin2bcd()`, to convert respectively from BCD-encoding to decimal, or from decimal to BCD. The next things you should pay attention to are some `rtc_time` fields, which have some boundaries requirements, and where some translation must be done. Data is read in BCD from the device, and should be converted using `bcd2bin()`.

Since the `struct rtc_time` structure is complex, the kernel provides the `rtc_valid_tm()` helper, in order to validate a given `rtc_time` structure, and which returns 0 on success, meaning the structure represents a valid date/time:

```
int rtc_valid_tm(struct rtc_time *tm);
```

RTC Drivers

The following sample describes an RTC-read operation callback:

```c
static int foo_rtc_read_time(struct device *dev, struct rtc_time *tm)
{
    struct foo_regs regs;
    int error;

    error = foo_device_read(dev, &regs, 0, sizeof(regs));
    if (error)
        return error;

    tm->tm_sec = bcd2bin(regs.seconds);
    tm->tm_min = bcd2bin(regs.minutes);
    tm->tm_hour = bcd2bin(regs.cent_hours);
    tm->tm_mday = bcd2bin(regs.date);
    /*
     * This device returns weekdays from 1 to 7
     * But rtc_time.wday expect days from 0 to 6.
     * So we need to substract 1 to the value returned by the chip
     */
    tm->tm_wday = bcd2bin(regs.day) - 1;

    /*
     * This device returns months from 1 to 12
     * But rtc_time.tm_month expect a months 0 to 11.
     * So we need to substract 1 to the value returned by the chip
     */
    tm->tm_mon = bcd2bin(regs.month) - 1;

    /*
     * This device's Epoch is 2000.
     * But rtc_time.tm_year expect years from Epoch 1900.
     * So we need to add 100 to the value returned by the chip
     */
    tm->tm_year = bcd2bin(regs.years) + 100;

    return rtc_valid_tm(tm);
}
```

The following header is necessary prior to using BCD-conversion functions:

```c
#include <linux/bcd.h>
```

RTC Drivers

When it comes to the `set_time` function, a pointer to a `struct rtc_time` is given as an input parameter. This parameter is already filled with values to be stored in the RTC chip. Unfortunately, these are decimal-encoded, and should be converted into BCD prior to being sent to the chip. `bin2bcd` does the conversion. The same attention should be paid to some fields of the `struct rtc_time` structure. The following is a pseudo-code describing a generic `set_time` function:

```
static int foo_rtc_set_time(struct device *dev, struct rtc_time *tm)
{
    regs.seconds = bin2bcd(tm->tm_sec);
    regs.minutes = bin2bcd(tm->tm_min);
    regs.cent_hours = bin2bcd(tm->tm_hour);

    /*
     * This device expects week days from 1 to 7
     * But rtc_time.wday contains week days from 0 to 6.
     * So we need to add 1 to the value given by rtc_time.wday
     */
    regs.day = bin2bcd(tm->tm_wday + 1);
    regs.date = bin2bcd(tm->tm_mday);
    /*
     * This device expects months from 1 to 12
     * But rtc_time.tm_mon contains months from 0 to 11.
     * So we need to add 1 to the value given by rtc_time.tm_mon
     */
    regs.month = bin2bcd(tm->tm_mon + 1);
    /*
     * This device expects year since Epoch 2000
     * But rtc_time.tm_year contains year since Epoch 1900.
     * We can just extract the year of the century with the
     * rest of the division by 100.
     */
    regs.cent_hours |= BQ32K_CENT;
    regs.years = bin2bcd(tm->tm_year % 100);

    return write_into_device(dev, &regs, 0, sizeof(regs));
}
```

RTC's epoch differs from the POSIX epoch, which is only used for the system clock. If the year according to the RTC's epoch and the year register is less than 1970, it is assumed to be 100 years later, that is, between 2000 and 2069.

Driver example

One can summarize the preceding concepts in a simple and fake driver, which simply registers an RTC device on the system:

```
#include <linux/platform_device.h>
#include <linux/module.h>
#include <linux/types.h>
#include <linux/time.h>
#include <linux/err.h>
#include <linux/rtc.h>
#include <linux/of.h>

static int fake_rtc_read_time(struct device *dev, struct rtc_time *tm)
{
    /*
     * One can update "tm" with fake values and then call
     */
    return rtc_valid_tm(tm);
}

static int fake_rtc_set_time(struct device *dev, struct rtc_time *tm)
{
    return 0;
}

static const struct rtc_class_ops fake_rtc_ops = {
    .read_time = fake_rtc_read_time,
    .set_time = fake_rtc_set_time
};

static const struct of_device_id rtc_dt_ids[] = {
    { .compatible = "packt,rtc-fake", },
    { /* sentinel */ }
};

static int fake_rtc_probe(struct platform_device *pdev)
{
    struct rtc_device *rtc;
    rtc = rtc_device_register(pdev->name, &pdev->dev,
                    &fake_rtc_ops, THIS_MODULE);

    if (IS_ERR(rtc))
        return PTR_ERR(rtc);

    platform_set_drvdata(pdev, rtc);
    pr_info("Fake RTC module loaded\n");
```

```
        return 0;
}

static int fake_rtc_remove(struct platform_device *pdev)
{
    rtc_device_unregister(platform_get_drvdata(pdev));
    return 0;
}

static struct platform_driver fake_rtc_drv = {
    .probe = fake_rtc_probe,
    .remove = fake_rtc_remove,
    .driver = {
        .name = KBUILD_MODNAME,
        .owner = THIS_MODULE,
        .of_match_table = of_match_ptr(rtc_dt_ids),
    },
};
module_platform_driver(fake_rtc_drv);

MODULE_LICENSE("GPL");
MODULE_AUTHOR("John Madieu <john.madieu@gmail.com>");
MODULE_DESCRIPTION("Fake RTC driver description");
```

Playing with alarms

RTC alarms are programmable events to be triggered by the device at a given time. An RTC alarm is represented as an instance of the `struct rtc_wkalarm` structure:

```
struct rtc_wkalrm {
unsigned char enabled;    /* 0 = alarm disabled, 1 = enabled */
unsigned char pending;    /* 0 = alarm not pending, 1 = pending */
struct rtc_time time;     /* time the alarm is set to */
};
```

The driver should provide `set_alarm()` and `read_alarm()` operations, to set and read time at which the alarm should occur, as well as `alarm_irq_enable()`, which is a function used to enable/disable the alarm. When the `set_alarm()` function is invoked, it is given as an input parameter, a pointer to a `struct rtc_wkalrm`, whose `.time` field contains the time the alarm must be set to. It is up to the driver to extract each value in a correct manner (using `bin2dcb()` if necessary), and write it into the device in appropriate registers. `rtc_wkalrm.enabled` tell if the alarm should be enabled right after it has been set. If true, the driver must enable the alarm in the chip. The same is true for `read_alarm()` that is given a pointer to `struct rtc_wkalrm`, but as an output parameter this time. The driver has to fill the structure with data read from the device.

{read | set}_alarm() and {read | set}_time() functions behave the same way, except that each pair of functions reads/stores data from/into different sets of registers in the device.

Prior to report alarm event to the system, it is mandatory to connect the RTC chip to an IRQ line of the SoC. It relies on INT line of the RTC driven low when the alarm occur. Depending on the manufacturer, the line remains low until a status register get read, or a special bit get cleared:

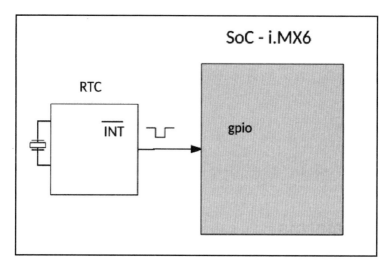

At this point we can use a generic IRQ API, such as request_threaded_irq(), in order to register the alarm IRQ's handler. From within the IRQ handler, it is important to inform the kernel about the RTC IRQ event, using the rtc_update_irq() function:

```
void rtc_update_irq(struct rtc_device *rtc,
            unsigned long num, unsigned long events)
```

- rtc: This is the rtc device that raised the IRQ
- num: This shows how many IRQs are being reported (usually one)
- events: This is a mask of RTC_IRQF with one or more of RTC_PF, RTC_AF, RTC_UF

```
/* RTC interrupt flags */
#define RTC_IRQF 0x80  /* Any of the following is active */
#define RTC_PF   0x40  /* Periodic interrupt */
#define RTC_AF   0x20  /* Alarm interrupt */
#define RTC_UF   0x10  /* Update interrupt for 1Hz RTC */
```

RTC Drivers

That function can be called from any context, atomic or not. The IRQ handler could look as follows:

```
static irqreturn_t foo_rtc_alarm_irq(int irq, void *data)
{
    struct foo_rtc_struct * foo_device = data;
    dev_info(foo_device ->dev, "%s:irq(%d)\n", __func__, irq);
    rtc_update_irq(foo_device ->rtc_dev, 1, RTC_IRQF | RTC_AF);

    return IRQ_HANDLED;
}
```

Keep in mind that RTC devices that have the alarm feature can be used as a wake-up source. That said, the system can be woken up from suspend mode whenever the alarm triggers. This feature relies on the interrupt raised by the RTC device. One declares a device as being wake-up source using the `device_init_wakeup()` function. The IRQ that actually wakes the system up must be registered with the power management core too, using the `dev_pm_set_wake_irq()` function:

```
int device_init_wakeup(struct device *dev, bool enable)
int dev_pm_set_wake_irq(struct device *dev, int irq)
```

We will not discuss power management in detail in this book. The idea is just to give you an overview of how RTC devices may improve your system. The driver `drivers/rtc/rtc-ds1343.c` may help to implement such functions. Let us put everything together by writing a fake `probe` function for an SPI foo RTC device:

```
static const struct rtc_class_ops foo_rtc_ops = {
    .read_time  = foo_rtc_read_time,
    .set_time   = foo_rtc_set_time,
    .read_alarm = foo_rtc_read_alarm,
    .set_alarm  = foo_rtc_set_alarm,
    .alarm_irq_enable = foo_rtc_alarm_irq_enable,
    .ioctl      = foo_rtc_ioctl,
};

static int foo_spi_probe(struct spi_device *spi)
{
    int ret;
    /* initialise and configure the RTC chip */
    [...]

foo_rtc->rtc_dev =
devm_rtc_device_register(&spi->dev, "foo-rtc",
&foo_rtc_ops, THIS_MODULE);
    if (IS_ERR(foo_rtc->rtc_dev)) {
        dev_err(&spi->dev, "unable to register foo rtc\n");
```

```
            return PTR_ERR(priv->rtc);
    }

    foo_rtc->irq = spi->irq;

    if (foo_rtc->irq >= 0) {
        ret = devm_request_threaded_irq(&spi->dev, spi->irq,
                        NULL, foo_rtc_alarm_irq,
                        IRQF_ONESHOT, "foo-rtc", priv);
        if (ret) {
            foo_rtc->irq = -1;
            dev_err(&spi->dev,
                "unable to request irq for rtc foo-rtc\n");
        } else {
            device_init_wakeup(&spi->dev, true);
            dev_pm_set_wake_irq(&spi->dev, spi->irq);
        }
    }

    return 0;
}
```

RTCs and user space

On Linux systems, there are two kernel options one needs to care about in order to properly manage RTCs from user space. These are CONFIG_RTC_HCTOSYS and CONFIG_RTC_HCTOSYS_DEVICE.

CONFIG_RTC_HCTOSYS includes the code file drivers/rtc/hctosys.c in kernel build process, which sets system time from the RTC on startup and resume. Once this option is enabled, the system time will be set using the value read from the specified RTC device. RTC devices should be specified in CONFIG_RTC_HCTOSYS_DEVICE:

```
CONFIG_RTC_HCTOSYS=y
CONFIG_RTC_HCTOSYS_DEVICE="rtc0"
```

In the preceding example, we tell the kernel to set the system time from the RTC, and we specify that the RTC to use is rtc0.

The sysfs interface

The kernel code responsible for instantiating RTC attributes in sysfs is defined in `drivers/rtc/rtc-sysfs.c`, in the kernel source tree. Once registered, an RTC device will create a `rtc<id>` directory under `/sys/class/rtc`. That directory contains a set of read-only attributes, among which the most important are:

- `date`: This file prints the current date of the RTC interface:

    ```
    $ cat /sys/class/rtc/rtc0/date
    2017-08-28
    ```

- `time`: This prints the current time of this RTC:

    ```
    $ cat /sys/class/rtc/rtc0/time
    14:54:20
    ```

- `hctosys`: This attribute indicates whether the RTC device is the one specified in `CONFIG_RTC_HCTOSYS_DEVICE`, meaning that this RTC is used to set system time on startup and resume. Read 1 as true, and 0 as false:

    ```
    $ cat /sys/class/rtc/rtc0/hctosys
    1
    ```

- `dev`: This attribute shows the device's major and minor. Read as major:minor:

    ```
    $ cat /sys/class/rtc/rtc0/dev
    251:0
    ```

- `since_epoch`: This attribute will print the number of seconds elapsed since the UNIX epoch (since January 1rst 1970):

    ```
    $ cat /sys/class/rtc/rtc0/since_epoch
    1503931738
    ```

The hwclock utility

Hardware clock (hwclock) is a tool used to access RTC devices. The `man hwclock` command will probably be much more meaningful than everything discussed in this section. That said, let us write some commands, to set hwclock RTC from the system clock:

```
$ sudo ntpd -q          # make sure system clock is set from network time
$ sudo hwclock --systohc # set rtc from the system clock
$ sudo hwclock --show   # check rtc was set
Sat May 17 17:36:50 2017 -0.671045 seconds
```

The preceding example assumes the host has a network connection on which it can access an NTP server. It is also possible to set the system time manually:

```
$ sudo date -s '2017-08-28 17:14:00' '+%s' #set system clock manually
$ sudo hwclock --systohc #synchronize rtc chip on system time
```

If not given as argument, `hwclock` assumes the RTC device file is `/dev/rtc`, which is actually a symbolic link to the real RTC device:

```
$ ls -l /dev/rtc
lrwxrwxrwx 1 root root 4 août  27 17:50 /dev/rtc -> rtc0
```

Summary

This chapter introduced you to the RTC framework and its API. Its reduced set of functions and data structures make it the most lightweight framework, and easy to master. Using skills described in this chapter, you will be able to develop a driver for most of the existing RTC chips, and even go further and handle such devices from the user space, easily setting up date and time, as well as alarms. The next chapter, PWM drivers, has nothing common with this one, but is a must-know for embedded engineers.

19
PWM Drivers

Pulse Wide Modulation (PWM) operates like a switch that constantly cycles on and off. It is a hardware feature used to control servomotors, for voltage regulation, and so on. The most well-known applications of PWM are:

- Motor speed control
- Light dimming
- Voltage regulation

Now, let us introduce PWM with a simple following figure:

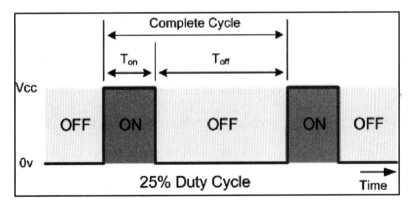

PWM Drivers

The preceding figure describes a complete PWM cycle, introducing some terms we need to clarify prior to getting deeper into the kernel PWM framework:

- `Ton`: This is the duration during which the signal is high.
- `Toff`: This is the duration during which the signal is low.
- `Period`: This is the duration of a complete PWM cycle. It represents the sum of `Ton` and `Toff` of the PWM signal.
- `Duty cycle`: It is represented as a percentage of the time signal that remains on during the period of the PWM signal.

Different formulas are detailed as follows:

- PWM period: $Ton + Toff$
- Duty cycle: $D = \frac{Ton}{Ton+Toff} \times 100 = \frac{Ton}{Period} \times 100$

You can find details about PWM at https://en.wikipedia.org/wiki/Pulse-width_modulation.

The Linux PWM framework has two interfaces:

1. **Controller interface**: The one that exposes the PWM line. It is the PWM chip, that is, the producer.
2. **Consumer interface**: The device consuming PWM lines exposed by the controller. Drivers of such devices use helper functions exported by the controller by means of a generic PWM framework.

Either the consumer or producer interface depends on the following header file:

```
#include <linux/pwm.h>
```

In this chapter, we will deal with:

- PWM driver architecture and data structures, for both controller and consumer, along with a dummy driver
- Instantiating PWM devices and controllers in the device tree
- Requesting and consuming PWM devices
- Using PWM from user space through sysfs interface

PWM controller driver

As you need `struct gpio_chip` when writing GPIO-controller drivers and `struct irq_chip` when writing IRQ-controller drivers, a PWM controller is represented in the kernel as an instance of `struct pwm_chip` structure.

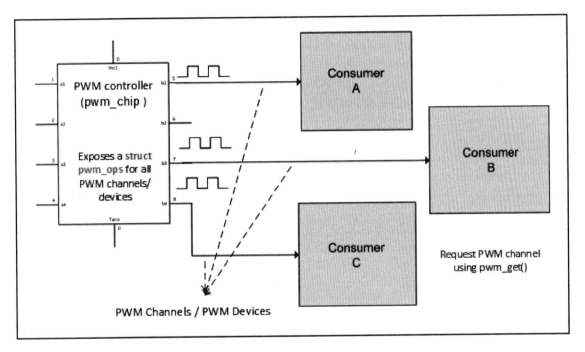

PWM controller and devices

```
struct pwm_chip {
    struct device *dev;
    const struct pwm_ops *ops;
    int base;
    unsigned int npwm;

    struct pwm_device *pwms;
    struct pwm_device * (*of_xlate)(struct pwm_chip *pc,
                    const struct of_phandle_args *args);
    unsigned int of_pwm_n_cells;
    bool can_sleep;
};
```

The following is the meaning of each elements in the structure:

- `dev`: This represents the device associated with this chip.
- `Ops`: This is a data structure providing callback functions this chip exposes to consumer drivers.
- `Base`: This is the number of the first PWM controlled by this chip. If `chip->base < 0` then, the kernel will dynamically assign a base.
- `can_sleep`: This should be set to `true` by the chip driver if `.config()`, `.enable()`, or `.disable()` operations of the ops field may sleep.
- `npwm`: This is the number of PWM channels (devices) this chip provide.
- `pwms`: This is an array of PWM devices of this chip, allocated by the framework, to consumer drivers.
- `of_xlate`: This is an optional callback to request a PWM device given a DT PWM specifier. If not defined, it will be set to `of_pwm_simple_xlate` by the PWM core, which will force `of_pwm_n_cells` to 2 as well.
- `of_pwm_n_cells`: This is the number of cells expected in the DT for a PWM specifier.

PWM controller/chip adding and removal rely on two basic functions, `pwmchip_add()` and `pwmchip_remove()`. Each function should be given a filled in `struct pwm_chip` structure as an argument. Their respective prototypes are as follows:

```
int pwmchip_add(struct pwm_chip *chip)
int pwmchip_remove(struct pwm_chip *chip)
```

Unlike other framework removal functions that do not have return values, `pwmchip_remove()` has a return value. It returns 0 on success, or -EBUSY if the chip has a PWM line still in use (still requested).

Each PWM driver must implement some hooks through the `struct pwm_ops` field, which is used by the PWM core or the consumer interface in order to configure and make full use of its PWM channels. Some of them are optional.

```
struct pwm_ops {
    int (*request)(struct pwm_chip *chip, struct pwm_device *pwm);
    void (*free)(struct pwm_chip *chip, struct pwm_device *pwm);
    int (*config)(struct pwm_chip *chip, struct pwm_device *pwm,
                  int duty_ns, int period_ns);
    int (*set_polarity)(struct pwm_chip *chip, struct pwm_device *pwm,
                        enum pwm_polarity polarity);
    int (*enable)(struct pwm_chip *chip, struct pwm_device *pwm);
    void (*disable)(struct pwm_chip *chip, struct pwm_device *pwm);
```

```
        void (*get_state)(struct pwm_chip *chip, struct pwm_device *pwm,
                    struct pwm_state *state); /* since kernel v4.7 */
        struct module *owner;
};
```

Let us see what each element in the structure means:

- `request`: This is an optional hook that, if provided, is executed during a PWM channel request.
- `free`: This is the same as request, ran during PWM freeing.
- `config`: This is the PMW configuration hook. It configures duty cycles and period length for this PWM.
- `set_polarity`: This hook configures the polarity of this PWM.
- `Enable`: This enables the PWM line, starting output toggling.
- `Disable`: This disables the PWM line, stopping output toggling.
- `Apply`: This atomically applies a new PWM config. The state argument should be adjusted with the real hardware config.
- `get_state`: This returns the current PWM state. This function is only called once per PWM device when the PWM chip is registered.
- `Owner`: This is the module that owns this chip, usually `THIS_MODULE`.

In the `probe` function of the PWM controller driver, it is good practice to retrieve DT resources, initialize hardware, fill a `struct pwm_chip` and its `struct pwm_ops`, and then, add the PWM chip with the `pwmchip_add` function.

Driver example

Now let us summarize things by writing a dummy driver for a PWM controller, which has three channels:

```c
#include <linux/module.h>
#include <linux/of.h>
#include <linux/platform_device.h>
#include <linux/pwm.h>

struct fake_chip {
    struct pwm_chip chip;
    int foo;
    int bar;
    /* put the client structure here (SPI/I2C) */
};
```

PWM Drivers

```c
static inline struct fake_chip *to_fake_chip(struct pwm_chip *chip)
{
    return container_of(chip, struct fake_chip, chip);
}

static int fake_pwm_request(struct pwm_chip *chip,
                            struct pwm_device *pwm)
{
    /*
     * One may need to do some initialization when a PWM channel
     * of the controller is requested. This should be done here.
     *
     * One may do something like
     *      prepare_pwm_device(struct pwm_chip *chip, pwm->hwpwm);
     */

    return 0;
}

static int fake_pwm_config(struct pwm_chip *chip,
                   struct pwm_device *pwm,
                   int duty_ns, int period_ns)
{

    /*
     * In this function, one ne can do something like:
     *      struct fake_chip *priv = to_fake_chip(chip);
     *
     *      return send_command_to_set_config(priv,
     *                  duty_ns, period_ns);
     */

    return 0;
}

static int fake_pwm_enable(struct pwm_chip *chip, struct pwm_device *pwm)
{
    /*
     * In this function, one ne can do something like:
     *   struct fake_chip *priv = to_fake_chip(chip);
     *
     * return foo_chip_set_pwm_enable(priv, pwm->hwpwm, true);
     */
    pr_info("Somebody enabled PWM device number %d of this chip",
            pwm->hwpwm);
    return 0;
}
```

```c
static void fake_pwm_disable(struct pwm_chip *chip,
                             struct pwm_device *pwm)
{
    /*
     * In this function, one ne can do something like:
     *   struct fake_chip *priv = to_fake_chip(chip);
     *
     * return foo_chip_set_pwm_enable(priv, pwm->hwpwm, false);
     */
     pr_info("Somebody disabled PWM device number %d of this chip",
             pwm->hwpwm);
}

static const struct pwm_ops fake_pwm_ops = {
    .request = fake_pwm_request,
    .config  = fake_pwm_config,
    .enable  = fake_pwm_enable,
    .disable = fake_pwm_disable,
    .owner   = THIS_MODULE,
};

static int fake_pwm_probe(struct platform_device *pdev)
{
    struct fake_chip *priv;

    priv = devm_kzalloc(&pdev->dev, sizeof(*priv), GFP_KERNEL);
    if (!priv)
          return -ENOMEM;

    priv->chip.ops  = &fake_pwm_ops;
    priv->chip.dev  = &pdev->dev;
    priv->chip.base = -1;    /* Dynamic base */
    priv->chip.npwm = 3;     /* 3 channel controller */

    platform_set_drvdata(pdev, priv);
    return pwmchip_add(&priv->chip);
}

static int fake_pwm_remove(struct platform_device *pdev)
{
    struct fake_chip *priv = platform_get_drvdata(pdev);
    return pwmchip_remove(&priv->chip);
}

static const struct of_device_id fake_pwm_dt_ids[] = {
    { .compatible = "packt,fake-pwm", },
    { }
};
```

```
MODULE_DEVICE_TABLE(of, fake_pwm_dt_ids);

static struct platform_driver fake_pwm_driver = {
   .driver = {
        .name = KBUILD_MODNAME,
.owner = THIS_MODULE,
        .of_match_table = of_match_ptr(fake_pwm_dt_ids),
   },
   .probe = fake_pwm_probe,
   .remove = fake_pwm_remove,
};
module_platform_driver(fake_pwm_driver);

MODULE_AUTHOR("John Madieu <john.madieu@gmail.com>");
MODULE_DESCRIPTION("Fake pwm driver");
MODULE_LICENSE("GPL");
```

PWM controller binding

While binding the PWM controller from within the DT, the most important property is `#pwm-cells`. It represents the number of cells used to represent a PWM device of this controller. If you remember, in the `struct pwm_chip` structure, the `of_xlate` hook is used to translate a given PWM specifier. If the hook has not been set, `pwm-cells` here must be set to 2, else, it should be set with the same value as `of_pwm_n_cells`. The following is an example of a PWM controller node in the DT, for an i.MX6 SoC.

```
pwm3: pwm@02088000 {
    #pwm-cells = <2>;
    compatible = "fsl,imx6q-pwm", "fsl,imx27-pwm";
    reg = <0x02088000 0x4000>;
    interrupts = <0 85 IRQ_TYPE_LEVEL_HIGH>;
    clocks = <&clks IMX6QDL_CLK_IPG>,
        <&clks IMX6QDL_CLK_PWM3>;
    clock-names = "ipg", "per";
    status = "disabled";
};
```

On the other hand, the node that corresponds to our fake-pwm driver looks like:

```
fake_pwm: pwm@0 {
    #pwm-cells = <2>;
    compatible = "packt,fake-pwm";
    /*
     * Our driver does not use resource
     * neither mem, IRQ, nor Clock)
     */
};
```

PWM consumer interface

A consumer is the device that actually uses PWM channels. A PWM channel is represented in the kernel as an instance of struct pwm_device structure:

```
struct pwm_device {
    const char *label;
    unsigned long flags;
    unsigned int hwpwm;
    unsigned int pwm;
    struct pwm_chip *chip;
    void *chip_data;

    unsigned int period;     /* in nanoseconds */
    unsigned int duty_cycle; /* in nanoseconds */
    enum pwm_polarity polarity;
};
```

- Label: This is the name of this PWM device
- Flags: This represents the flags associated with the PWM device
- hwpw: This is a relative index of the PWM device, local to the chip
- pwm: This is a system global index of the PWM device
- chip: This is a PWM chip, the controller providing this PWM device
- chip_data: This is chip-private data associated with this PWM device

Since kernel v4.7, the structure changed into:

```
struct pwm_device {
    const char *label;
    unsigned long flags;
    unsigned int hwpwm;
    unsigned int pwm;
    struct pwm_chip *chip;
    void *chip_data;

    struct pwm_args args;
    struct pwm_state state;
};
```

- `args`: This represents the board-dependent PWM arguments attached to this PWM device, which are usually retrieved from the PWM lookup table or device tree. PWM arguments represent the initial configuration that users want to use on this PWM device rather than the current PWM hardware state.
- `state`: This represents the current PWM channel state.

```
struct pwm_args {
    unsigned int period; /* Device's nitial period */
    enum pwm_polarity polarity;
};

struct pwm_state {
    unsigned int period; /* PWM period (in nanoseconds) */
    unsigned int duty_cycle; /* PWM duty cycle (in nanoseconds) */
    enum pwm_polarity polarity; /* PWM polarity */
    bool enabled; /* PWM enabled status */
}
```

Over Linux evolutions, the PWM framework faced several changes. These changes concern the way one requests PWM devices from within the consumer side. We can split the consumer interface into two parts, or more precisely into two versions.

The legacy version, where you use `pwm_request()` and `pwm_free()` in order to request a PWM device and free it after usage.

The new and recommended API, using `pwm_get()` and `pwm_put()` functions. The former is given the consumer device along with the channel name as arguments to request the PWM device, and the second is given the PWM device to be freed as a parameter. Managed variants of these functions, `devm_pwm_get()` and `devm_pwm_put()`, also exist.

```
struct pwm_device *pwm_get(struct device *dev, const char *con_id)
void pwm_put(struct pwm_device *pwm)
```

`pwm_request()`/`pwm_get()` and `pwm_free()`/`pwm_put()` cannot be called from an atomic context, since the PWM core make use of mutexes, which may sleep.

After being requested, a PWM has to be configured using:

```
int pwm_config(struct pwm_device *pwm, int duty_ns, int period_ns);
```

To start/stop toggling the PWM output, use `pwm_enable()`/`pwm_disable()`. Both functions take a pointer to a `struct pwm_device` as a parameter, and are all wrappers around hooks exposed by the controller through the `pwm_chip.pwm_ops` field.

```
int pwm_enable(struct pwm_device *pwm)
void pwm_disable(struct pwm_device *pwm)
```

`pwm_enable()` returns 0 on success, or a negative error code on failure. A good example of a PWM consumer driver is `drivers/leds/leds-pwm.c` in the kernel source tree. The following is an example of consumer code, driving a PWM led:

```
static void pwm_led_drive(struct pwm_device *pwm,
                struct private_data *priv)
{
    /* Configure the PWM, applying a period and duty cycle */
    pwm_config(pwm, priv->duty, priv->pwm_period);

    /* Start toggling */
    pwm_enable(pchip->pwmd);

    [...] /* Do some work */

    /* And then stop toggling*/
    pwm_disable(pchip->pwmd);
}
```

PWM clients binding

PWM devices can be assigned to the consumer from:

- Device tree
- ACPI
- Static lookup tables, in board `init` file.

PWM Drivers

This book will only deal with DT binding, as it is the recommended method. When binding a PWM consumer (client) to its driver, you need to provide the phandle of the controller to which it is linked.

It is recommended you give the name `pwms` to PWM properties; since PWM devices are named resources, you can provide an optional property `pwm-names`, containing a list of strings to name each of the PWM devices listed in the `pwms` property. In case no `pwm-names` property is given, the name of the user node will be used as fallback.

Drivers for devices that use more than a single PWM device can use the `pwm-names` property to map the name of the PWM device requested by the `pwm_get()` call to an index into the list given by the `pwms` property.

The following example describes a PWM-based backlight device, which is an excerpt from the kernel documentation on PWM device binding (see *Documentation/devicetree/bindings/pwm/pwm.txt*):

```
pwm: pwm {
    #pwm-cells = <2>;
};

[...]

bl: backlight {
pwms = <&pwm 0 5000000>;
    pwm-names = "backlight";
};
```

The PWM-specifier typically encodes the chip-relative PWM number and the PWM period in nanoseconds. With the line as follows:

```
pwms = <&pwm 0 5000000>;
```

0 corresponds to the PWM index relative to the controller, and 5000000 represents the period in nanoseconds. Note that in the preceding example, specifying the `pwm-names` is redundant because the name `backlight` would be used as a fallback anyway. Therefore, the driver would have to call:

```
static int my_consummer_probe(struct platform_device *pdev)
{
    struct pwm_device *pwm;

    pwm = pwm_get(&pdev->dev, "backlight");
    if (IS_ERR(pwm)) {
        pr_info("unable to request PWM, trying legacy API\n");
        /*
```

```
         * Some drivers use the legacy API as fallback, in order
         * to request a PWM ID, global to the system
         * pwm = pwm_request(global_pwm_id, "pwm beeper");
         */
    }

    [...]
    return 0;
}
```

 The PWM-specifier typically encodes the chip-relative PWM number and the PWM period in nanoseconds.

Using PWMs with the sysfs interface

The PWM core `sysfs` root path is `/sys/class/pwm/`. It is the user space way to manage PWM device. Each PWM controller/chip added to the system creates a `pwmchipN` directory entry under the `sysfs` root path, where N is the base of the PWM chip. The directory contains the following files:

- `npwm`: This is a reads only file printing the number of PWM channels that this chip supports
- `Export`: This is a write-only file allowing to export a PWM channel for use with `sysfs` (this functionality is equivalent to GPIO sysfs interface)
- `Unexport`: Unexports a PWM channel from `sysfs` (write-only)

The PWM channels are numbered using an index from 0 to `pwm<n-1>`. These numbers are local to the chip. Each PWM channel exportation creates a `pwmX` directory in the `pwmchipN`, which is the same directory as the one containing the `export` file used. X is the number of the channel that was exported. Each channel directory contains the following files:

- `Period`: This is a readable/writable file to get/set the total period of the PWM signal. Value is in nanoseconds.
- `duty_cycle`: This is a readable/writable file to get/set the duty cycle of the PWM signal. It represents the active time of the PWM signal. Value is in nanoseconds and must always be less than the period.

- `Polarity`: This is a readable/writable file to use only if the chip of this PWM device supports polarity inversion. It is better to change the polarity only when this PWM is not enabled. Accepted values are string *normal* or *inversed*.
- `Enable`: This is a readable/writable file, to enable (start toggling)/disable (stop toggling) the PWM signal. Accepted values are:

- 0: disabled
- 1: enabled

The following is an example of using PWM from user space through the `sysfs` interface:

1. Enable PWM:

   `# echo 1 > /sys/class/pwm/pwmchip<pwmchipnr>/pwm<pwmnr>/enable`

2. Set PWM period:

   `# echo <value in nanoseconds> > /sys/class/pwm/pwmchip<pwmchipnr>/pwm<pwmnr>/period`

3. Set PWM duty cycle: The value of the duty cycle must be less than the value of PWM period:

   `# echo <value in nanoseconds> > /sys/class/pwm/pwmchip<pwmchipnr>/pwm<pwmnr>/duty_cycle`

4. Disable PWM:

   `# echo 0 > /sys/class/pwm/pwmchip<pwmchipnr>/pwm<pwmnr>/enable`

The complete PWM framework API and sysfs description is available in the *Documentation/pwm.txt* file, in the kernel source tree.

Summary

By the end of this chapter, you are armed for any PWM controller, whether it is memory mapped, or externally sitting on a bus. The API described in this chapter will be sufficient to write as well as to enhance a controller driver as a consumer device driver. If you are not comfortable with the PWM kernel side yet, you can fully use the user space sysfs interface. That said, in the next chapter, we will discuss about regulators, which sometimes are driven by PWM. So, please hold on, we are almost done.

20
Regulator Framework

A regulator is an electronic device that supplies power to other devices. Devices powered by regulators are called consumers. One said they consume power provided by regulators. Most regulators can enable and disable their output and some can also control their output voltage or current. The driver should expose those capabilities to consumers by means of specific functions and data structures, which we will discuss in this chapter.

The chip that physically provides regulators is called a **Power Management Integrated Circuit (PMIC)**:

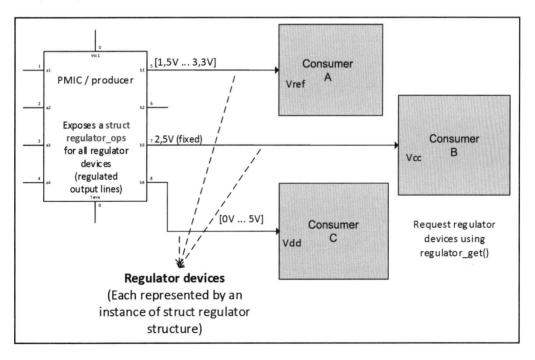

The Linux regulator framework has been designed to interface and control voltage and current regulators. It is divided into four separate interfaces as follows:

- A regulator drivers interface for regulator PMIC drivers. The structure of this interface can be found in `include/linux/regulator/driver.h`.
- A consumer interface for device drivers.
- A machine interface for board configuration.
- A sysfs interface for user space.

In this chapter, we will cover the following topics:

- Introducing the PMIC/producer driver interface, driver methods and data structures
- A case study with ISL6271A MIC driver, as well as a dummy regulator for testing purpose
- A regulator consumer interface along with its API
- Regulator (producer/consumer) binding in DT

PMIC/producer driver interface

The producer is the device generating the regulated voltage or current. The name of such a device is PMIC and it can be used for power sequencing, battery management, DC-to-DC conversion, or simple power switches (on/off). It regulates the output power from the input power, with the help of (and under) software control.

It deals with regulator drivers, and especially the producer PMIC side, which requires a few headers:

```
#include <linux/platform_device.h>
#include <linux/regulator/driver.h>
#include <linux/regulator/of_regulator.h>
```

Driver data structures

We will start with a short walkthrough of data structures used by the regulator framework. Only the producer interface is described in this section.

Description structure

The kernel describes every regulator provided by a PMIC by means of a `struct regulator_desc` structure, which characterizes a regulator. By regulator, I mean any independent regulated output. For example, the ISL6271A from Intersil is a PMIC with three independent regulated outputs. There should then be three instances of `regulator_desc` in its driver. This structure, which contains the fixed properties of a regulator, looks like the following:

```
struct regulator_desc {
    const char *name;
    const char *of_match;

    int id;
    unsigned n_voltages;
    const struct regulator_ops *ops;
    int irq;
    enum regulator_type type;
    struct module *owner;

    unsigned int min_uV;
    unsigned int uV_step;
};
```

Let us omit some fields for simplicity reasons. Full structure definition is available in `include/linux/regulator/driver.h`:

- `name` holds the name of the regulator.
- `of_match` holds the name used to identify the regulator in DT.
- `id` is a numerical identifier for the regulator.
- `owner` represents the module providing the regulator. Set this field to `THIS_MODULE`.
- `type` indicates if the regulator is a voltage regulator or a current regulator. It can either be `REGULATOR_VOLTAGE` or `REGULATOR_CURRENT`. Any other value will result in a regulator registering failure.
- `n_voltages` indicates the number of selectors available for this regulator. It represents the numerical value that the regulator can output. For fixed output voltage, `n_voltages` should be set to 1.
- `min_uV` indicates the minimum voltage value this regulator can provide. It is the voltage given by the lowest selector.
- `uV_step` represents the voltage increase with each selector.

- `ops` represents the regulator operations table. It is a structure pointing to a set of operation callbacks that the regulator can support. This field is discussed later.
- `irq` is the interrupt number of the regulator.

Constraints structure

When a PMIC exposes a regulator to consumers, it has to impose some nominal limits for this regulator with the help of the `struct regulation_constraints` structure. It is a structure gathering security limit of the regulator and defines boundaries the consumers cannot cross. It is a kind of a contract between the regulator driver and the consumer driver:

```
struct regulation_constraints {
    const char *name;

    /* voltage output range (inclusive) - for voltage control */
    int min_uV;
    int max_uV;

    int uV_offset;

    /* current output range (inclusive) - for current control */
    int min_uA;
    int max_uA;

    /* valid regulator operating modes for this machine */
    unsigned int valid_modes_mask;

    /* valid operations for regulator on this machine */
    unsigned int valid_ops_mask;

    struct regulator_state state_disk;
    struct regulator_state state_mem;
    struct regulator_state state_standby;
    suspend_state_t initial_state; /* suspend state to set at init */

    /* mode to set on startup */
    unsigned int initial_mode;

    /* constraint flags */
    unsigned always_on:1;    /* regulator never off when system is on */
    unsigned boot_on:1;      /* bootloader/firmware enabled regulator */
    unsigned apply_uV:1;     /* apply uV constraint if min == max */
};
```

Let us describe each element in the structure:

- `min_uV`, `min_uA`, `max_uA`, and `max_uV` are the smallest voltage/current values that the consumers may set.
- `uV_offset` is the offset applied to voltages from the consumer to compensate for voltage drops.
- `valid_modes_mask` and `valid_ops_mask` respectively are masks of modes/operations which may be configured/performed by consumers.
- `always_on` should be set if the regulator should never be disabled.
- `boot_on` should be set if the regulator is enabled when the system is initially started. If the regulator is not enabled by the hardware or bootloader then it will be enabled when the constraints are applied.
- `name` is a descriptive name for the constraints used for display purposes.
- `apply_uV` applies the voltage constraint when initializing.
- `input_uV` represents the input voltage for this regulator when it is supplied by another regulator.
- `state_disk`, `state_mem`, and `state_standby` define the state for the regulator when the system is suspended in the disk mode, mem mode, or in standby.
- `initial_state` indicates the suspended state is set by default.
- `initial_mode` is the mode to set at startup.

init data structure

There are two ways to pass `regulator_init_data` to a driver; this can be done by platform data in the board initialization file or by a node in the device tree using the `of_get_regulator_init_data` function:

```
struct regulator_init_data {
    struct regulation_constraints constraints;

    /* optional regulator machine specific init */
    int (*regulator_init)(void *driver_data);
    void *driver_data;      /* core does not touch this */
};
```

The following are the meanings of elements in the structure:

- `constraints` represents the regulator constraints
- `regulator_init` is an optional callback invoked at a given moment when the core registers the regulator
- `driver_data` represents the data passed to `regulator_init`

As one can see, the `struct constraints` structure is part of the `init data`. This is explained by the fact that at the initialization of the regulator, its constraint is directly applied to it, far before any consumer can use it.

Feeding init data into a board file

This method consists of filling an array of constraints, either from within the driver, or in the board file, and using it as part of the platform data. The following is the sample based on the device from the case study, the ISL6271A from Intersil:

```
static struct regulator_init_data isl_init_data[] = {
    [0] = {
            .constraints = {
                .name            = "Core Buck",
                .min_uV          = 850000,
                .max_uV          = 1600000,
                .valid_modes_mask = REGULATOR_MODE_NORMAL
                                 | REGULATOR_MODE_STANDBY,
                .valid_ops_mask   = REGULATOR_CHANGE_MODE
                                 | REGULATOR_CHANGE_STATUS,
            },
    },
    [1] = {
            .constraints = {
                .name            = "LDO1",
                .min_uV          = 1100000,
                .max_uV          = 1100000,
                .always_on       = true,
                .valid_modes_mask = REGULATOR_MODE_NORMAL
                                 | REGULATOR_MODE_STANDBY,
                .valid_ops_mask   = REGULATOR_CHANGE_MODE
                                 | REGULATOR_CHANGE_STATUS,
            },
    },
    [2] = {
            .constraints = {
                .name            = "LDO2",
                .min_uV          = 1300000,
```

```
                    .max_uV          = 1300000,
                    .always_on       = true,
                    .valid_modes_mask = REGULATOR_MODE_NORMAL
                                      | REGULATOR_MODE_STANDBY,
                    .valid_ops_mask   = REGULATOR_CHANGE_MODE
                                      | REGULATOR_CHANGE_STATUS,
            },
        },
};
```

This method is now depreciated, though it is presented here for your information. The new and recommended approach is the DT, which is described in the next section.

Feeding init data into the DT

In order to extract init data passed from within the DT, there is a new data type that we need to introduce, `struct of_regulator_match`, which looks like this:

```
struct of_regulator_match {
    const char *name;
    void *driver_data;
    struct regulator_init_data *init_data;
    struct device_node *of_node;
    const struct regulator_desc *desc;
};
```

Prior to making any use of this data structure, we need to figure out how to achieve the regulator binding of a DT file.

Every PMIC node in the DT should have a sub-node named `regulators`, in which we have to declare each of the regulators this PMIC provides as a dedicated sub-node. In other words, every regulator of a PMIC is defined as a sub-node of the `regulators` node, which in turn is a child of the PMIC node in the DT.

There are standardized properties you can define in a regulator node:

- `regulator-name`: This is a string used as a descriptive name for regulator outputs
- `regulator-min-microvolt`: This is the smallest voltage that consumers may set
- `regulator-max-microvolt`: This is the largest voltage consumers may set
- `regulator-microvolt-offset`: This is the offset applied to voltages to compensate for voltage drops

Regulator Framework

- `regulator-min-microamp`: This is the smallest current consumers may set
- `regulator-max-microamp`: This is the largest current consumers may set
- `regulator-always-on`: This is a Boolean value, indicated if the regulator should never be disabled
- `regulator-boot-on`: This is a bootloader/firmware enabled regulator
- `<name>-supply`: This is a phandle to the parent supply/regulator node
- `regulator-ramp-delay`: This is the ramp delay for the regulator (in uV/uS)

Those properties really look like fields in `struct regulator_init_data`. Back with the `ISL6271A` driver, its DT entry could look like this:

```
isl6271a@3c {
    compatible = "isl6271a";
    reg = <0x3c>;
    interrupts = <0 86 0x4>;

    /* supposing our regulator is powered by another regulator */
    in-v1-supply = <&some_reg>;
    [...]

    regulators {
        reg1: core_buck {
            regulator-name = "Core Buck";
            regulator-min-microvolt = <850000>;
            regulator-max-microvolt = <1600000>;
        };

        reg2: ldo1 {
            regulator-name = "LDO1";
            regulator-min-microvolt = <1100000>;
            regulator-max-microvolt = <1100000>;
            regulator-always-on;
        };

        reg3: ldo2 {
            regulator-name = "LDO2";
            regulator-min-microvolt = <1300000>;
            regulator-max-microvolt = <1300000>;
            regulator-always-on;
        };
    };
};
```

Using the kernel helper function `of_regulator_match()`, given the `regulators` sub-node as the parameter, the function will walk through each regulator device node and build a `struct init_data` structure for each of them. There is an example in the `probe()` function, discussed in the driver methods section.

Configuration structure

Regulator devices are configured by means of the `struct regulator_config` structure, which holds variable elements of the regulator description. This structure is passed to the framework when it comes to registering a regulator with the core:

```
struct regulator_config {
    struct device *dev;
    const struct regulator_init_data *init_data;
    void *driver_data;
    struct device_node *of_node;
};
```

- `dev` represents the struct device structure the regulator belongs to.
- `init_data` is the most important field of the structure, since it contains an element holding the regulator constraints (a machine specific structure).
- `driver_data` holds the regulator's private data.
- `of_node` is for DT capable drivers. It is the node to parse for DT bindings. It is up to the developer to set this field. It may be `NULL` also.

Device operation structure

The `struct regulator_ops` structure is a list of callbacks representing all operations a regulator can perform. These callbacks are helpers and are wrapped by generic kernel functions:

```
struct regulator_ops {
    /* enumerate supported voltages */
    int (*list_voltage) (struct regulator_dev *,
                         unsigned selector);

    /* get/set regulator voltage */
    int (*set_voltage) (struct regulator_dev *,
                        int min_uV, int max_uV,
                        unsigned *selector);
    int (*map_voltage)(struct regulator_dev *,
                       int min_uV, int max_uV);
```

```c
        int (*set_voltage_sel) (struct regulator_dev *,
                        unsigned selector);
        int (*get_voltage) (struct regulator_dev *);
        int (*get_voltage_sel) (struct regulator_dev *);

        /* get/set regulator current */
        int (*set_current_limit) (struct regulator_dev *,
                        int min_uA, int max_uA);
        int (*get_current_limit) (struct regulator_dev *);

        int (*set_input_current_limit) (struct regulator_dev *,
                        int lim_uA);
        int (*set_over_current_protection) (struct regulator_dev *);
        int (*set_active_discharge) (struct regulator_dev *,
                        bool enable);

        /* enable/disable regulator */
        int (*enable) (struct regulator_dev *);
        int (*disable) (struct regulator_dev *);
        int (*is_enabled) (struct regulator_dev *);

        /* get/set regulator operating mode (defined in consumer.h) */
        int (*set_mode) (struct regulator_dev *, unsigned int mode);
        unsigned int (*get_mode) (struct regulator_dev *);
};
```

Callback names explain quite well what they do. There are other callbacks that are not listed here, for which you must enable the appropriate mask in `valid_ops_mask` or `valid_modes_mask` of the regulator's constraints before the consumer can use them. Available operation mask flags are defined in `include/linux/regulator/machine.h`.

Therefore, given a `struct regulator_dev` structure, one can get the ID of the corresponding regulator by calling the `rdev_get_id()` function:

```c
int rdev_get_id(struct regulator_dev *rdev)
```

Driver methods

Driver methods consist of `probe()` and `remove()` functions. Please refer to the preceding data structure if this section seems unclear to you.

Probe function

The `probe` function of a PMIC driver can be split into a few steps, enumerated as follows:

1. Define an array of `struct regulator_desc` objects for all the regulators provided by this PMIC. In this step, you should have defined a valid `struct regulator_ops` to be linked to the appropriate `regulator_desc`. It could be the same `regulator_ops` for all, assuming they all support the same operations.
2. Now in the `probe` function, for each regulator:

    - Fetch the appropriate `struct regulator_init_data` either from the platform data, which must already contain a valid `struct regulation_constraints` or build a `struct regulation_constraints` from DT, in order to build a new `struct regulator_init_data` object.
    - Use the previous `struct regulator_init_data` to set up a `struct regulator_config` structure. If the driver supports DT, one can make `regulator_config.of_node` point to the node used to extract the regulator properties.
    - Call `regulator_register()` (or the managed version `devm_regulator_register()`) to register the regulator with the core, giving the previous `regulator_desc` and `regulator_config` as parameters.

A regulator is registered with the kernel using the `regulator_register()` function, or `devm_regulator_register()`, which is the managed version:

```
struct regulator_dev * regulator_register(const struct regulator_desc
*regulator_desc, const struct regulator_config *cfg)
```

This function returns a data type we have not discussed so far: a `struct regulator_dev` object, defined in `include/linux/regulator/driver.h`. That structure represents an instance of a regulator device from the producer side (it is different in the consumer side). Instances of the `struct regulator_dev` structure should not be used directly by anything except the regulator core and notification injection (which should take the mutex and not other direct access). That being said, to keep track of the registered regulator from within the driver, one should hold references for each `regulator_dev` object returned by the registering function.

Remove function

The `remove()` function is where every operation performed earlier during the `probe`. Therefore, the essential function you should keep in mind is `regulator_unregister()`, when it comes to removing a regulator from the system:

```
void regulator_unregister(struct regulator_dev *rdev)
```

This function accepts a pointer to a `struct regulator_dev` structure as a parameter. This is another reason a reference for each registered regulator should be kept. The following is the `remove` function of the ISL6271A driver:

```
static int __devexit isl6271a_remove(struct i2c_client *i2c)
{
    struct isl_pmic *pmic = i2c_get_clientdata(i2c);
    int i;

    for (i = 0; i < 3; i++)
        regulator_unregister(pmic->rdev[i]);

    kfree(pmic);
    return 0;
}
```

Case study: Intersil ISL6271A voltage regulator

As a recall, this PMIC provides three regulator's devices, among which only one can have its output value changed. The two others provide fixed voltages:

```
struct isl_pmic {
    struct i2c_client   *client;
    struct regulator_dev    *rdev[3];
    struct mutex        mtx;
};
```

Regulator Framework

First we define ops callbacks, to set up a `struct regulator_desc`:

1. Callback to handle a `get_voltage_sel` operation:

    ```
    static int isl6271a_get_voltage_sel(struct regulator_dev *rdev)
    {
        struct isl_pmic *pmic = rdev_get_drvdata(dev);
        int idx = rdev_get_id(rdev);
        idx = i2c_smbus_read_byte(pmic->client);
        if (idx < 0)
                [...] /* handle this error */

        return idx;
    }
    ```

    The following is the callback to handle a `set_voltage_sel` operation:

    ```
    static int isl6271a_set_voltage_sel(
    struct regulator_dev *dev, unsigned selector)
    {
        struct isl_pmic *pmic = rdev_get_drvdata(dev);
        int err;

        err = i2c_smbus_write_byte(pmic->client, selector);
        if (err < 0)
                [...] /* handle this error */

        return err;
    }
    ```

2. Since we are done with the callback definition, we can build a `struct regulator_ops`:

    ```
    static struct regulator_ops isl_core_ops = {
        .get_voltage_sel = isl6271a_get_voltage_sel,
        .set_voltage_sel = isl6271a_set_voltage_sel,
        .list_voltage    = regulator_list_voltage_linear,
        .map_voltage     = regulator_map_voltage_linear,
    };

    static struct regulator_ops isl_fixed_ops = {
        .list_voltage    = regulator_list_voltage_linear,
    };
    ```

Regulator Framework

 You can ask yourself where the `regulator_list_voltage_linear` and `regulator_list_voltage_linear` functions come from. As with many other regulator helper functions, they are also defined in `drivers/regulator/helpers.c`. The kernel provides helper functions for linear output regulators, as is the case for the ISL6271A.

It is time to build an array of `struct regulator_desc` for all regulators:

```
static const struct regulator_desc isl_rd[] = {
    {
        .name       = "Core Buck",
        .id         = 0,
        .n_voltages = 16,
        .ops        = &isl_core_ops,
        .type       = REGULATOR_VOLTAGE,
        .owner          = THIS_MODULE,
        .min_uV     = ISL6271A_VOLTAGE_MIN,
        .uV_step    = ISL6271A_VOLTAGE_STEP,
    }, {
        .name       = "LDO1",
        .id         = 1,
        .n_voltages = 1,
        .ops        = &isl_fixed_ops,
        .type       = REGULATOR_VOLTAGE,
        .owner          = THIS_MODULE,
        .min_uV     = 1100000,
    }, {
        .name       = "LDO2",
        .id         = 2,
        .n_voltages = 1,
        .ops        = &isl_fixed_ops,
        .type       = REGULATOR_VOLTAGE,
        .owner          = THIS_MODULE,
        .min_uV     = 1300000,
    },
};
```

LDO1 and LDO2 have a fixed output voltage. It is why their `n_voltages` properties are set to 1, and their ops only provide `regulator_list_voltage_linear` mapping.

3. Now we are in the `probe` function, the place where we need to build our `struct init_data` structures. If you remember, we will use the `struct of_regulator_match` introduced previously. We should declare an array of that type, in which we should set the `.name` property of each regulator, for which we need to fetch `init_data`:

```
static struct of_regulator_match isl6271a_matches[] = {
    { .name = "core_buck", },
    { .name = "ldo1",      },
    { .name = "ldo2",      },
};
```

Looking a bit closer, you will notice that the `.name` property is set with exactly the same value as the label of the regulator in the device tree. This is a rule you should care about and respect.

Now let us look at the probe function. The ISL6271A provides three regulator outputs, which means that the `regulator_register()` function should be called three times:

```
static int isl6271a_probe(struct i2c_client *i2c,
                          const struct i2c_device_id *id)
{
struct regulator_config config = { };
struct regulator_init_data *init_data    = dev_get_platdata(&i2c->dev);
struct isl_pmic *pmic;
int i, ret;

    struct device *dev = &i2c->dev;
    struct device_node *np, *parent;

    if (!i2c_check_functionality(i2c->adapter,
                I2C_FUNC_SMBUS_BYTE_DATA))
        return -EIO;

    pmic = devm_kzalloc(&i2c->dev,
sizeof(struct isl_pmic), GFP_KERNEL);
    if (!pmic)
        return -ENOMEM;

    /* Get the device (PMIC) node */
    np = of_node_get(dev->of_node);
```

```c
        if (!np)
              return -EINVAL;

        /* Get 'regulators' subnode */
        parent = of_get_child_by_name(np, "regulators");
        if (!parent) {
              dev_err(dev, "regulators node not found\n");
              return -EINVAL;
        }

        /* fill isl6271a_matches array */
        ret = of_regulator_match(dev, parent, isl6271a_matches,
                              ARRAY_SIZE(isl6271a_matches));

        of_node_put(parent);
        if (ret < 0) {
              dev_err(dev, "Error parsing regulator init data: %d\n",
                    ret);
              return ret;
        }

        pmic->client = i2c;
        mutex_init(&pmic->mtx);

        for (i = 0; i < 3; i++) {
              struct regulator_init_data *init_data;
              struct regulator_desc *desc;
              int val;

              if (pdata)
                    /* Given as platform data */
                    config.init_data = pdata->init_data[i];
              else
                    /* Fetched from device tree */
                    config.init_data = isl6271a_matches[i].init_data;

              config.dev = &i2c->dev;
config.of_node = isl6271a_matches[i].of_node;
config.ena_gpio = -EINVAL;

              /*
               * config is passed by reference because the kernel
               * internally duplicate it to create its own copy
               * so that it can override some fields
               */
              pmic->rdev[i] = devm_regulator_register(&i2c->dev,
                                      &isl_rd[i], &config);
              if (IS_ERR(pmic->rdev[i])) {
```

```
                        dev_err(&i2c->dev, "failed to register %s\n",
id->name);
                        return PTR_ERR(pmic->rdev[i]);
                }
        }
        i2c_set_clientdata(i2c, pmic);
        return 0;
}
```

 init_data can be NULL for a fixed regulator. It means that for the ISL6271A, only the regulator whose voltage output may change may be assigned an init_data.

```
/* Only the first regulator actually need it */
if (i == 0)
    if(pdata)
            config.init_data = init_data; /* pdata */
        else
            isl6271a_matches[i].init_data; /* DT */
else
    config.init_data = NULL;
```

The preceding driver does not fill every field of the struct regulator_desc. It greatly depends on the type of device for which we write a driver. Some drivers leave the whole job to the regulator core, and only provide the chip's register address, which the regulator core needs to work with. Such drivers use **regmap** API, which is a generic I2C and SPI register map library. drivers/regulator/max8649.c is an example.

Driver example

Let's summarize things discussed previously in a real driver, for a dummy PMIC with two regulators, where the first one has a voltage range of 850000 µV to 1600000 µV with a step of 50000 µV, and the second regulator has a fixed voltage of 1300000 µV:

```
#include <linux/init.h>
#include <linux/module.h>
#include <linux/kernel.h>
#include <linux/platform_device.h>    /* For platform devices */
#include <linux/interrupt.h>          /* For IRQ */
#include <linux/of.h>                 /* For DT*/
#include <linux/err.h>
#include <linux/regulator/driver.h>
#include <linux/regulator/machine.h>
```

```c
#define DUMMY_VOLTAGE_MIN    850000
#define DUMMY_VOLTAGE_MAX    1600000
#define DUMMY_VOLTAGE_STEP   50000

struct my_private_data {
    int foo;
    int bar;
    struct mutex lock;
};

static const struct of_device_id regulator_dummy_ids[] = {
    { .compatible = "packt,regulator-dummy", },
    { /* sentinel */ }
};

static struct regulator_init_data dummy_initdata[] = {
    [0] = {
        .constraints = {
            .always_on = 0,
            .min_uV = DUMMY_VOLTAGE_MIN,
            .max_uV = DUMMY_VOLTAGE_MAX,
        },
    },
    [1] = {
        .constraints = {
            .always_on = 1,
        },
    },
};

static int isl6271a_get_voltage_sel(struct regulator_dev *dev)
{
    return 0;
}

static int isl6271a_set_voltage_sel(struct regulator_dev *dev,
                    unsigned selector)
{
    return 0;
}

static struct regulator_ops dummy_fixed_ops = {
    .list_voltage   = regulator_list_voltage_linear,
};

static struct regulator_ops dummy_core_ops = {
```

```c
    .get_voltage_sel = isl6271a_get_voltage_sel,
    .set_voltage_sel = isl6271a_set_voltage_sel,
    .list_voltage    = regulator_list_voltage_linear,
    .map_voltage     = regulator_map_voltage_linear,
};


static const struct regulator_desc dummy_desc[] = {
    {
        .name       = "Dummy Core",
        .id         = 0,
        .n_voltages = 16,
        .ops        = &dummy_core_ops,
        .type       = REGULATOR_VOLTAGE,
        .owner      = THIS_MODULE,
        .min_uV     = DUMMY_VOLTAGE_MIN,
        .uV_step    = DUMMY_VOLTAGE_STEP,
    }, {
        .name       = "Dummy Fixed",
        .id         = 1,
        .n_voltages = 1,
        .ops        = &dummy_fixed_ops,
        .type       = REGULATOR_VOLTAGE,
        .owner      = THIS_MODULE,
        .min_uV     = 1300000,
    },
};

static int my_pdrv_probe (struct platform_device *pdev)
{
   struct regulator_config config = { };
   config.dev = &pdev->dev;
   struct regulator_dev *dummy_regulator_rdev[2];
    int ret, i;
    for (i = 0; i < 2; i++){
        config.init_data = &dummy_initdata[i];
        dummy_regulator_rdev[i] = \
            regulator_register(&dummy_desc[i], &config);
        if (IS_ERR(dummy_regulator_rdev)) {
            ret = PTR_ERR(dummy_regulator_rdev);
            pr_err("Failed to register regulator: %d\n", ret);
            return ret;
        }
    }

    platform_set_drvdata(pdev, dummy_regulator_rdev);
    return 0;
}
```

Regulator Framework

```
static void my_pdrv_remove(struct platform_device *pdev)
{
    int i;
    struct regulator_dev *dummy_regulator_rdev = \
                        platform_get_drvdata(pdev);
    for (i = 0; i < 2; i++)
        regulator_unregister(&dummy_regulator_rdev[i]);
}

static struct platform_driver mypdrv = {
    .probe      = my_pdrv_probe,
    .remove     = my_pdrv_remove,
    .driver     = {
        .name           = "regulator-dummy",
        .of_match_table = of_match_ptr(regulator_dummy_ids),
        .owner          = THIS_MODULE,
    },
};
module_platform_driver(mypdrv);
MODULE_AUTHOR("John Madieu <john.madieu@gmail.com>");
MODULE_LICENSE("GPL");
```

Once the module is loaded and the device matched, the kernel will print something like this:

Dummy Core: at 850 mV
Dummy Fixed: 1300 mV

One can then check what happened under the hood:

ls /sys/class/regulator/
regulator.0 regulator.11 regulator.14 regulator.4 regulator.7
regulator.1 regulator.12 regulator.2 regulator.5 regulator.8
regulator.10 regulator.13 regulator.3 regulator.6 regulator.9

regulator.13 and regulator.14 have been added by our driver. Let us now check their properties:

```
# cd /sys/class/regulator
# cat regulator.13/name
Dummy Core
# cat regulator.14/name
Dummy Fixed
# cat regulator.14/type
voltage
# cat regulator.14/microvolts
1300000
# cat regulator.13/microvolts
850000
```

Regulators consumer interface

The consumer interface only requires the driver to include one header:

```
#include <linux/regulator/consumer.h>
```

A consumer can be static or dynamic. A static one requires only a fixed supply, whereas a dynamic one requires active management of the regulator at runtime. From the consumer point side, a regulator device is represented in the kernel as an instance of a struct regulator structure, defined in drivers/regulator/internal.h and shown as follows:

```
/*
 * struct regulator
 *
 * One for each consumer device.
 */
struct regulator {
    struct device *dev;
    struct list_head list;
    unsigned int always_on:1;
    unsigned int bypass:1;
    int uA_load;
    int min_uV;
    int max_uV;
    char *supply_name;
    struct device_attribute dev_attr;
    struct regulator_dev *rdev;
    struct dentry *debugfs;
};
```

This structure is meaningful enough and does not need us to add any comments. To see how easy it is to consume a regulator, here is a little example of how a consumer acquires a regulator:

```
[...]
int ret;
struct regulator *reg;
const char *supply = "vdd1";
int min_uV, max_uV;
reg = regulator_get(dev, supply);
[...]
```

Regulator device requesting

Prior to gaining access to a regulator, the consumer has to request the kernel by means of the `regulator_get()` function. It is also possible to use the managed version, the `devm_regulator_get()` function:

```
struct regulator *regulator_get(struct device *dev,
const char *id)
```

An example of using this function is:

```
reg = regulator_get(dev, "Vcc");
```

The consumer passes in its `struct device` pointer and power supply ID. The core will try to find the correct regulator by consulting the DT or a machine-specific lookup table. If we focus only on the device tree, `*id` should match the `<name>` pattern of the regulator supply in the device tree. If the lookup is successful then this call will return a pointer to the `struct regulator` that supplies this consumer.

To release the regulator, the consumer driver should call:

```
void regulator_put(struct regulator *regulator)
```

Prior to calling this function, the driver should ensure that all `regulator_enable()` calls made on this regulator source are balanced by `regulator_disable()` calls.

More than one regulator can supply a consumer, for example, codec consumers with analog and digital supplies:

```
digital = regulator_get(dev, "Vcc");   /* digital core */
analog = regulator_get(dev, "Avdd");   /* analog */
```

Consumer `probe()` and `remove()` functions are an appropriate place to grab and release regulators.

Controlling the regulator device

Regulator control consists of enabling, disabling, and setting output values for a regulator.

Regulator output enable and disable

A consumer can enable its power supply by calling the following:

```
int regulator_enable(regulator);
```

This function returns 0 on success. The reverse operation consists of disabling the power supply, by calling this:

```
int regulator_disable(regulator);
```

To check whether a regulator is already enabled or not, the consumer should call this:

```
int regulator_is_enabled(regulator);
```

This function returns a value greater than 0 if the regulator is enabled. Since the regulator may be enabled early by the bootloader or shared with another consumer, one can use the `regulator_is_enabled()` function to check the regulator state.

Here is an example,

```
printk (KERN_INFO "Regulator Enabled = %d\n",
                    regulator_is_enabled(reg));
```

> For a shared regulator, `regulator_disable()` will actually disable the regulator only when the enabled reference count is zero. That said, you can force disabling in case of an emergency, for example, by calling `regulator_force_disable()`:

```
int regulator_force_disable(regulator);
```

Each of the functions that we will discuss in the sections that follows is actually a wrapper around a `regulator_ops` operation. For example, `regulator_set_voltage()` internally calls `regulator_ops.set_voltage` after checking the corresponding mask allowing this operation is set, and so on.

Voltage control and status

For consumers that need to adapt their power supplies according to their operating modes, the kernel provides this:

```
int regulator_set_voltage(regulator, min_uV, max_uV);
```

`min_uV` and `max_uV` are the minimum and maximum acceptable voltages in microvolts.

If called when the regulator is disabled, this function will change the voltage configuration so that the voltage is physically set when the regulator is next enabled. That said, consumers can get the regulator configured voltage output by calling `regulator_get_voltage()`, which will return the configured output voltage whether the regulator is enabled or not:

```
int regulator_get_voltage(regulator);
```

Here is an example,

```
printk (KERN_INFO "Regulator Voltage = %d\n",
regulator_get_voltage(reg));
```

Current limit control and status

What we have discussed in the voltage section also applies here. For example, USB drivers may want to set the limit to 500 mA when supplying power.

Consumers can control their supply current limit by calling:

```
int regulator_set_current_limit(regulator, min_uA, max_uA);
```

`min_uA` and `max_uA` are the minimum and maximum acceptable current limits in microamps.

In the same way, consumers can get the regulator configured to the current limit by calling `regulator_get_current_limit()`, which will return the current limit whether the regulator is enabled or not:

```
int regulator_get_current_limit(regulator);
```

Operating mode control and status

For efficient power management, some consumers may change the operating mode of their supply when their (consumers) operating state changes. Consumer drivers can request a change in their supply regulator operating mode by calling:

```
int regulator_set_optimum_mode(struct regulator *regulator,
int load_uA);
int regulator_set_mode(struct regulator *regulator,
unsigned int mode);
unsigned int regulator_get_mode(struct regulator *regulator);
```

Consumers should use `regulator_set_mode()` on a regulator only when it knows about the regulator and does not share the regulator with other consumers. This is known as **direct mode**. `regulator_set_uptimum_mode()` causes the core to undertake some background work in order to determine what operating mode is best for the requested current. This is called the **indirect mode**.

Regulator binding

This section only deals with consumer interface binding. Because PMIC binding consists of providing `init data` for regulators that this PMIC provides, you should refer to the section *Feeding init data into the DT* to understand producer binding.

Consumer nodes can reference one or more of its supplies/regulators using the following bindings:

```
<name>-supply: phandle to the regulator node
```

It is the same principle as PWM consumer binding. `<name>` should be meaningful enough, so that the driver can easily refer to it when requesting the regulator. That said, `<name>` must match the `*id` parameter of the `regulator_get()` function:

```
twl_reg1: regulator@0 {
    [...]
};

twl_reg2: regulator@1 {
    [...]
};

mmc: mmc@0x0 {
    [...]
    vmmc-supply = <&twl_reg1>;
```

```
        vmmcaux-supply = <&twl_reg2>;
};
```

The consumer code (which is the MMC driver) that actually requests its supplies could look like this:

```
struct regulator *main_regulator;
struct regulator *aux_regulator;
int ret;
main_regulator = devm_regulator_get(dev, "vmmc");

/*
 * It is a good practive to apply the config before
 * enabling the regulator
 */
if (!IS_ERR(io_regulator)) {
    regulator_set_voltage(main_regulator,
                    MMC_VOLTAGE_DIGITAL,
                    MMC_VOLTAGE_DIGITAL);
    ret = regulator_enable(io_regulator);
}
[...]
aux_regulator = devm_regulator_get(dev, "vmmcaux");
[...]
```

Summary

With the wide range of devices that need to be smartly and smoothly supplied, this chapter can be relied on to take care of their power supply management. PMIC devices usually sit on SPI or I2C buses. Having already dealt with these buses in previous chapters, you should be able to write any PMIC driver. Let's now jump to the next chapter, which deals with framebuffer drivers, which is a completely different and no less interesting topic.

21
Framebuffer Drivers

Video cards always have a certain amount of RAM. This RAM is where the bitmap of image data is buffered for display. From the software point of view, the framebuffer is a character device providing access to this RAM.

That said, a framebuffer driver provides an interface for:

- Display mode setting
- Memory access to the video buffer
- Basic 2D acceleration operations (for example, scrolling)

To provide this interface, the framebuffer driver generally talks to the hardware directly. There are well-known framebuffer drivers such as:

- **intelfb**, which is a framebuffer for various Intel 8xx/9xx compatible graphic devices
- **vesafb**, which is a framebuffer driver that uses the VESA standard interface to talk to the video hardware
- **mxcfb**, the framebuffer driver for i.MX6 chip series

Framebuffer drivers are the simplest form of graphics drivers under Linux, not to confuse them with X.org drivers, which implement advanced features such as 3D acceleration and so on, or Kernel mode setting (KMS) drivers, which expose both framebuffer and GPU functionalities (like X.org drivers do).

i.MX6 X.org driver is a closed source and called **vivante**.

Framebuffer Drivers

Back to our framebuffer drivers, they are very simple API drivers that expose video card functionalities by means of character devices, accessible from the user space through /dev/fbX entries. One can find more information on Linux graphical stack in the comprehensive talk *Linux Graphics Demystified* by Martin Fiedler: http://keyj.emphy.de/files/linuxgraphics_en.pdf.

In this chapter, we cover the following topics:

- Framebuffer driver data structures and methods, thus covering the whole driver architecture
- Framebuffer device operations, accelerated and non-accelerated
- Accessing framebuffer from user space

Driver data structures

The framebuffer drivers depend heavily on four data structures, all defined in include/linux/fb.h, which is also the header you should include in your code in order to deal with framebuffer driver:

```
#include <linux/fb.h>
```

These structures are fb_var_screeninfo, fb_fix_screeninfo, fb_cmap, and fb_info. The first three are made available to and from user space code. Now let us describe the purpose of each structure, their meaning, and what they are used for.

1. The kernel use an instance of struct struct fb_var_screeninfo to hold variable properties of the video card. These values are those defined by the user, such as resolution depth:

```
struct fb_var_screeninfo {
    __u32 xres; /* visible resolution */
    __u32 yres;

    __u32 xres_virtual; /* virtual resolution */
    __u32 yres_virtual;

    __u32 xoffset; /* offset from virtual to visible resolution */
    __u32 yoffset;

    __u32 bits_per_pixel; /* # of bits needed to hold a pixel */
    [...]

    /* Timing: All values in pixclocks, except pixclock (of course)
```

```
    */
    __u32 pixclock;    /* pixel clock in ps (pico seconds) */
    __u32 left_margin;    /* time from sync to picture */
    __u32 right_margin; /* time from picture to sync */
    __u32 upper_margin; /* time from sync to picture */
    __u32 lower_margin;
    __u32 hsync_len;   /* length of horizontal sync */
    __u32 vsync_len;   /* length of vertical sync */
    __u32 rotate; /* angle we rotate counter clockwise */
};
```

This can be summarized into a figure shown as follows:

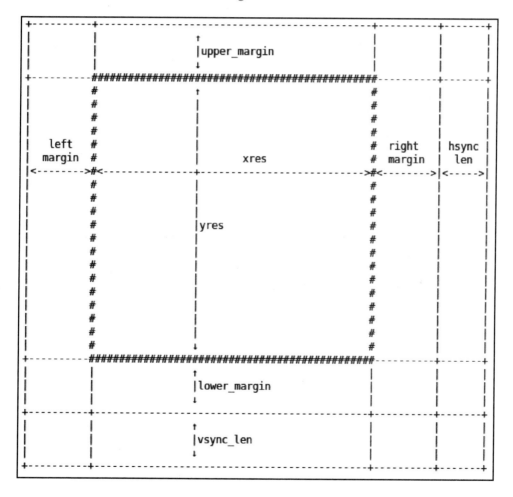

2. There are properties of video card that are fixed, either by the manufacturer, or applied when a mode is set, and can't be changed otherwise. This is generally hardware information. A good example of this is the start of the framebuffer memory, which cannot change, even by user program. The kernel holds such information in an instance of struct fb_fix_screeninfo structure:

```
struct fb_fix_screeninfo {
    char id[16];         /* identification string eg "TT Builtin" */
    unsigned long smem_start;       /* Start of frame buffer mem */
                         /* (physical address) */
    __u32 smem_len;/* Length of frame buffer mem */
    __u32 type;      /* see FB_TYPE_*            */
    __u32 type_aux;  /* Interleave for interleaved Planes */
    __u32 visual;    /* see FB_VISUAL_* */
    __u16 xpanstep;  /* zero if no hardware panning  */
    __u16 ypanstep;    /* zero if no hardware panning  */
    __u16 ywrapstep;   /* zero if no hardware ywrap    */
    __u32 line_length;   /* length of a line in bytes    */
    unsigned long mmio_start; /* Start of Memory Mapped I/O
 *(physical address)
 */
    __u32 mmio_len;   /* Length of Memory Mapped I/O  */
    __u32 accel;      /* Indicate to driver which     */
                      /* specific chip/card we have */
    __u16 capabilities; /* see FB_CAP_* */
};
```

3. The struct fb_cmap structure specifies the color map, which is used to store the user's definition of colors in a manner the kernel can understand, in order to send it to the underlying video hardware. One can use this structure to define the RGB ratio that you desire for different colors:

```
struct fb_cmap {
    __u32 start;    /* First entry */
    __u32 len;      /* Number of entries */
    __u16 *red;     /* Red values */
    __u16 *green;   /* Green values */
    __u16 *blue;    /* Blue values */
    __u16 *transp;  /* Transparency. Discussed later on */
};
```

4. The `struct fb_info` structure, which represents the framebuffer itself, is the main data structure of framebuffer drivers. Unlike other preceding structure discussed, `fb_info` exists only in the kernel, and is not part of the user space framebuffer API:

```
struct fb_info {
    [...]
    struct fb_var_screeninfo var; /* Variable screen information.
                                     Discussed earlier. */
    struct fb_fix_screeninfo fix; /* Fixed screen information. */
    struct fb_cmap cmap;          /* Color map. */
    struct fb_ops *fbops;         /* Driver operations.*/
    char __iomem *screen_base;    /* Frame buffer's
                                     virtual address */
    unsigned long screen_size;    /* Frame buffer's size */
    [...]
    struct device *device;        /* This is the parent */
struct device *dev;               /* This is this fb device */
#ifdef CONFIG_FB_BACKLIGHT
    /* assigned backlight device */
    /* set before framebuffer registration,
       remove after unregister */
    struct backlight_device *bl_dev;

    /* Backlight level curve */
    struct mutex bl_curve_mutex;
    u8 bl_curve[FB_BACKLIGHT_LEVELS];
#endif
    [...]
    void *par; /* Pointer to private memory */
};
```

`struct fb_info` structure should always be allocated dynamically, using `framebuffer_alloc()`, which is a kernel (framebuffer core) helper functions to allocate memory for instance of framebuffer devices, along with their private data memory:

```
struct fb_info *framebuffer_alloc(size_t size, struct device *dev)
```

In this prototype, `size` represents the size of the private area as an argument and appends that to the end of the allocated `fb_info`. This private area can be referenced using the `.par` pointer in the `fb_info` structure. `framebuffer_release()` does the reverse operation:

```
void framebuffer_release(struct fb_info *info)
```

Framebuffer Drivers

Once set up, a framebuffer should be registered with the kernel using `register_framebuffer()`, which returns negative `errno` on error, or `zero` for success:

```
int register_framebuffer(struct fb_info *fb_info)
```

Once registered, one can unregister the framebuffer with the `unregister_framebuffer()` function, which also returns a negative `errno` on error, or `zero` for success:

```
int unregister_framebuffer(struct fb_info *fb_info)
```

Allocation and registering should be done during the device probing, whereas unregistering and deallocation (release) should be done from within the driver's `remove()` function.

Device methods

In the `struct fb_info` structure, there is a `.fbops` field, which is an instance of `struct fb_ops` structure. This structure contains a collection of functions that need to perform some operations on the framebuffer device. These are entry points for `fbdev` and `fbcon` tools. Some methods in that structure are mandatory, the minimum required for a framebuffer to work, whereas others are optional, and depend on the features the driver needs to expose, assuming the device itself supports those features.

The following is the definition of the `struct fb_ops` structure:

```
struct fb_ops {
/* open/release and usage marking */
struct module *owner;
int (*fb_open)(struct fb_info *info, int user);
int (*fb_release)(struct fb_info *info, int user);

/* For framebuffers with strange nonlinear layouts or that do not
 * work with normal memory mapped access
 */
ssize_t (*fb_read)(struct fb_info *info, char __user *buf,
         size_t count, loff_t *ppos);
ssize_t (*fb_write)(struct fb_info *info, const char __user *buf,
         size_t count, loff_t *ppos);

/* checks var and eventually tweaks it to something supported,
 * DO NOT MODIFY PAR */
int (*fb_check_var)(struct fb_var_screeninfo *var, struct fb_info
*info);
```

```c
    /* set the video mode according to info->var */
    int (*fb_set_par)(struct fb_info *info);

    /* set color register */
    int (*fb_setcolreg)(unsigned regno, unsigned red, unsigned green,
                unsigned blue, unsigned transp, struct fb_info *info);

    /* set color registers in batch */
    int (*fb_setcmap)(struct fb_cmap *cmap, struct fb_info *info);

    /* blank display */
    int (*fb_blank)(int blank_mode, struct fb_info *info);

    /* pan display */
    int (*fb_pan_display)(struct fb_var_screeninfo *var, struct fb_info
*info);

    /* Draws a rectangle */
    void (*fb_fillrect) (struct fb_info *info, const struct fb_fillrect
*rect);
    /* Copy data from area to another */
    void (*fb_copyarea) (struct fb_info *info, const struct fb_copyarea
*region);
    /* Draws a image to the display */
    void (*fb_imageblit) (struct fb_info *info, const struct fb_image
*image);

    /* Draws cursor */
    int (*fb_cursor) (struct fb_info *info, struct fb_cursor *cursor);

    /* wait for blit idle, optional */
    int (*fb_sync)(struct fb_info *info);

    /* perform fb specific ioctl (optional) */
    int (*fb_ioctl)(struct fb_info *info, unsigned int cmd,
            unsigned long arg);

    /* Handle 32bit compat ioctl (optional) */
    int (*fb_compat_ioctl)(struct fb_info *info, unsigned cmd,
            unsigned long arg);

    /* perform fb specific mmap */
    int (*fb_mmap)(struct fb_info *info, struct vm_area_struct *vma);

    /* get capability given var */
    void (*fb_get_caps)(struct fb_info *info, struct fb_blit_caps *caps,
                struct fb_var_screeninfo *var);
```

```
        /* teardown any resources to do with this framebuffer */
        void (*fb_destroy)(struct fb_info *info);
        [...]
};
```

Different callbacks can be set depending on what functionality one wishes to implement.

In Chapter 4, *Character Device Drivers*, we learned that character devices, by means of `struct file_operations` structure, can export a collection of file operations, which are entry points for file-related system calls such as `open()`, `close()`, `read()`, `write()`, `mmap()`, `ioctl()`, and so on.

That being said, do not confuse `fb_ops` with `file_operations` structure. `fb_ops` offers an abstraction of low-level operations, while `file_operations` is for an upper-level system call interface. The kernel implements framebuffer file operations in `drivers/video/fbdev/core/fbmem.c`, which internally call methods we defined in `fb_ops`. In this manner, one can implement the low-level hardware operations according to the need of the system call interface, namely the `file_operations` structure. For example, when the user `open()` the device, the core's open file operation method will perform some core operations, and execute `fb_ops.fb_open()` method if set, same for `release`, `mmap`, and so on.

Framebuffer devices support some ioctl commands defined in `include/uapi/linux/fb.h`, that user programs can use to operate on hardware. These commands are all handled by the core's `fops.ioctl` method. For some of those commands, the core's ioctl method may internally execute methods defined in `fb_ops` structure.

One may wonder what the `fb_ops.ffb_ioctl` is used for. The framebuffer core executes `fb_ops.fb_ioctl` only when the given ioctl command is not known to the kernel. In other words, `fb_ops.fb_ioctl` is executed in the default statement of the framebuffer core's `fops.ioctl` method.

Driver methods

Drivers methods consist of `probe()` and `remove()` functions. Prior to going further in these method descriptions, let us set up our `fb_ops` structure:

```
        static struct fb_ops myfb_ops = {
           .owner         = THIS_MODULE,
           .fb_check_var  = myfb_check_var,
           .fb_set_par    = myfb_set_par,
```

```
    .fb_setcolreg  = myfb_setcolreg,
    .fb_fillrect   = cfb_fillrect,  /* Those three hooks are */
    .fb_copyarea   = cfb_copyarea,  /* non accelerated and   */
    .fb_imageblit  = cfb_imageblit, /* are provided by kernel */
    .fb_blank      = myfb_blank,
};
```

- Probe: Driver `probe` function is in charge of initializing the hardware, creating the `struct fb_info` structure using `framebuffer_alloc()` function, and `register_framebuffer()` on it. The following sample assumes the device is memory mapped. Therefore, your nonmemory map can exist, such as screen sitting on SPI buses. In this case, bus specific routines should be used:

```
static int myfb_probe(struct platform_device *pdev)
{
    struct fb_info *info;
    struct resource *res;
    [...]

    dev_info(&pdev->dev, "My framebuffer driver\n");

/*
 * Query resource, like DMA channels, I/O memory,
 * regulators, and so on.
 */
    res = platform_get_resource(pdev, IORESOURCE_MEM, 0);
    if (!res)
        return -ENODEV;
    /* use request_mem_region(), ioremap() and so on */
    [...]
    pwr = regulator_get(&pdev->dev, "lcd");

    info = framebuffer_alloc(sizeof(
struct my_private_struct), &pdev->dev);
    if (!info)
        return -ENOMEM;

    /* Device init and default info value*/
    [...]
    info->fbops = &myfb_ops;

    /* Clock setup, using devm_clk_get() and so on */
    [...]

    /* DMA setup using dma_alloc_coherent() and so on*/
    [...]
```

```
            /* Register with the kernel */
            ret = register_framebuffer(info);

            hardware_enable_controller(my_private_struct);
            return 0;
        }
```

- Remove: The `remove()` function should release whatever was acquired in `probe()`, and call:

```
        static int myfb_remove(struct platform_device *pdev)
        {
            /* iounmap() memory and release_mem_region() */
            [...]
            /* Reverse DMA, dma_free_*(); */
            [...]

            hardware_disable_controller(fbi);

             /* first unregister, */
            unregister_framebuffer(info);
             /* and then free the memory */
            framebuffer_release(info);

            return 0;
        }
```

- Assuming you used the manager version for resource allocations, you'll just need to use `unregister_framebuffer()` and `framebuffer_release()`. Everything else will be done by the kernel.

Detailed fb_ops

Let us describe some of the hooks declared in `fb_ops` structure. That being said, for an idea on writing framebuffer drivers, you can have a look at `drivers/video/fbdev/vfb.c`, which is a simple virtual framebuffer driver in the kernel. One can also have a look at other specific framebuffer drivers, like i.MX6 one, at `drivers/video/fbdev/imxfb.c`, or at the kernel documentation about framebuffer driver API at `Documentation/fb/api.txt`.

Checking information

The hook `fb_ops->fb_check_var` is responsible for checking framebuffer parameters. Its prototype is as follows:

```
int (*fb_check_var)(struct fb_var_screeninfo *var,
struct fb_info *info);
```

This function should check framebuffer variable parameters and adjust to valid values. `var` represents the framebuffer variable parameters, which should be checked and adjusted:

```
static int myfb_check_var(struct fb_var_screeninfo *var,
struct fb_info *info)
{
    if (var->xres_virtual < var->xres)
        var->xres_virtual = var->xres;

    if (var->yres_virtual < var->yres)
        var->yres_virtual = var->yres;

    if ((var->bits_per_pixel != 32) && (var->bits_per_pixel != 24) && (var->bits_per_pixel != 16) && (var->bits_per_pixel != 12) &&
        (var->bits_per_pixel != 8))
            var->bits_per_pixel = 16;

    switch (var->bits_per_pixel) {
    case 8:
        /* Adjust red*/
        var->red.length = 3;
        var->red.offset = 5;
        var->red.msb_right = 0;
        /*adjust green*/
        var->green.length = 3;
        var->green.offset = 2;
        var->green.msb_right = 0;

        /* adjust blue */
        var->blue.length = 2;
        var->blue.offset = 0;
        var->blue.msb_right = 0;
        /* Adjust transparency */
        var->transp.length = 0;
        var->transp.offset = 0;
        var->transp.msb_right = 0;
        break;
    case 16:
```

Framebuffer Drivers

```
                [...]
                break;
        case 24:
                [...]
                break;
        case 32:
                var->red.length = 8;
                var->red.offset = 16;
                var->red.msb_right = 0;

                var->green.length = 8;
                var->green.offset = 8;
                var->green.msb_right = 0;

                var->blue.length = 8;
                var->blue.offset = 0;
                var->blue.msb_right = 0;

                var->transp.length = 8;
                var->transp.offset = 24;
                var->transp.msb_right = 0;
                break;
        }
        /*
         * Any other field in *var* can be adjusted
         * like var->xres,     var->yres, var->bits_per_pixel,
         * var->pixclock and so on.
         */
           return 0;
    }
```

The preceding code adjusts variable framebuffer properties according to the configuration chosen by user.

Set controller's parameters

The hook `fp_ops->fb_set_par` is another hardware specific hook, responsible for sending parameters to the hardware. It programs the hardware based on user settings (`info->var`):

```
    static int myfb_set_par(struct fb_info *info)
    {
        struct fb_var_screeninfo *var = &info->var;

        /* Make some compute or other sanity check */
        [...]
```

```
    /*
     * This function writes value to the hardware,
     * in the appropriate registers
     */
    set_controller_vars(var, info);

    return 0;
}
```

Screen blanking

The hook `fb_ops->fb_blank` is a hardware specific hook, responsible for screen blanking. Its prototype is as follows:

```
int (*fb_blank)(int blank_mode, struct fb_info *info)
```

`blank_mode` parameter is always one of the following values:

```
enum {
    /* screen: unblanked, hsync: on,  vsync: on */
    FB_BLANK_UNBLANK       = VESA_NO_BLANKING,

    /* screen: blanked,   hsync: on,  vsync: on */
    FB_BLANK_NORMAL        = VESA_NO_BLANKING + 1,

    /* screen: blanked,   hsync: on,  vsync: off */
    FB_BLANK_VSYNC_SUSPEND = VESA_VSYNC_SUSPEND + 1,

    /* screen: blanked,   hsync: off, vsync: on */
    FB_BLANK_HSYNC_SUSPEND = VESA_HSYNC_SUSPEND + 1,

    /* screen: blanked,   hsync: off, vsync: off */
    FB_BLANK_POWERDOWN     = VESA_POWERDOWN + 1
};
```

The usual way of blank display is to do a `switch case` on the `blank_mode` parameter as follows:

```
static int myfb_blank(int blank_mode, struct fb_info *info)
{
    pr_debug("fb_blank: blank=%d\n", blank);

    switch (blank) {
    case FB_BLANK_POWERDOWN:
    case FB_BLANK_VSYNC_SUSPEND:
    case FB_BLANK_HSYNC_SUSPEND:
    case FB_BLANK_NORMAL:
```

```
                myfb_disable_controller(fbi);
                break;

        case FB_BLANK_UNBLANK:
                myfb_enable_controller(fbi);
                break;
        }
        return 0;
}
```

Blanking operation should disable the controller, stop its clocks and power it down. Unblanking should perform the reverse operations.

Accelerated methods

Users video operations such as blending, stretching, moving bitmaps, or dynamic gradient generation are all heavy-duty tasks. They require graphics acceleration to obtain acceptable performance. One can implement framebuffer accelerated methods using the following fields of `struct fp_ops` structure:

- `.fb_imageblit()`: This method draws an image on the display and is very useful
- `.fb_copyarea()`: This method copies a rectangular area from one screen region to another
- `.fb_fillrect()`: This method fills in an optimized manner a rectangle with pixel lines

Therefore, kernel developers thought of controllers that did not have hardware acceleration, and provided a software-optimized method. This makes acceleration implementation optional, since software fall-back exists. That said, if the framebuffer controller does not provide any acceleration mechanism, one must populate these methods using the kernel generic routines.

These are respectively:

- `cfb_imageblit()`: This is a kernel-provided fallback for imageblit. The kernel uses it to output a logo to the screen during boot up.
- `cfb_copyarea()`: This is for area copy operations.
- `cfb_fillrect()`: This is the framebuffer core non-accelerated method to achieve operations of the same name.

Putting it all together

In this section, let us summarize things discussed in the preceding section. In order to write framebuffer driver, one has to:

- Fill a `struct fb_var_screeninfo` structure in order to provide information on framebuffer variable properties. Those properties can be changed by user space.
- Fill a `struct fb_fix_screeninfo` structure, to provide fixed parameters.
- Set up a `struct fb_ops` structure, providing necessary callback functions, which will used by the framebuffer subsystem in response to user actions.
- Still in the `struct fb_ops` structure, one has to provide accelerated functions callback, if supported by the device.
- Set up a `struct fb_info` structure, feeding it with structures filled in previous steps, and call `register_framebuffer()` on it in order to have it registered with the kernel.

For an idea on writing a simple framebuffer driver, one can have a look at drivers/video/fbdev/vfb.c, which is a virtual framebuffer driver in kernel. One can enable this in the kernel by means of the CONGIF_FB_VIRTUAL option.

Framebuffer from user space

One usually accesses framebuffer memory by means of mmap() command in order to map the framebuffer memory to the part of system RAM, so that drawing pixels on the screen becomes a simple matter affecting memory value. Screen parameters (variable and fixed) are extracted by means of ioctl commands, especially FBIOGET_VSCREENINFO and FBIOGET_FSCREENINFO. The complete list is available at include/uapi/linux/fb.h in the kernel source.

The following is a sample code to draw a 300*300 square on the framebuffer:

```
#include <stdlib.h>
#include <unistd.h>
#include <stdio.h>
#include <fcntl.h>
#include <linux/fb.h>
#include <sys/mman.h>
#include <sys/ioctl.h>

#define FBCTL(_fd, _cmd, _arg)                \
    if(ioctl(_fd, _cmd, _arg) == -1) {        \
```

Framebuffer Drivers

```c
            ERROR("ioctl failed");         \
            exit(1); }
int main()
{
    int fd;
    int x, y, pos;
    int r, g, b;
    unsigned short color;
    void *fbmem;

    struct fb_var_screeninfo var_info;
    struct fb_fix_screeninfo fix_info;

    fd = open(FBVIDEO, O_RDWR);
    if (tfd == -1 || vfd == -1) {
        exit(-1);
    }

    /* Gather variable screen info (virtual and visible) */
    FBCTL(fd, FBIOGET_VSCREENINFO, &var_info);

    /* Gather fixed screen info */
    FBCTL(fd, FBIOGET_FSCREENINFO, &fix_info);

    printf("****** Frame Buffer Info ******\n");
    printf("Visible: %d,%d \nvirtual: %d,%d \n  line_len %d\n",
            var_info.xres, this->var_info.yres,
            var_info.xres_virtual, var_info.yres_virtual,
            fix_info.line_length);
    printf("dim %d,%d\n\n", var_info.width, var_info.height);

    /* Let's mmap frame buffer memory */
    fbmem = mmap(0, v_var.yres_virtual * v_fix.line_length, \
                    PROT_WRITE | PROT_READ, \
                    MAP_SHARED, fd, 0);

    if (fbmem == MAP_FAILED) {
        perror("Video or Text frame bufer mmap failed");
        exit(1);
    }

    /* upper left corner (100,100). The square is 300px width */
    for (y = 100; y < 400; y++) {
        for (x = 100; x < 400; x++) {
            pos = (x + vinfo.xoffset) * (vinfo.bits_per_pixel / 8)
                +  (y + vinfo.yoffset) * finfo.line_length;
```

```
            /* if 32 bits per pixel */
            if (vinfo.bits_per_pixel == 32) {
                /* We prepare some blue color */
                *(fbmem + pos) = 100;

                /* adding a little green */
                *(fbmem + pos + 1) = 15+(x-100)/2;

                /* With lot of read */
                *(fbmem + pos + 2) = 200-(y-100)/5;

                /* And no transparency */
                *(fbmem + pos + 3) = 0;
            } else  { /* This assume 16bpp */
                r = 31-(y-100)/16;
                g = (x-100)/6;
                b = 10;
                /* Compute color */
                color = r << 11 | g << 5 | b;
                *((unsigned short int*)(fbmem + pos)) = color;
            }
        }
    }

    munmap(fbp, screensize);
    close(fbfd);
    return 0;
}
```

One can also dump the framebuffer memory into a raw image, using cat or dd command:

> # cat /dev/fb0 > my_image

Write it back using:

> # cat my_image > /dev/fb0

It is possible to blank/unblank the screen through a special /sys/class/graphics/fb<N>/blank sysfs file, where <N> is the framebuffer index. Writing a 1 will blank the screen, whereas 0 will unblank it:

> # echo 0 > /sys/class/graphics/fb0/blank
> # echo 1 > /sys/class/graphics/fb0/blank

Summary

The framebuffer drivers are the simplest form of Linux graphics drivers, requiring little implementation work. They heavily abstract hardware. At this stage, you should be able to enhance either an existing driver (with graphical acceleration functions for example), or write a fresh one from scratch. However, it is recommended to rely on an existing driver whose hardware shares as many characteristics as possible with the one you need to write the driver for. Let us jump to the next and last chapter, dealing with network devices.

22
Network Interface Card Drivers

We all know that networking is inherent to the Linux kernel. Some years ago, Linux was only used for its network performances, but things have changed now; Linux is much more than a server, and runs on billions of embedded devices. Through the years, Linux gained the reputation of being the best network operating system. In spite of all this, Linux cannot do everything. Given the huge variety of Ethernet controllers that exist, Linux has found no other way than to expose an API to developers who need a writing driver for their network device, or who need to perform kernel networking development in a general manner. This API offers a sufficient abstraction layer, allowing for gauranteeing the generosity of the developed code, as well as porting on other architectures. This chapter will simply walk through the part of this API that deals with **Network Interface Card** (**NIC**) driver development, and discuss its data structures and methods.

In this chapter, we will cover the following topics:

- NIC driver data structure and a walk through its main socket buffer structure
- NIC driver architecture and methods description, as well as packets transmission and reception
- Developing a dummy NIC driver for testing purposes

Driver data structures

When you deal with NIC devices, there are two data structures that you need to play with:

- The `struct sk_buff` structure, defined in `include/linux/skbuff.h`, which is the fundamental data structure in the Linux networking code, and which should be included in your code:

    `#include <linux/skbuff.h>`

- Each packet sent or received is handled using this data structure.
- The `struct net_device` structure; this is the structure by which any NIC device is represented in the kernel. It is the interface by which data transit takes place. It is defined in `include/linux/netdevice.h`, which should also be included in your code:

    `#include <linux/netdevice.h>`

Other files that one should include in the code are `include/linux/etherdevice.h` for MAC and Ethernet-related functions (such as `alloc_etherdev()`) and `include/linux/ethtool.h` for ethtools support:

```
#include <linux/ethtool.h>
#include <linux/etherdevice.h>
```

The socket buffer structure

This structure wraps any packet that transits through an NIC:

```
struct sk_buff {
  struct sk_buff * next;
  struct sk_buff * prev;
  ktime_t tstamp;
  struct rb_node      rbnode; /* used in netem & tcp stack */
  struct sock * sk;
  struct net_device * dev;
  unsigned int        len;
  unsigned int        data_len;
  __u16               mac_len;
  __u16               hdr_len;
  unsigned int len;
  unsigned int data_len;
  __u16 mac_len;
  __u16 hdr_len;
```

```
    __u32 priority;
    dma_cookie_t dma_cookie;
    sk_buff_data_t tail;
    sk_buff_data_t end;
    unsigned char * head;
    unsigned char * data;
    unsigned int truesize;
    atomic_t users;
};
```

The following is the meanings of the elements in the structure:

- next and prev : This represents the next and previous buffer in the list.
- sk: This is the socket associated with this packet.
- tstamp: This is the time when the packet arrived/left.
- rbnode: This is an alternative to next/prev represented in a red-black tree.
- dev: This represents the device this packet arrived on/is leaving by. This field is associated with two other fields not listed here. These are input_dev and real_dev. They track devices associated with the packet. Therefore, input_dev always refers to a device the packet is received from.
- len: This is the total number of bytes in the packet. Socket Buffers (SKBs) are composed of a linear data buffer and, optionally, a set of one or more regions called **rooms**. In case there are such rooms, data_len will hold the total number of bytes of the data area.
- mac_len: This holds the length of the MAC header.
- csum: This contains the checksum of the packet.
- Priority: This represents the packet priority in QoS.
- truesize: This keeps track of how many bytes of system memory are consumed by a packet, including the memory occupied by the struct sk_buff structure itself.
- users: This is used for reference counting for the SKB objects.
- Head: Head, data, tail are pointers to different regions (rooms) in the socket buffer.
- end: This points to the end of the socket buffer.

Only a few fields of this structure have been discussed here. A full description is available in include/linux/skbuff.h., which is the header file you should include to deal with socket buffers.

Socket buffer allocation

Allocation of a socket buffer is a bit tricky, since it needs at least three different functions:

- First of all, the whole memory allocation should be done using the `netdev_alloc_skb()` function
- Increase and align header room with the `skb_reserve()` function
- Extend the used data area of the buffer (which will contain the packet) using the `skb_put()` function.

Let us have a look at the following figure:

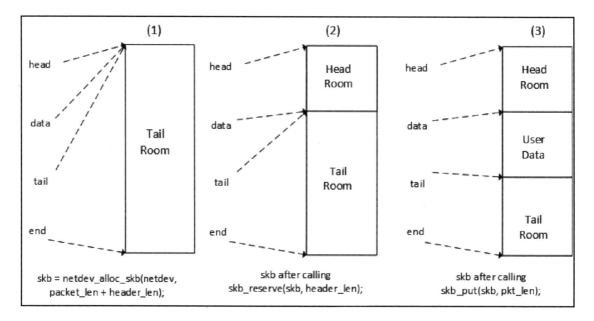

Socket buffers allocation process

1. We allocate a buffer large enough to contain a packet along with the Ethernet header by means of the `netdev_alloc_skb()` function:

   ```
   struct sk_buff *netdev_alloc_skb(struct net_device *dev,
                                    unsigned int length)
   ```

   This function returns `NULL` on failure. Therefore, even if it allocates memory, `netdev_alloc_skb()` can be called from an atomic context.

Since the Ethernet header is 14 bytes long, it needs to have some alignment so that the CPU does not encounter any performance issues while accessing that part of the buffer. The appropriate name of the `header_len` parameter should be `header_alignment`, since this parameter is used for alignment. The usual value is 2, and it is the reason why the kernel defined a dedicated macro for this purpose, `NET_IP_ALIGN`, in `include/linux/skbuff.h`:

```
#define NET_IP_ALIGN 2
```

2. The second step reserves aligned memory for the header by reducing the tail room. The function that does is `skb_reserve()`:

   ```
   void skb_reserve(struct sk_buff *skb, int len)
   ```

3. The last step consists of extending the used data area of the buffer to as large as the packet size, by means of the `skb_put()` function. This function returns a pointer to the first byte of the data area:

   ```
   unsigned char *skb_put(struct sk_buff *skb, unsigned int len)
   ```

   The allocated socket buffer should be forwarded to the kernel-networking layer. This is the last step of the socket buffer's lifecycle. One should use the `netif_rx_ni()` function for that:

   ```
   int netif_rx_ni(struct sk_buff *skb)
   ```

We will discuss how to use the preceding steps in the section of this chapter that deals with packet reception.

Network interface structure

A network interface is represented in the kernel as an instance of `struct net_device` structure, defined in `include/linux/netdevice.h`:

```
struct net_device {
    char name[IFNAMSIZ];
    char *ifalias;
    unsigned long mem_end;
    unsigned long mem_start;
    unsigned long base_addr;
    int irq;
    netdev_features_t features;
    netdev_features_t hw_features;
    netdev_features_t wanted_features;
```

```
        int ifindex;
        struct net_device_stats stats;
        atomic_long_t rx_dropped;
        atomic_long_t tx_dropped;
        const struct net_device_ops *netdev_ops;
        const struct ethtool_ops *ethtool_ops;
        unsigned int flags;
        unsigned int priv_flags;
        unsigned char link_mode;
           unsigned char if_port;
        unsigned char dma;
        unsigned int mtu;
        unsigned short type;
        /* Interface address info. */
        unsigned char perm_addr[MAX_ADDR_LEN];
        unsigned char addr_assign_type;
        unsigned char addr_len;
        unsigned short neigh_priv_len;
        unsigned short dev_id;
        unsigned short dev_port;
        unsigned long last_rx;
        /* Interface address info used in eth_type_trans() */
        unsigned char *dev_addr;

        struct device dev;
        struct phy_device *phydev;
};
```

The `struct net_device` structure belongs to the kernel data structures that need to be allocated dynamically, having their own allocation function. An NIC is allocated in the kernel by means of the `alloc_etherdev()` function.

```
struct net_device *alloc_etherdev(int sizeof_priv);
```

The function returns `NULL` on failure. The `sizeof_priv` parameter represents the memory size to be allocated for a private data structure, attached to this NIC, and which can be extracted with the `netdev_priv()` function:

```
void *netdev_priv(const struct net_device *dev)
```

Given the `struct priv_struct`, which is our private structure, the following is an implementation of how you allocate a network device along with the private data structure:

```
struct net_device *net_dev;
struct priv_struct *priv_net_struct;
net_dev = alloc_etherdev(sizeof(struct priv_struct));
my_priv_struct = netdev_priv(dev);
```

Unused network devices should be freed with the `free_netdev()` function, which also frees memory allocated for private data. You should call this method only after the device has been unregistered from the kernel:

```
void free_netdev(struct net_device *dev)
```

After your `net_device` structure has been completed and filled, you should call `register_netdev()` on it. This function is explained later in this chapter in the section *Driver Methods*. Just keep in mind this function registers our network device with the kernel, so that it can be used. That being said, you should make sure the device really can process network operations before calling this function.

```
int register_netdev(struct net_device *dev)
```

The device methods

Network devices fall into the category of devices not appearing in the /dev directory (unlike block, input, or char devices). Therefore, like all of those kinds of devices, the NIC driver exposes a set of facilities in order to perform. The kernel exposes operations that can be performed on the network interfaces by means of the `struct net_device_ops` structure, which is a field of the `struct net_device` structure, representing the network device (dev->netdev_ops). The `struct net_device_ops` fields are described as follows:

```
struct net_device_ops {
    int (*ndo_init)(struct net_device *dev);
    void (*ndo_uninit)(struct net_device *dev);
    int (*ndo_open)(struct net_device *dev);
    int (*ndo_stop)(struct net_device *dev);
    netdev_tx_t (*ndo_start_xmit) (struct sk_buff *skb,
                        struct net_device *dev);
    void (*ndo_change_rx_flags)(struct net_device *dev, int flags);
    void (*ndo_set_rx_mode)(struct net_device *dev);
    int (*ndo_set_mac_address)(struct net_device *dev, void *addr);
    int (*ndo_validate_addr)(struct net_device *dev);
    int (*ndo_do_ioctl)(struct net_device *dev,
                        struct ifreq *ifr, int cmd);
    int (*ndo_set_config)(struct net_device *dev, struct ifmap *map);
    int (*ndo_change_mtu)(struct net_device *dev, int new_mtu);
    void (*ndo_tx_timeout) (struct net_device *dev);

    struct net_device_stats* (*ndo_get_stats)(
    struct net_device *dev);
};
```

Let us see what the meaning of each element in the structure is:

- `int (*ndo_init)(struct net_device *dev)` and `void(*ndo_uninit)(struct net_device *dev);` They are extra initialization/unitialization functions, respectively executed when the driver calls `register_netdev()`/`unregister_netdev()` in order to register/unregister the network device with the kernel. Most drivers do not provide those functions, since the real job is done by `ndo_open()` and `ndo_stop()` functions.
- `int (*ndo_open)(struct net_device *dev);` Prepares and opens the interface. The interface is opened whenever `ip` or `ifconfig` utilities activate it. In this method, the driver should request/map/register any system resource it needs (I/O ports, IRQ, DMA, and so on), turn on the hardware, and perform any other setup the device requires.
- `int (*ndo_stop)(struct net_device *dev)`:The kernel executes this function when the interface is brought down (For example, `ifconfig <name> down` and so on). This function should perform reverse operations of what has been done in `ndo_open()`.
- `int (*ndo_start_xmit) (struct sk_buff *skb, struct net_device *dev)`: This method is called whenever the kernel wants to send a packet through this interface.
- `void (*ndo_set_rx_mode)(struct net_device *dev)`: This method is called to change the interface address list filter mode, multicast or promiscuous. It is recommended to provide this function.
- `void (*ndo_tx_timeout)(struct net_device *dev)`: The kernel calls this method when a packet transmission fails to complete within a reasonable period, usually for `dev->watchdog` ticks. The driver should check what happened, handle the problem, and resume packet transmission.
- `struct net_device_stats *(*get_stats)(struct net_device *dev)`: This method returns the device statistic. It is what one can see when `netstat -i` or `ifconfig` is run.

The preceding descriptions miss a lot of fields. The complete structure description is available in the `include/linux/netdevice.h` file. Actually, only `ndo_start_xmit` is mandatory, but it is a good practice to provide as many helper hooks as your device has features.

Opening and closing

The `ndo_open()` function is called by the kernel whenever this network interface is configured by authorized users (admin for example) who make use of any user space utilities like `ifconfig` or `ip`.

Like other network device operations, the `ndo_open()` function receives a `struct net_device` object as its parameter, from which the driver should get the device-specific object stored in the `priv` field at the time of allocating the `net_device` object.

The network controller usually raises an interrupt whenever it receives or completes a packet transmission. The driver needs to register an interrupt handler that will be called whenever the controller raises an interrupt. The driver can register the interrupt handler either in the `init()`/`probe()` routine or in the `open` function. Some devices need the interrupt to be enabled by setting this in a special register in the hardware. In this case, one can request the interrupt in the `probe` function and just set/clear the enable bit in the open/close method.

Let us summarize what the `open` function should do:

1. Update the interface MAC address (in case the user changed it and if your device allows this).
2. Reset the hardware if necessary, and take it out of the low-power mode.
3. Request any resources (I/O memory, DMA channels, IRQ).
4. Map IRQ and register interrupt handlers.
5. Check the interface link status.
6. Call `net_if_start_queue()` on the device in order to let the kernel know that your device is ready to transmit packets.

An example of `open` function is follows:

```
/*
 * This routine should set everything up new at each open, even
 * registers that should only need to be set once at boot, so that
 * there is non-reboot way to recover if something goes wrong.
 */
static int enc28j60_net_open(struct net_device *dev)
{
    struct priv_net_struct *priv = netdev_priv(dev);

    if (!is_valid_ether_addr(dev->dev_addr)) {
        [...] /* Maybe print a debug message ? */
        return -EADDRNOTAVAIL;
```

```
    }
    /*
     * Reset the hardware here and take it out of low
     * power mode
     */
    my_netdev_lowpower(priv, false);

    if (!my_netdev_hw_init(priv)) {
            [...] /* handle hardware reset failure */
            return -EINVAL;
    }

    /* Update the MAC address (in case user has changed it)
     * The new address is stored in netdev->dev_addr field
     */
    set_hw_macaddr_registers(netdev, MAC_REGADDR_START,
    netdev->addr_len, netdev->dev_addr);

    /* Enable interrupts */
    my_netdev_hw_enable(priv);

    /* We are now ready to accept transmit requests from
     * the queueing layer of the networking.
     */
    netif_start_queue(dev);

    return 0;
}
```

`netif_start_queue()` simply allows upper layers to call the device `ndo_start_xmit` routine. In other words, it informs the kernel that the device is ready to handle transmit requests.

The closing method on the other side just has to do the reverse of the operations done when the device was opened:

```
    /* The inverse routine to net_open(). */
    static int enc28j60_net_close(struct net_device *dev)
    {
        struct priv_net_struct *priv = netdev_priv(dev);

        my_netdev_hw_disable(priv);
        my_netdev_lowpower(priv, true);

         /**
          *     netif_stop_queue - stop transmitted packets
          *
          *     Stop upper layers calling the device ndo_start_xmit routine.
```

```
 *      Used for flow control when transmit resources are unavailable.
 */
    netif_stop_queue(dev);

    return 0;
}
```

`netif_stop_queue()` simply does the reverse of `netif_start_queue()`, telling the kernel to stop calling the device `ndo_start_xmit` routine. We can't handle transmit request anymore.

Packet handling

Packet handling consists of transmission and reception of packets. This is the main task of any network interface driver. Transmission refers only to sending outgoing frames, whereas reception refers to frames coming in.

There are two ways to drive networking data exchange: by polling or by interrupt. Polling, which is a kind of timer-driven interrupt, consists of a kernel continuously checking at given intervals for any change from the device. On the other hand, interrupt mode consists of the kernel doing nothing, listening to an IRQ line, and waiting for the device to notify a change, by means of the IRQ. Interrupt-driven data exchange can increase system overhead during time of high traffic. That is why some drivers mix the two methods. The part of the kernel that allows mixing of the two methods is called **New API** (**NAPI**), which consists of using polling during times of high traffic and using interrupt IRQ-driven management when the traffic becomes normal. New drivers should use NAPI if the hardware can support it. However, NAPI is not discussed in this chapter, which will focus on the interrupt-driven method.

Packet reception

When a packet arrives into the network interface card, the driver must build a new socket buffer around it, and copy the packet into the `sk_ff->data` field. The kind of copy does not really matter, and DMA can be used too. The driver is generally aware of new data arrivals by means of interrupts. When the NIC receives a packet, it raises an interrupt, which will be handled by the driver, which has to check the interrupt status register of the device and check the real reason why this interrupt was raised (in could be RX ok, RX error, and so on). Bit(s) that correspond to the event that raised the interrupt will be set in the status register.

Network Interface Card Drivers

The tricky part will be in allocating and building the socket buffer. But fortunately, we already discussed that in the first section of this chapter. So let's not waste time and let's jump to a sample RX handler. The driver has to perform as many `sk_buff` allocations as the number of packets it received:

```c
/*
 * RX handler
 * This function is called in the work responsible of packet
 * reception (bottom half) handler. We use work because access to
 * our device (which sit on a SPI bus) may sleep
 */
static int my_rx_interrupt(struct net_device *ndev)
{
    struct priv_net_struct *priv = netdev_priv(ndev);
    int pk_counter, ret;

    /* Let's get the number of packet our device received */
    pk_counter = my_device_reg_read(priv, REG_PKT_CNT);
    if (pk_counter > priv->max_pk_counter) {
        /* update statistics */
        priv->max_pk_counter = pk_counter;
    }
    ret = pk_counter;

    /* set receive buffer start */
    priv->next_pk_ptr = KNOWN_START_REGISTER;
    while (pk_counter-- > 0)
        /*
     * By calling this internal helper function in a "while"
     * loop, packets get extracted one by one from the device
     * and forwarder to the network layer.
     */
        my_hw_rx(ndev);

    return ret;
}
```

The following helper is responsible for getting one packet from the device, forwarding it to the kernel network, and decrementing the packet counter:

```c
/*
 * Hardware receive function.
 * Read the buffer memory, update the FIFO pointer to
 * free the buffer.
 * This function decrements the packet counter.
 */
static void my_hw_rx(struct net_device *ndev)
```

[538]

```c
{
    struct priv_net_struct *priv = netdev_priv(ndev);
    struct sk_buff *skb = NULL;
    u16 erxrdpt, next_packet, rxstat;
    u8 rsv[RSV_SIZE];
    int packet_len;

    packet_len = my_device_read_current_packet_size();
    /* Can't cross boundaries */
    if ((priv->next_pk_ptr > RXEND_INIT)) {
            /* packet address corrupted: reset RX logic */
            [...]
            /* Update RX errors stats */
            ndev->stats.rx_errors++;
            return;
    }
    /* Read next packet pointer and rx status vector
     * This is device-specific
     */
    my_device_reg_read(priv, priv->next_pk_ptr, sizeof(rsv), rsv);

    /* Check for errors in the device RX status reg,
     * and update error stats accordingly
     */
    if(an_error_is_detected_in_device_status_registers())
            /* Depending on the error,
             * stats.rx_errors++;
             * ndev->stats.rx_crc_errors++;
             * ndev->stats.rx_frame_errors++;
             * ndev->stats.rx_over_errors++;
             */
    } else {
            skb = netdev_alloc_skb(ndev, len + NET_IP_ALIGN);
            if (!skb) {
                    ndev->stats.rx_dropped++;
            } else {
                    skb_reserve(skb, NET_IP_ALIGN);
                    /*
                      * copy the packet from the device' receive buffer
                      * to the socket buffer data memory.
                      * Remember skb_put() return a pointer to the
                      * beginning of data region.
                      */
                    my_netdev_mem_read(priv,
                        rx_packet_start(priv->next_pk_ptr),
                        len, skb_put(skb, len));
                    /* Set the packet's protocol ID */
                    skb->protocol = eth_type_trans(skb, ndev);
```

```
                /* update RX statistics */
                ndev->stats.rx_packets++;
                ndev->stats.rx_bytes += len;

                /* Submit socket buffer to the network layer */
                netif_rx_ni(skb);
            }
        }
        /* Move the RX read pointer to the start of the next
         * received packet.
         */
        priv->next_pk_ptr = my_netdev_update_reg_next_pkt();
    }
```

Of course the only reason we call the RX handler from within a deferred work is because we sit on an SPI bus. All of the preceding operations could be performed from within the hwriq in case of an MMIO device. Have a look at the NXP FEC driver, in drivers/net/ethernet/freescale/fec.c to see how this is achieved.

Packet transmission

When the kernel needs to send packets out of the interface, it calls the driver's ndo_start_xmit method, which should return NETDEV_TX_OK on success, or NETDEV_TX_BUSY on failure, and in this case you can't do anything to the socket buffer since it is still owned by the network queuing layer when the error is returned. This means you cannot modify any SKB fields, or free the SKB, and so on. This function is protected from the concurrent call by a spinlock.

Packet transmission is done asynchronously in most cases. The sk_buff of the transmitted packet is filled by the upper layers. Its data field contains packets to be sent. Drivers should extract packet from sk_buff->data and write it into the device hardware FIFO, or put it into a temporary TX buffer (if the device needs a certain size of data before sending it) before writing it into the device hardware FIFO. Data is really only sent once the FIFO reaches a threshold value (usually defined by the driver, or provided in a device datasheet) or when the driver intentionally starts the transmission, by setting a bit (a kind of trigger) in a special register of the device. That being said, the driver needs to inform the kernel not to start any transmissions until the hardware is ready to accept new data. This notification is done by means of the netif_stop_queue() function.

```
    void netif_stop_queue(struct net_device *dev)
```

After sending the packet, the network interface card will raise an interrupt. The interrupt handler should check why the interrupt has occurred. In case of transmission interrupt, it should update its statistics (`net_device->stats.tx_errors` and `net_device->stats.tx_packets`), and notify the kernel that the device is free for sending new packets. This notification is done by means of `netif_wake_queue()`:

```
void netif_wake_queue(struct net_device *dev)
```

To summarize, packet transmission is split into two parts:

- `ndo_start_xmit` operation, which notifies the kernel that the device is busy, set up everything, and starts the transfer.
- The TX interrupt handler, which updates TX statistics and notifies the kernel that the device is available again.

The `ndo_start_xmit` function must roughly contain the following steps:

1. Call `netif_stop_queue()` on the network device in order to inform the kernel that the device will be busy in data transmission.
2. Write `sk_buff->data` content into the device FIFO.
3. Trigger the transmission (instruct the device to start transmission).

Operations (2) and (3) may lead to sleep for devices sitting on slow buses (SPI for example) and may need to be deferred to the work structure. This is the case for our sample.

Once the packet is transferred, the TX interrupt handler should perform the following steps:

4. Depending on the device being memory mapped or sitting on a bus whose access functions may sleep, the following operations should be performed directly in the hwirq handler or scheduled in a work (or threaded IRQ):

    1. Check if the interrupt is a transmission interrupt.

    2. Read the transmission descriptor status register and see what the status of the packet is.

    3. Increment error statistics if there are any problems in the transmission.

    4. Increment statistics of successful transmitted packets.

[541]

5. Start the transmission queue allowing the kernel to call the driver's `ndo_start_xmit` method again, by means of the `netif_wake_queue()` function.

Let us summarize in a short sample code:

```
/* Somewhere in the code */
INIT_WORK(&priv->tx_work, my_netdev_hw_tx);

static netdev_tx_t my_netdev_start_xmit(struct sk_buff *skb,
                        struct net_device *dev)
{
   struct priv_net_struct *priv = netdev_priv(dev);

   /* Notify the kernel our device will be busy */
   netif_stop_queue(dev);

   /* Remember the skb for deferred processing */
   priv->tx_skb = skb;

   /* This work will copy data from sk_buffer->data to
    * the hardware's FIFO and start transmission
    */
   schedule_work(&priv->tx_work);

   /* Everything is OK */
   return NETDEV_TX_OK;
}
```

The work is described below:

```
/*
 * Hardware transmit function.
 * Fill the buffer memory and send the contents of the
 * transmit buffer onto the network
 */
static void my_netdev_hw_tx(struct priv_net_struct *priv)
{
   /* Write packet to hardware device TX buffer memory */
   my_netdev_packet_write(priv, priv->tx_skb->len,
priv->tx_skb->data);

   /*
    * does this network device support write-verify?
    * Perform it
    */
   [...];

      /* set TX request flag,
```

```
 * so that the hardware can perform transmission.
 * This is device-specific
 */
    my_netdev_reg_bitset(priv, ECON1, ECON1_TXRTS);
}
```

TX interrupt management will be discussed in the next section.

Driver example

We can summarize the concepts discussed above in the following fake Ethernet driver:

```c
#include <linux/module.h>
#include <linux/kernel.h>
#include <linux/errno.h>
#include <linux/init.h>
#include <linux/netdevice.h>
#include <linux/etherdevice.h>
#include <linux/ethtool.h>
#include <linux/skbuff.h>
#include <linux/slab.h>
#include <linux/of.h>                    /* For DT*/
#include <linux/platform_device.h>       /* For platform devices */

struct eth_struct {
    int bar;
    int foo;
    struct net_device *dummy_ndev;
};

static int fake_eth_open(struct net_device *dev) {
    printk("fake_eth_open called\n");
    /* We are now ready to accept transmit requests from
     * the queueing layer of the networking.
     */
    netif_start_queue(dev);
    return 0;
}

static int fake_eth_release(struct net_device *dev) {
    pr_info("fake_eth_release called\n");
    netif_stop_queue(dev);
    return 0;
}
```

```c
static int fake_eth_xmit(struct sk_buff *skb, struct net_device *ndev) {
    pr_info("dummy xmit called...\n");
    ndev->stats.tx_bytes += skb->len;
    ndev->stats.tx_packets++;

    skb_tx_timestamp(skb);
    dev_kfree_skb(skb);
    return NETDEV_TX_OK;
}

static int fake_eth_init(struct net_device *dev)
{
    pr_info("fake eth device initialized\n");
    return 0;
};


static const struct net_device_ops my_netdev_ops = {
    .ndo_init = fake_eth_init,
    .ndo_open = fake_eth_open,
    .ndo_stop = fake_eth_release,
    .ndo_start_xmit = fake_eth_xmit,
    .ndo_validate_addr    = eth_validate_addr,
    .ndo_validate_addr    = eth_validate_addr,
};


static const struct of_device_id fake_eth_dt_ids[] = {
    { .compatible = "packt,fake-eth", },
    { /* sentinel */ }
};


static int fake_eth_probe(struct platform_device *pdev)
{
    int ret;
    struct eth_struct *priv;
    struct net_device *dummy_ndev;

    priv = devm_kzalloc(&pdev->dev, sizeof(*priv), GFP_KERNEL);
    if (!priv)
        return -ENOMEM;

    dummy_ndev = alloc_etherdev(sizeof(struct eth_struct));
    dummy_ndev->if_port = IF_PORT_10BASET;
    dummy_ndev->netdev_ops = &my_netdev_ops;
    /* If needed, dev->ethtool_ops = &fake_ethtool_ops; */
```

```
        ret = register_netdev(dummy_ndev);
        if(ret) {
            pr_info("dummy net dev: Error %d initalizing card ...", ret);
            return ret;
        }

        priv->dummy_ndev = dummy_ndev;
        platform_set_drvdata(pdev, priv);
        return 0;
}

static int fake_eth_remove(struct platform_device *pdev)
{
    struct eth_struct *priv;
    priv = platform_get_drvdata(pdev);
    pr_info("Cleaning Up the Module\n");
    unregister_netdev(priv->dummy_ndev);
    free_netdev(priv->dummy_ndev);

    return 0;
}

static struct platform_driver mypdrv = {
    .probe      = fake_eth_probe,
    .remove     = fake_eth_remove,
    .driver     = {
        .name       = "fake-eth",
        .of_match_table = of_match_ptr(fake_eth_dt_ids),
        .owner      = THIS_MODULE,
    },
};
module_platform_driver(mypdrv);

MODULE_LICENSE("GPL");
MODULE_AUTHOR("John Madieu <john.madieu@gmail.com>");
MODULE_DESCRIPTION("Fake Ethernet driver");
```

Once the module is loaded and a device matched, an Ethernet interface will be created on the system. First, let us see what the `dmesg` command shows us:

```
# dmesg
[...]
[146698.060074] fake eth device initialized
[146698.087297] IPv6: ADDRCONF(NETDEV_UP): eth0: link is not ready
```

If one runs the `ifconfig -a` command, the interface will be printed on the screen:

```
# ifconfig -a
[...]
eth0 Link encap:Ethernet HWaddr 00:00:00:00:00:00
BROADCAST MULTICAST MTU:1500 Metric:1
RX packets:0 errors:0 dropped:0 overruns:0 frame:0
TX packets:0 errors:0 dropped:0 overruns:0 carrier:0
collisions:0 txqueuelen:1000
RX bytes:0 (0.0 B) TX bytes:0 (0.0 B)
```

One can finally configure the interface, assigning an IP address, so that it can be shown by using `ifconfig`:

```
# ifconfig eth0 192.168.1.45
# ifconfig
[...]
eth0 Link encap:Ethernet HWaddr 00:00:00:00:00:00
inet addr:192.168.1.45 Bcast:192.168.1.255 Mask:255.255.255.0
BROADCAST MULTICAST MTU:1500 Metric:1
RX packets:0 errors:0 dropped:0 overruns:0 frame:0
TX packets:0 errors:0 dropped:0 overruns:0 carrier:0
collisions:0 txqueuelen:1000
RX bytes:0 (0.0 B) TX bytes:0 (0.0 B)
```

Status and control

Device control refers to a situation where the kernel needs to change properties of the interface on its own initiative, or in response to a user action. It can then use either operations exposed through the `struct net_device_ops` structure, as discussed, or use another control tool, **ethtool**, which requires the driver to introduce a new set of hooks that we will discuss in the next section. Conversely, status consists of reporting the state of the interface.

The interrupt handler

So far, we have only dealt with two different interrupts: when a new packet has arrived or when the transmission of an outgoing packet is complete; but now-a-days hardware interfaces are becoming smart, and they able to report their status either for sanity purposes, or for data transfer purposes. This way, network interfaces can also generate interrupts to signal errors, link status changes, and so on. They should all be handled in the interrupt handler.

This is what our hwrirq handler looks like:

```
static irqreturn_t my_netdev_irq(int irq, void *dev_id)
{
    struct priv_net_struct *priv = dev_id;

    /*
     * Can't do anything in interrupt context because we need to
     * block (spi_sync() is blocking) so fire of the interrupt
     * handling workqueue.
     * Remember, we access our netdev registers through SPI bus
     * via spi_sync() call.
     */
    schedule_work(&priv->irq_work);

    return IRQ_HANDLED;
}
```

Because our device sits on an SPI bus, everything is deferred into a work_struct, which is defined as follows:

```
static void my_netdev_irq_work_handler(struct work_struct *work)
{
    struct priv_net_struct *priv =
            container_of(work, struct priv_net_struct, irq_work);
    struct net_device *ndev = priv->netdev;
    int intflags, loop;

    /* disable further interrupts */
    my_netdev_reg_bitclear(priv, EIE, EIE_INTIE);

    do {
        loop = 0;
        intflags = my_netdev_regb_read(priv, EIR);
        /* DMA interrupt handler (not currently used) */
        if ((intflags & EIR_DMAIF) != 0) {
            loop++;
            handle_dma_complete();
            clear_dma_interrupt_flag();
        }
        /* LINK changed handler */
        if ((intflags & EIR_LINKIF) != 0) {
            loop++;
            my_netdev_check_link_status(ndev);
            clear_link_interrupt_flag();
        }
        /* TX complete handler */
        if ((intflags & EIR_TXIF) != 0) {
```

```
                    bool err = false;
                    loop++;
                    priv->tx_retry_count = 0;
                    if (locked_regb_read(priv, ESTAT) & ESTAT_TXABRT)
                            clear_tx_interrupt_flag();
            /* TX Error handler */
            if ((intflags & EIR_TXERIF) != 0) {
                    loop++;
                    /*
                     * Reset TX logic by setting/clearing appropriate
                     * bit in the right register
                     */
                    [...]

                    /* Transmit Late collision check for retransmit */
                    if (my_netdev_cpllision_bit_set())
                            /* Handlecollision */
                            [...]
            }
            /* RX Error handler */
            if ((intflags & EIR_RXERIF) != 0) {
                    loop++;
                    /* Check free FIFO space to flag RX overrun */
                    [...]
            }
            /* RX handler */
            if (my_rx_interrupt(ndev))
                    loop++;
    } while (loop);

    /* re-enable interrupts */
    my_netdev_reg_bitset(priv, EIE, EIE_INTIE);
}
```

Ethtool support

Ethtool is a small utility for examining and tuning the settings of Ethernet-based network interfaces. With ethtool, it is possible to control various parameters like:

- Speed
- Media type
- Duplex operation
- Get/set eeprom register content
- Hardware check summing
- Wake-on-LAN, and so on.

Drivers that need support from ethtool should include `<linux/ethtool.h>`. It relies on the `struct ethtool_ops` structure which is the core of this feature, and contains a set of methods for ethtool operations support. Most of these methods are relatively straightforward; see `include/linux/ethtool.h` for the details.

For ethtool support to be fully part of the driver, the driver should fill an `ethtool_ops` structure and assign it to the `.ethtool_ops` field of the `struct net_device` structure.

```
my_netdev->ethtool_ops = &my_ethtool_ops;
```

The macro `SET_ETHTOOL_OPS` can be used for this purpose too. Do note that your ethtool methods can be called even when the interface is down.

For example, the following drivers implement ethtool support:

- `drivers/net/ethernet/microchip/enc28j60.c`
- `drivers/net/ethernet/freescale/fec.c`
- `drivers/net/usb/rtl8150.c`

Driver methods

Driver methods are the `probe()` and `remove()` functions. They are responsible for (un)registering the network device with the kernel. The driver has to provide its functionalities to the kernel through the device methods by means of the `struct net_device` structure. These are the operations that can be performed on the network interface:

```
static const struct net_device_ops my_netdev_ops = {
    .ndo_open           = my_netdev_open,
    .ndo_stop           = my_netdev_close,
    .ndo_start_xmit     = my_netdev_start_xmit,
    .ndo_set_rx_mode    = my_netdev_set_multicast_list,
    .ndo_set_mac_address    = my_netdev_set_mac_address,
    .ndo_tx_timeout     = my_netdev_tx_timeout,
    .ndo_change_mtu     = eth_change_mtu,
    .ndo_validate_addr      = eth_validate_addr,
};
```

The preceding are the operations that most drivers implement.

The probe function

The `probe` function is quite basic, and only needs to perform a device's early `init`, and then register our network device with the kernel.

In other words, the `probe` function has to:

1. Allocate the network device along with its private data using the `alloc_etherdev()` function (helped by `netdev_priv()`).
2. Initialize private data fields (mutexes, spinlock, work_queue, and so on). One should use work queues (and mutexes) in case the device sits on a bus whose access functions may sleep (SPI for example). In this case, the hwirq just has to acknowledge the kernel code, and schedule the work that will perform operations on the device. The alternative solution is to use threaded IRQs. If the device is MMIO, one can use spinlock to protect critical sections and get rid of work queues.
3. Initialize bus-specific parameters and functionalities (SPI, USB, PCI, and so on).
4. Request and map resources (I/O memory, DMA channel, IRQ).
5. If necessary, generate a random MAC address and assign it to the device.
6. Fill the mandatories (or useful) netdev properties: `if_port`, `irq`, `netdev_ops`, `ethtool_ops`, and so on.
7. Put the device into the low-power state (the `open()` function will remove it from this mode).
8. Finally, call `register_netdev()` on the device.

With an SPI network device, the `probe` function can look like this:

```
static int my_netdev_probe(struct spi_device *spi)
{
    struct net_device *dev;
    struct priv_net_struct *priv;
    int ret = 0;

    /* Allocate network interface */
    dev = alloc_etherdev(sizeof(struct priv_net_struct));
    if (!dev)
            [...] /* handle -ENOMEM error */

    /* Private data */
    priv = netdev_priv(dev);

    /* set private data and bus-specific parameter */
    [...]
```

Network Interface Card Drivers

```
    /* Initialize some works */
    INIT_WORK(&priv->tx_work, data_tx_work_handler);
    [...]
    /* Devicerealy init, only few things */
    if (!my_netdev_chipset_init(dev))
            [...] /* handle -EIO error */

    /* Generate and assign random MAC address to the device */
    eth_hw_addr_random(dev);
    my_netdev_set_hw_macaddr(dev);

    /* Board setup must set the relevant edge trigger type;
     * level triggers won't currently work.
     */
    ret = request_irq(spi->irq, my_netdev_irq, 0, DRV_NAME, priv);
    if (ret < 0)
            [...]; /* Handle irq request failure */

    /* Fill some netdev mandatory or useful properties */
    dev->if_port = IF_PORT_10BASET;
    dev->irq = spi->irq;
    dev->netdev_ops = &my_netdev_ops;
    dev->ethtool_ops = &my_ethtool_ops;

    /* Put device into sleep mode */
    My_netdev_lowpower(priv, true);

    /* Register our device with the kernel */
    if (register_netdev(dev))
            [...]; /* Handle registration failure error */

    dev_info(&dev->dev, DRV_NAME " driver registered\n");

    return 0;
}
```

 This whole chapter is heavily inspired by the enc28j60 from Microchip. You may have a look into its code in `drivers/net/ethernet/microchip/enc28j60.c`.

The `register_netdev()` function takes a completed `struct net_device` object and adds it to the kernel interfaces; 0 is returned on success and a negative error code is returned on failure. The `struct net_device` object should be stored in your bus device structure so that it can be accessed later. That being said, if your net device is part of a global private structure, it is that structure that you should register.

[551]

Network Interface Card Drivers

 Do note that the duplicate device name may lead to registration failure.

Module unloading

This is the cleaning function, which relies on two functions. Our driver release function could look like this:

```
static int my_netdev_remove(struct spi_device *spi)
{
    struct priv_net_struct *priv = spi_get_drvdata(spi);

    unregister_netdev(priv->netdev);
    free_irq(spi->irq, priv);
    free_netdev(priv->netdev);

    return 0;
}
```

The `unregister_netdev()` function removes the interface from the system, and the kernel can no longer call its methods; `free_netdev()` frees the memory used by the `struct net_device` structure itself along with the memory allocated for private data, as well as any internally allocated memory related to the network device. Do note that you should never free `netdev->priv` by yourself.

Summary

This chapter has explained everything needed to write an NIC device driver. Even if the chapter relies on a network interface sitting on an SPI bus, the principle is the same for USB or PCI network interfaces. One can also use the dummy driver provided for testing purposes. After this chapter, it is obvious NIC drivers will no longer be mystery to you.

Index

A

ACPI match 136
analogic to digitals converters (ADC) 225
application-specific data
   Boolean 155
   cells 154
   extracting 153
   parsing 155
   sub-nodes, extracting 155
   text strings 153
   unsigned 32-bit integers 154
architecture, I2C client drivers
   about 170
   i2c_driver structure 170
architecture, SPI device drivers
   about 186
   initialization and registration 191
   provisioning 192
   spi_driver structure 189
   structure 186
attributes, sysfs files
   bus attributes 364
   class attributes 366
   device attributes 362
   device drivers attributes 365
   polling, enabling 367
attributes
   about 357
   group 358

B

bits 249
block started by symbol (bss) 27
BMA220
   about 251, 256
   reference 251
bottom halves
   about 86
   as solution 86
   interrupt handler design limitations issue 86
   softirqs, using as 89
   tasklets, using as 87
   workqueue, using as 88
buddies 286
buffer data access
   about 258
   capturing, using sysfs trigger 258
buffer support
   triggering 241, 244
bus matching 132

C

cache 308
caching algorithm
   write back cache 312
   write-aroundcache 311
   write-throughcache 311
chaining IRQ
   about 423
   chained interrupts 423
   nested interrupts 424
channel
   distinguishing 236
character device
   about 96
   allocating 101
   registering 101
Chip Select (CS) 185
coherent mapping 317
coherent systems 316
completion 322, 323
container_of macro 47, 49
content-addressable memory (CAM) 280

copy-on-write (CoW) case 297
CPU cache (memory caching)
  L1 cache 309
  L2 cache 309
  L3 cache 309
Cyclic Redundancy Code (CRC) 11

D

data structures, Linux Device Model (LDM)
  bus 342, 343, 346
  bus registration 347
  device driver 348
  device driver registration 349
  Kobject structure 352
  struct device 350
  struct device registration 351
data structures, Network Interface Card (NIC) driver
  socket buffer structure 528
data structures, producer driver
  configuration structure 491
  constraints structure 486
  description structures 485
  device operation structure 491
  init data structure 487
data writing delay, to disks
  reasons 310
  write caching strategies 311
dedicated work queue
  about 78
  programming syntax 78
descriptor-based GPIO interface
  GPIO descriptor mapping 383
  GPIO, allocating 384
  GPIO, using 385
  summarizing 387
device access functions, regmap
  other functions 219
  regmap_multi_reg_write function 218
  regmap_update_bits function 217
device file operations
  about 98
  file representation, in kernel 99
device methods, Network Interface Card (NIC) driver
  open function 535

packet, handling 537
device model
  and sysfs 359
device number
  allocating 97
  allocating dynamically 97
  allocating statically 97
device provisioning
  new method 131
  old method 126
  platform data 129
  resources 127, 128
Device Tree (DT) mechanism
  about 141
  aliases 143
  compiler 144
  labels 143
  naming convention 142
  OF match style 156
  pointer handle (phandle) 143
  reference 141
device tree compiler (dtc) 144
devices
  about 132
  addressing 145
  platform device addressing 147
  representing 145
  SPI and I2C addressing 145
devres 312
digital to analogic converters (DAC) 225
Direct Memory Access (DMA) 198
DMA DT binding
  about 337
  consumer binding 337, 338
DMA engine AP
  slave and controller specific parameters, setting 326
DMA engine API
  about 324
  callback notification, waiting 331
  descriptor, obtaining for transaction 329
  DMA slave channel, allocating 325
  pending DMA requests, issuing 331
  slave and controller specific parameters, setting 328

transaction, submitting 330
DMA mappings
  about 317
  coherent mappings 317
  setting up 316
  streaming DMA mappings 317
DMA
  about 315
  and cache coherency 316
driver
  about 132
  examples 445, 447, 451
dynamic tick kernel
  about 64
  atomic context 64
  delays 64
  nonatomic context 65
  sleep 64

E

end of interrupt (eoi) 412
environment setup, Linux
  about 10
  kernel configuration 12
  kernel, building 13
error codes
  printing 31
error
  handling 31
  null pointer errors, handling 34
exceptions
  about 408
  processor-detected exception 408
  programmed exception 408
Executable and Linkable Format (ELF) 26

F

fb_ops structure
  about 518
  accelerated methods 522
  controller's parameters, setting 520
  information, checking 519
  screen blanking 521
file operations
  file_operations structure, filling 120

ioctl method 116
llseek method 110
open method 104
poll method 112
read method 108
release method 105
structure, filling 120
write method 106
writing 102
flusher threads 312
framebuffer drivers
  data structures 510, 514
  device methods 514, 516
  intelfb 509
  methods 516
  mxcfb 509
  vesafb 509
  writing 523
framebuffer
  using, from user space 523

G

General Purpose Input Output (GPIO) 371
Global Interrupt Controller (GIC) 151
GPIO and IRQ chip 424, 426
GPIO controller driver
  architecture 399
  data structures 399
GPIO controllers
  and DT 405
GPIO interface
  legacy integer-based interface, and device tree 390
  mapping, to IRQ in device tree 393
GPIO subsystem
  about 376
  descriptor-based GPIO interface 382
  GPIO interface, and device tree 389
  integer-based GPIO interface 377
GPIO
  exporting, from kernel 396
gpiolib irqchip API 427

H

hard fault 296
hardware clock (hwclock) utility 466
heap memory allocation
   reference 292
helloworld module
   __init attribute 25
   _exit attribute 25
   about 24
   authors 30
   describing 31
   entry and exit point 25
   information 27
   licensing 29
high resolution timers (HRTs)
   about 62
   HRT API 62
   setup initialization 62
hrtimer trigger interface 247

I

I/O memory
   __iomem cookie 300
   Memory Mapped Input Output (MMIO) 298
   MMIO devices access 299
   PIO devices access 298
   through I/O ports 298
   working with 298
I2C client drivers, accessing
   I2C devices, instantiating in board configuration file 178
   plain I2C communication 175
   System Management Bus (SMBus) compatible functions, using 177
I2C client drivers
   accessing 175
   device, provisioning 174
   initialization and registration 174
   writing, steps 182
I2C devices
   instantiating, in device tree 182
I2C driver
   defining 180
   kernel version 181
   registering 180
I2C specification
   reference 170
I2C
   and device tree 179
i2c_driver structure, I2C client drivers
   probe() function 171
   remove() function 173
i2c_driver structure
   about 170
   probe () function, per-device data 172
ID table matching
   about 136, 138
   per device-specific data 138
IIO buffer
   about 248
   setup 249, 250
   sysfs interface 248
IIO channels
   channel attribute naming conventions 234
IIO data access
   about 257
   buffer data access 258
   hrtimer trigger, used for capturing 260
   one-shot capture 258
IIO data structures
   about 227
   IIO channels 232
   iio_dev structure 227, 230
   iio_info structure 231
   voltage channels 238
IIO tools 261
IIO trigger
   and sysfs (user space) 245
Industrial I/O (IIO) 225
init data structure
   elements 487
   init data, feeding in board file 488
   init data, feeding in DT 489
input device
   allocating 434
   drivers, writing 443
   polled input devices, sub-class 435, 438
   registering 434
   structures 431, 434

input event
  generating 439, 441
  reporting 439, 441
input/output control(ioctl) commands 203
  about 116
  reference 118
integer-based GPIO interface
  about 377
  accessing 378
  configuring 377
  GPIOs, mapped to IRQ 379
  modifying 377
  value, getting/setting 378
Inter-Process Communication (IPC) 11
interrupt controller code property 152
interrupt controller
  and DT 429
  multiplexing 410, 419
interrupt handler
  about 151
  registering 83
  using, with lock 85
interrupt requests 422
interrupt trigger interface 246
interrupts
  about 408
  multiplexing 410, 419
  propagation 422
ioctl method
  about 116
  implementing, steps 118
  used, for generating ioctl numbers 117

K

kernel addresses
  high memory 267, 268
  low memory 268
kernel interruption mechanism
  about 82
  bottom halves 86
  interrupt handler, registering 82
kernel locking mechanism
  about 65
  mutual exclusion (mutex) 66
  spinlock 68

kernel memory, mapping to user space
  io_remap_pfn_range, using 305
  mmap file operation 305
  remap_pfn_range, using 303
kernel object (kobject)
  about 11
  structure 352
kernel object sets (ksets) 356
kernel object type 354
kernel sleeping mechanism
  about 55
  wait queue 55, 58
kernel space
  about 20, 265
  and data space, data exchange 102
  single value copy 103
kernel timers
  high resolution timers (HRTs) 62
  standard timers 59
kernel version
  reference 181
kernel
  about 14
  classes 16
  coding style 14
  objects 16
  OOP 16
  structures allocation/initialization 15
  user-space applications, invoking 92
ksoftirqd 89

L

legacy version 476
linked list
  about 50
  creating 51
  dynamic method, using 51
  initializing 51
  list node, adding 53
  list node, creating 52
  node, deleting 54
  static method, using 52
  traversal 54
linker definition file (LDF) 27
linker definition script (LDS) 27

Linux caching system 308
Linux Device Model (LDM)
  about 341
  attribute 357
  data structures 342
  exploring 352
  kernel object sets (ksets) 356
  Kobj_type 354
Linux graphical stack
  reference 510
Linux page cache (disk caching) 310
Linux PWM framework
  controller interface 468
Linux regulator 484
Linux
  about 9
  advantages 9
  environmental setup 10
  source organization 11
  sources, obtaining 10
llseek method
  about 110
  implementing, steps 111

M

major 96
manual loading
  insmod 22
  modprobe 22
mapping, kernel
  Advanced Configuration and Power Interface
    mapping (ACPI) 383
  device tree 383
  platform data mapping 383
Master Input Slave Output (MISO) 185
Master Output Slave Input (MOSI) 185
match style mixing
  about 162
  platform resources 164
MCP23016 I2C I/O expander
  reference 402
memory (re)mapping
  about 302
  kernel memory, mapping to user space 303
  kmap 302

memory allocation mechanism
  about 282
  kmalloc family allocation 291
  page allocator 283
  processing, under hood 296
  reference 282
  slab allocator 286
  vmalloc allocator 294
Memory Management Unit (MMU)
  about 274
  address translation 274
message
  printing 31
  printing, with printk() 35
methods, framebuffer driver
  detailed fb_ops 518
methods, Network Interface Card (NIC) driver
  about 549
  module, unloading 552
  probe function 550
methods, producer driver
  about 492, 499, 503
  Intersil ISL6271A voltage regulator 494
  probe function 493
minor 96
mmap file
  implementing, in kernel 307
  operation 305
module
  about 21
  auto-loading 23
  building 39, 44
  building, in kernel tree 41, 44
  dependencies 21
  depmod utility 21
  external module, building 44
  file, creating 22
  loading 22
  makefile 39
  manual loading 22
  parameters 37
  unload 23
  unloading 22
most significant bits (MSBs) 279
mutual exclusion (mutex)

about 66
acquire and release 67
declare 66
mutex API 66

N

Network Interface Card (NIC) driver
  about 527
  close function 535
  control 546
  data structures 528
  device methods 533
  ethtool, using 548
  example 543, 546
  interrupt handler 546
  methods 549
  network interface structure 531, 533
  status 546
new and recommended API 476
New API (NAPI) 537
non-coherent systems 316
non-maskable interrupts (NMI) 408
NXP SDMA (i.MX6) 332, 335, 337

O

OF match style
  about 156, 157
  match style mixing 162
  multiple hardware, supporting with per device-specific data 160
  non-device tree platforms, dealing with 159
OF style 136
Open Firmware (OF) 134
open method
  about 104
  per-device data 104
operations (ops) structure 16

P

packet
  handling 537
  reception 537, 540
  transmission 540, 543
page allocator
  about 283

conversion function 285
  page allocation API 283
page fault 296
page frame number (PFN) 266
Page Global Directory (PGD) 277
page look up 280
Page Table (PTE) 278
page table base register (PTBR) 279
Page Upper Directory (PUD) 277
peripheral IRQs
  advanced management 419, 421
Pin control (pinctrl)
  about 371
  and device tree 372, 375, 376
pin controller
  guidelines 404
platform data
  versus DT 166
platform devices
  about 126
  and platform drivers match, in kernel 135
  and platform drivers, matching factors 134
  declaring 131
  device provisioning 126
  name matching 140
  platform data 126
  resources 126
platform drivers
  about 122, 123
  working, with DT 156
pointer handle (phandle) 143
poll method
  about 112
  implementing, steps 113
Port Input Output (PIO) 298
Power Management Integrated Circuit (PMIC) 484
private peripheral interrupt (PPI) 152
producer driver
  data structures 484
  interface 484
  methods 492
programmable interrupt timer (PIT) 59
programmed exception
  reference 408
publisher 367

Pulse Wide Modulation (PWM) consumer interface
  about 475
  clients binding 477
Pulse Wide Modulation (PWM)
  about 467
  consumer interface 475
  controller binding 474
  controller driver 469
  controller driver example 471
  reference 468
  using, with sysfs interface 479

R

read method
  about 108
  implementing, steps 109
real-time clock (RTC)
  about 58, 453
  and use space 464
  API 456
  framework data structures 454
regmap API
  about 210, 499
  device access functions 216
  regmap_config structure 211
  used, for programming 210
regmap
  caching mechanism 219
  example 221
  I2C initialization 215
  initializing 214
  SPI initialization 214
  subsystem, setting up 221
regmap_config structure 211
regulator 483
regulator device
  binding 507
  controlling 505
  current limit control 506
  operating mode control 507
  output, disabling 505
  output, enabling 505
  requesting 504
  status 506
  voltage control 506

regulators consumer interface 503
release method 105
request_threaded_irq() function
  @handler function 90
  @thread_fn function 90
resources
  application-specific data, extracting 153
  handling 148
  interrupts, handling 151
  named resources 149
  registers, accessing 150
RTC API
  about 456
  alarm, used for planning 461
  driver example 460
  time, setting 457
RTC framework data structures 454

S

Serial Clock (SCK) 185
Serial Clock (SCL) 169
Serial Data (SDA) 169
Serial Peripheral Interface (SPI) 152, 185
slab allocator
  about 286
  buddy algorithm 286
  exploring 289
slab states
  empty 289
  full 289
  partial 289
socket buffer structure
  about 528
  allocation 530
soft fault 296
software buffer structure
  allocation 531
Software IRQ (softirq)
  ksoftirqd 71
specialized caches (user space caching)
  libc (user-app cache) 310
  web browser cache 310
SPI device drivers
  accessing 197, 201
  and device tree 194

architecture 186
defining 196
instantiating, in board configuration file 193
instantiating, in device tree 196
registering 196
writing, steps 202
SPI user mode driver
  about 202
  with IOCTL 204, 205
spi_driver structure
  about 189
  probe() function 189
  probe() function, per-device data 190
  remove() function 191
spinlock
  about 68
  versus mutexes 70
standard timers
  HZ 59
  jiffies 59
  timers API 59
storagebits 249
streaming DMA mapping
  constraints 318
  scatter/gather mapping 319, 322
  single buffer mapping 319
subscriber 367
sysfs director 225
sysfs files
  and attributes 361
  current interfaces 362
sysfs GPIO interface 394
sysfs interface
  about 465
  using, with PWMs 479
sysfs trigger interface
  about 245
  add_trigger file 245
  device, tying with trigger 246
  remove_trigger file 246
system memory layout
  kernel space 265
  user space 265
system on chip (SoC) 10

T

tasklet
  about 72
  declaring 73
  disabling 73
  enabling 73
  scheduling 74
threaded IRQs
  about 89
  threaded bottom half 91
tickless kernel
  about 64
  atomic context 64
  delays 64
  nonatomic context 65
  sleep 64
timer management 58
timers API
  about 59
  setup initialization 60
  standard example 60
TLB miss event
  about 280
  hardware handling 280
  software handling 280
translation lookaside buffer (TLB)
  about 279
  working 280
Translation Table Base Register 0 (TTBR0) 279

U

user space addresses
  about 269
  VMA 272, 274
user space interface 441
user space
  about 21, 265
  and kernel space, data exchange 102
  framebuffer, using 523
user-space applications
  invoking, from kernel 92

V

virtual memory area (VMA) 271

W

work deferring mechanism
　about 70, 72, 73, 75
　kernel threads 82
　Software IRQ (softirq) 70

work queues
　about 75
　dedicated work queue 78
　kernel-global workqueue 75
　predefined (shared) workqueue 81
　standard workqueue functions 81
write method
　implementing, steps 106
writeback 308